Modern Diesel Technology: Heating, Ventilation, Air Conditioning & Refrigeration

2ND EDITION

John Dixon

Centennial College, Toronto, Ontario, Canada

DELMAR
CENGAGE Learning·

Australia • Brazil • Japan • Korea • Mexico • Singapore • Spain • United Kingdom • United States

Modern Diesel Technology: Heating, Ventilation, Air Conditioning & Refrigeration, 2nd Edition
John Dixon

Vice President, Careers & Computing:
 Dave Garza

Director of Learning Solutions: Sandy Clark

Executive Editor: Dave Boelio

Director, Development-Career
 & Computing: Marah Bellegarde

Managing Editor: Larry Main

Senior Product Manager: Sharon Chambliss

Editorial Assistant: Courtney Troeger

Brand Manager: Kristin McNary

Market Development Manager: Erin Brennan

Senior Production Director: Wendy Troeger

Production Manager: Mark Bernard

Content Project Manager: Christopher Chien

Art Director: Jackie Bates/GEX

Cover Image: Courtesy of Navistar, Inc.

Cover Inset Image: © 2014 Cengage Learning;
 Photo courtesy of John Dixon

For product information and technology assistance, contact us at
Cengage Learning Customer & Sales Support, 1-800-354-9706

For permission to use material from this text or product,
submit all requests online at **cengage.com/permissions**.
Further permissions questions can be e-mailed to
permissionrequest@cengage.com

Library of Congress Control Number: 2012948308

ISBN-13: 978-1-1337-1625-9

ISBN-10: 1-1337-1625-3

Delmar
5 Maxwell Drive
Clifton Park, NY 12065-2919
USA

Cengage Learning is a leading provider of customized learning solutions with office locations around the globe, including Singapore, the United Kingdom, Australia, Mexico, Brazil, and Japan. Locate your local office at: **international.cengage.com/region**

Cengage Learning products are represented in Canada by Nelson Education, Ltd.

To learn more about Delmar, visit **www.cengage.com/delmar**

Purchase any of our products at your local college store or at our preferred online store **www.cengagebrain.com**

Notice to the Reader
Publisher does not warrant or guarantee any of the products described herein or perform any independent analysis in connection with any of the product information contained herein. Publisher does not assume, and expressly disclaims, any obligation to obtain and include information other than that provided to it by the manufacturer. The reader is expressly warned to consider and adopt all safety precautions that might be indicated by the activities described herein and to avoid all potential hazards. By following the instructions contained herein, the reader willingly assumes all risks in connection with such instructions. The publisher makes no representations or warranties of any kind, including but not limited to, the warranties of fitness for particular purpose or merchantability, nor are any such representations implied with respect to the material set forth herein, and the publisher takes no responsibility with respect to such material. The publisher shall not be liable for any special, consequential, or exemplary damages resulting, in whole or part, from the readers' use of, or reliance upon, this material.

Printed in the United States of America
4 5 6 7 8 9 10 21 20 19 18 17

Table of Contents

Preface for Series . . . . *xvi*

Preface . . . . *xvii*

CHAPTER 1   Heating, Ventilation, and Air Conditioning 1

    Introduction 2

    System Overview 2

    History of Air Conditioning 2

    Today's Air-Conditioning Systems 2

    Vehicle Heat and Cold Sources 3

    Purpose of the HVAC System 3

    Air-Conditioning Components 4

        Compressor *4*

        Condenser *4*

        Pressure Regulating Devices *5*

        Evaporator *6*

        Receiver-Drier *6*

        Accumulator *6*

    Special Air-Conditioning Tools 7

        Manifold Gauge Set *7*

        Safety Eyewear *8*

        Leak Detectors *8*

        Thermometers *9*

        Shop Specialty Tools *9*

        Vacuum Pump *9*

        Refrigerant Recovery and Recycling Equipment *10*

        Antifreeze Recovery and Recycling Equipment *10*

        Electronic Weigh Scales *11*

        Scan Tools/Onboard Diagnostics *11*

        Refrigerant Identifier *11*

        Compressor Servicing Tools *12*

    Summary 12

    Review Questions 13

CHAPTER 2   Environmental and Safety Practices 15

    Introduction 16

    System Overview 16

    Stratospheric Ozone Depletion 16

        The Montreal Protocol *17*

The Clean Air Act 17
    EPA Penalties 18
Greenhouse Effect 18
    Greenhouse Gases 19
Refrigerants 19
    CFCs 19
    HCFCs 20
    HFCs 20
    Alternative Refrigerants 20
Disposable Refrigerant Cylinders 20
Refillable Cylinders 21
    Cylinder Color Code 22
Refrigerant Safety 22
    Health Hazards 22
    First Aid 22
    Poisonous Gas 22
General Workplace Precautions 22
Handling Refrigerant Cylinders 23
Summary 24
Online Research Tasks 24
Review Questions 24

CHAPTER 3   Thermodynamics 27
Introduction 28
System Overview 28
Heat 28
    Heat Transfer 29
    Thermal Equilibrium 29
    Rate of Heat Transfer 30
Temperature 30
    Temperature Scales 30
    British Thermal Unit 30
    Example 1 30
    Example 2 30
    Types of Heat 31
    Superheat 31
    Subcooling 31
Change of State 32
    Latent Heat of Fusion, Vaporization, and Condensation 32
Pressure 33
Atmospheric Pressure 33
Pressure Gauges 34
    Compound Gauge 34
Vacuum 35
Pressure/Temperature Relationship 35
    Raising the Boiling Point 36
    Lowering the Boiling Point 36
    Compressing a Vapor 36
Humidity 36
    Effects of Humidity 37
    Wet Bulb Temperature 37

Summary 37
Review Questions 38

CHAPTER 4 Air-Conditioning Components: Compressor, Condenser, and Receiver-Drier . . . 41

Introduction 41
System Overview 42
Compressor 42
Two-Piston-Type Compressors 43
Two-Piston Compressor Operation 43
Two-Piston Compressor Maintenance 45
Swash Plate Compressors 45
Swash Plate Compressor Operation 45
Swash Plate Compressor Maintenance 46
Rotary Vane Compressors 46
Rotary Vane Compressor Operation 46
Variable Displacement Compressors 47
Variable Displacement Compressor Operation 49
Scotch Yoke Compressors 49
Scotch Yoke Compressor Operation 49
Scroll Compressor 49
Scroll Compressor Operation 49
Lubrication 50
Condenser 52
Condenser Service 52
Receiver-Drier 53
Filter 53
Moisture Removal 54
Refrigerant Storage 54
Receiver-Drier Location 54
Receiver-Drier Service and Installation 55
Sight Glass 55
Maintenance Procedures 57
Summary 57
Review Questions 57

CHAPTER 5 Air-Conditioning Components: Metering Devices, Evaporator, Accumulator . . . 61

Introduction 61
Metering Devices 62
Thermostatic Expansion Valve 62
Internally Equalized Thermostatic Expansion Valve 62
Remote Bulb 63
Capillary Tube 63
Thermostatic Expansion Valve Operation 63
Throttling 63
Modulation 64
Controlling Action 64
Externally Equalized Thermostatic Expansion Valves 65
The H Valve 66
Fixed Orifice Tube 66
Variable Orifice Valve 68

    Superheat 68
        Determining Superheat *68*
    Evaporator 70
        Evaporator Service *71*
    Accumulator 71
    Maintenance Procedures 72
    Summary 73
    Review Questions 73

CHAPTER 6   The Refrigeration System 75
    Introduction 76
    System Overview 76
    The Thermostatic Expansion Valve System 76
    The Fixed Orifice Tube System 78
    Refrigerant Pressure and States 80
    Refrigeration Capacity—Performance Ratings 80
    Electromagnetic Clutch 81
        Description *81*
        Operation *81*
    Evaporator Temperature Control 82
        Compressor Operating Controls *82*
        Thermostatic Control Switch (Cold Switch) *82*
        Pressure Cycling Switch *84*
        Low-Pressure Switch *85*
    Compressor Protection Devices 85
        Low Pressure Cut-Off Switch *85*
        High Pressure Cut-Off Switch *85*
        Binary Pressure Switch *85*
        Trinary Switch *86*
        Fan Cycling Switch *86*
        Fan Timers *87*
        High-Pressure Relief Valve *87*
    Maintenance Procedures 88
    Performance Tasks 88
    Summary 88
    Review Questions 89

CHAPTER 7   Service Procedures 91
    Introduction 92
    System Overview 92
    Manifold Gauges 92
        Manifold Gauge Calibration *94*
        Manifold Service Hoses *95*
    Refrigerant Lines, Hoses, and Couplers 95
        Refrigerant Lines *95*
        Suction Line *96*
        Discharge Line *96*
        Liquid Line *96*
    Hose and Line Repair 96
        Simple Hose Repair *96*
        Finger-Style Crimp *96*

Beadlock Fittings 97
Crimping Tools 98
Alternate Method 98
Evaporator Inlet Repair 98
Aluminum Line Repair 98
Service Valves 98
Stem-Type Service Valve 99
Schrader-Type Service Valve 99
R-134 Service Valve 100
Leak Detection 102
Leak Detection Methods 102
Servicing Air-Conditioning Systems 104
Refrigerant Identification 104
Vacuum Pump 105
Correct Size 105
Correct Oil 105
Vacuum Pump Maintenance 106
Evacuating Procedure 106
Thermistor Vacuum Gauge 107
Refrigerant Charging 107
Charging Procedure 107
Partial Charge 108
Charging Cylinder 108
Refrigerant Recovery/Recycle 109
Refrigerant Management Center 109
Online Research Tasks 110
Summary 110
Review Questions 111

CHAPTER 8   Truck Engine Cooling Systems 113
Introduction 114
System Overview 114
Coolant 115
Testing Coolant Strength 117
Scaling 118
Testing Supplemental Coolant Additives 118
Mixing Heavy-Duty Coolant 120
High Silicate Antifreeze 120
Extended Life Coolants 121
Coolant Filters 121
Coolant Recycler 122
Cooling System Components 122
Radiators 122
Radiator Components 124
Radiator Servicing 124
Radiator Testing 125
Radiator Cap 125
Radiator Cap Testing 127
Water Pump 128
Water Pump Replacement, Inspection 128

Thermostats 128
    Thermostat Operation *129*
    By-Pass Circuit *129*
    Operating Without a Thermostat *129*
    Thermostat Testing *130*
Heater Core 130
Heater Control Valve 131
Bunk Heater and Air Conditioning 131
Shutters 132
Winter Fronts 132
Cooling Fans 132
    On/Off Fan Hubs *134*
    Thermatic Viscous Drive Fan Hubs/Thermo-Modulated Fans . . . *134*
    Fan Shrouds *134*
    Fan Belts and Pulleys *135*
    Cooling System Leaks *135*
    Testing for Leaks *135*
Cooling System Management 136
Summary 136
Review Questions 137

CHAPTER 9   Cab Climate Control/Supplemental Truck Heating and Cooling 141
Introduction 142
System Overview 142
    The Blend Air HVAC System *142*
    Water-Valve Controlled System *142*
    Supplemental Heating and Cooling Systems *142*
Water-Valve Controlled Systems 143
    The Fan Switch *143*
    Air Selection Switch *143*
    Temperature Control Switch *145*
    Air-Conditioning Switch *145*
    Recirculation *145*
    Optional Bunk Override Switch *146*
    Air Outlet Vents *147*
Sleeper Climate Control Panel 147
    Fan Switch *147*
    Temperature Control Switch *147*
Manual Water-Valve Controlled HVAC System 148
    Sleeper Climate Control Panel *148*
    Temperature Control *149*
Blend Air System 150
    Stepper Motor *150*
    Ventilation *150*
HVAC General Information 151
    Temperature Sensors *151*
    Operator Maintenance *151*
    General Maintenance *152*
Supplemental Cab Climate Control 152
    Fuel-Fired Interior Heaters *152*
    Operation *153*

Engine Coolant Heaters with Truck Cab or Bus Interior Heating 153
Operation 154
Auxiliary Power Units 154
Truck Stop Electrification 155
Stand-Alone Systems 156
Onboard or Shore Power Systems 156
Summary 156
Review Questions 157

CHAPTER 10 Troubleshooting and Performance Testing 159
Introduction 160
System Overview 160
Servicing 161
Performance Test 161
Gauge Testing 162
Some Air and Moisture 163
Symptoms 164
Cause 164
Cure 164
Excessive Air and Moisture 164
Symptoms 164
Cause 164
Cure 164
Condenser Air Flow Obstruction or Overcharged 165
Symptoms 165
Cause 165
Cure 166
Low Refrigerant Charge 166
Symptoms 166
Cause 166
Cure 166
Very Low Refrigerant Charge 166
Symptoms 166
Cause 167
Cure 167
Restriction in the High Side of the System 167
Symptoms 168
Cause 168
Cure 168
Expansion Valve Not Opening Enough 168
Symptoms 168
Cause 168
Cure 168
Expansion Valve Held Open 169
Symptoms 169
Cause 169
Cure 169
Defective Thermostatic Switch 169
Symptoms 169
Cause 170
Cure 170

Defective Compressor 170
   Symptoms *170*
   Cause *170*
   Cure *170*
Purging and Flushing 170
   Purging *170*
   Flushing *171*
Guidelines for Purging and Flushing 171
Purging and Flushing Procedures 172
   Purging *172*
   Flushing with HFCF-141b *172*
   How to Pop Components Dry *172*
Summary 173
Review Questions 173

CHAPTER 11   APAD/ACPU A/C Control Systems 177
Introduction 177
System Overview 178
Common Air-Conditioning Problems 178
The APAD System 178
Electrical I/O Definition 179
Inputs for ACPU CM-813 Controller 179
Outputs for ACPU CM-813 Controller 180
APADs Rules for Compressor Control (CM-813) 180
   Engine Fan Control *181*
Description of Diagnostic Faults 181
Blink Codes 181
   Clearing Blink Codes *181*
Fault Code Table 182
Testing the CM-813 Module 182
Troubleshooting 182
   Blink Codes *182*
   To Clear Fault Codes *187*
ACPU Control Functions CM-820 187
   Engine Fan Trigger *188*
Pinout Definition 188
Inputs for ACPU CM-820 Controller 189
   Low-Pressure Input *189*
   High-Pressure Input *189*
   Evaporator Thermostat (TStat) *189*
Outputs for ACPU CM-820 Controller 189
   A/C Drive (Compressor Clutch Drive) *189*
   DATA+ and DATA- *189*
   Fan (Fan Actuator) *189*
Diagnostics 189
Troubleshooting 191
Summary 194
Review Questions 194

CHAPTER 12   Coach Air Conditioning 197
Introduction 197
System Overview 197

Refrigeration Schematic 198
Service Procedures 202
   Triple Evacuation 204
   One-Time Evacuation Procedure 205
Air-Conditioning System Pressure 206
   Compressor Discharge Pressure 206
   Compressor Suction Pressure 206
Checking Refrigerant Charge 206
   Refrigerant Charging Procedures for Large Bus 207
   Partial Charging 207
Refrigerant Recovery 207
Checking Compressor Oil Level 208
   Adding Compressor Oil 208
   Removing Compressor Oil 208
Superheat Test Procedures 209
   Superheat Checklist 209
Air-Conditioning Troubleshooting Tips 210
Low-Side Pump-Down Procedures 211
Summary 211
Review Questions 212

CHAPTER 13 Truck-Trailer Refrigeration Equipment 215
Introduction 215
System Overview 216
System Components 216
   Engine 216
   Compressor 217
   Condenser 217
   Thermostatic Expansion Valve 217
   Evaporator 217
   Microprocessor 218
   Box Temperature 218
   Set Point 218
   Thermostat 218
Refrigerant 218
Reefer Van Construction 219
   Truck-Trailer Flooring 219
Multi-Temperature Refrigeration Units 219
Loading Factors 219
   Precooling the Product 219
   Precooling of the Controlled Space 220
   Air Circulation 220
   Pallet Positioning 220
Loading Procedures 221
   Proper Loading 221
   Side Spacing 221
   Roof Spacing 221
   Rear Door Spacing 221
   Front Bulkhead Spacing 222
Short Cycling 222
Auto Stop/Start 223

Maintenance Procedures 223
Performance Tasks 224
    Engine Maintenance 224
    Refrigeration Maintenance 224
Summary 224
Review Questions 225

CHAPTER 14 Refrigeration Components 227
Introduction 228
System Overview 228
The Compressor 228
    Compressor Operation 228
Service Valves 228
    Schrader Service Valves 230
    Vibrasorbers 230
The Condenser 230
The Receiver Tank 231
Filter Drier 232
    Drier Materials 232
    Liquid Line Installation 233
    Filtration 233
    Vapor Line Installation or Low-Side Installation 233
Moisture Indicators 233
Heat Exchanger 234
    Heat Exchanger Operation 234
The Thermostatic Expansion Valve 234
    Operation 235
    The Equalizer Line 235
    Valve Superheat 235
    Overview of Determining Superheat 236
    Sensing Element Charges 236
    Sensing Bulb Location 236
Distributor Tube 236
Evaporator 237
    Evaporator Construction 237
Accumulator 237
    Operation 237
Pressure Regulating Devices 238
    Evaporator Pressure Regulator 238
    Suction Pressure Regulator 238
    Operation 238
Safety Valves 239
Performance Tasks 239
Summary 239
Review Questions 240

CHAPTER 15 Refrigerant Flow Control 243
Introduction 243
System Overview 243

Refrigerant Cycle Control Valves 244
    Three-Way Valve *244*
    Pilot Solenoid *245*
    Three-Way Valve Operation *245*
Condenser Pressure Bypass Valve 246
Check Valves 247
Refrigerant Flow for Three-Way Valve Systems (Thermo King Units) 248
    Cool Cycle *248*
    Heat Cycle *248*
    Defrost Cycle *249*
Solenoid Control System (Carrier) 250
    Operation of the Solenoid Control System (Carrier) *250*
    Cooling Cycle *251*
    Heating Cycle *252*
    Defrost Cycle *253*
Four-Way Valve (Trane/Arctic Traveler) 253
    Four-Way Valve Operation *253*
    Cool Cycle *254*
    Heat Cycle *255*
    Four-Way Valve Defrost Cycle *256*
Summary 256
Review Questions 256

CHAPTER 16   Truck-Trailer Refrigeration Electrical Components 259
Introduction 260
System Overview 260
Storage Batteries 260
    Battery Construction *260*
    Cell Operation *261*
    Cell Voltage *261*
Battery Safety 261
Batteries 262
    Dry Charged Batteries *262*
    Wet Charged Batteries *262*
Battery Types 262
    Conventional Batteries *262*
    Low Maintenance Batteries *263*
    Maintenance-Free Batteries *263*
Battery Ratings 263
    Cold Cranking Amps *263*
    Reserve Capacity *263*
    Battery Council International (BCI) Group Dimensional Number *264*
Battery Maintenance 264
    Battery Storage *264*
    Truth or Urban Legend *265*
Battery Testing 265
    Hydrometer Testing *265*
    Open Circuit Voltage Test *266*
    Load Test *266*
    Using a Commercial Battery Load Tester *266*
    Using the Reefer Unit's Engine *268*

Battery Charging 268
  Slow Charging 268
  Fast Charging 268
Jump-Starting a Unit 269
  Battery Removal and Installation 269
Charging Systems 269
Alternator Components 270
  Stator 270
  Rotor 270
  Rectifier Diodes 271
  Field Diode 271
  Voltage Regulator 271
  Alternator Output Test 272
  Alternator Removal and Installation 272
Starters 273
  Starter Motor Types 273
  Conventional Starter Motors 273
  Gear Reduction Starter Motors 274
  Overrunning Clutch 274
Starter Testing 274
  Test Results 275
Refrigeration Unit Safety Switches 275
  Low Engine Oil Pressure Safety Switch 275
  High Engine Coolant Temperature 276
  High Compressor Discharge Pressure 276
  Low Compressor Oil Pressure Switch 276
Performance Tasks 276
Summary 277
Review Questions 277

CHAPTER 17 Truck and Trailer Refrigeration Maintenance 281
Introduction 281
System Overview 282
Engine Lubrication System 282
  Engine Oil Change 282
  Oil Filter Replacement 282
  Fuel Filter Replacement 282
Bleeding the Fuel System 283
  Bleeding Fuel System with Electric Fuel Pump 284
Air Filter Service/Replacement 284
  Oil Bath Air Cleaner 284
  Dry-Type Air Cleaner 285
Drive Belts 285
Glow Plugs 286
  Glow Plug Test 286
Engine Cooling System 286
  Coolant Replacement 286
  Flushing the Cooling System 287
Defrost System 287
  Defrost Air Switch Check 288
  Defrost Termination Switch 289

Refrigeration Unit Pre-Trip 289
External Leak Checking 289
Testing Refrigerant Level 289
    Recharging of the Refrigeration System 289
    Partial Recharging of the Refrigeration System 292
Compressor Oil Level Check 292
Compressor Pump Down 293
    Placing Compressor in Service 293
Compressor Oil Change 293
Low-Side Pump Down 294
    Preparing for Back in Service/Filter Drier Replacement 294
Refrigerant Removal 294
Evacuation Procedures 294
Soldering and Silver Brazing 295
Inert Gas Brazing 295
    Silver Brazing 296
    Vertical Down Joint Technique 298
    Vertical Up Joint Technique 298
    Horizontal Joint Technique 298
    Brazed Joint Disassembly 299
    Soft Soldering 299
Structural Maintenance 300
    Mounting Bolts 300
    Unit Visual Inspection 300
    Condenser 300
    Defrost Drain Hoses 300
    Evaporator 300
    Defrost Damper Door 300
Summary 301
Review Questions 301

Glossary **303**

Index **309**

Preface for Series

The Modern Diesel Technology (MDT) series of textbooks debuted in 2007 as a means of addressing the learning requirements of schools and colleges whose syllabi used a modular approach to curricula. The initial intent was to provide comprehensive coverage of the subject matter of each title using ASE/NATEF learning outcomes and thus provide educators in programs that directly target a single certification field with a little more flexibility. In some cases, an MDT textbook exceeds the certification competency standards. An example would be Joe Bell's *MDT: Electricity and Electronics* in which the approach is to challenge the student to attain a higher level of understanding than that required by the general service technician but suited to one specializing in the key areas of chassis electrical and electronics systems.

The MDT series now boasts nine textbooks. As the series has evolved, it has expanded in scope with the introduction of books addressing a much broader spectrum of commercial vehicles. Titles now include *Heavy Equipment Systems; Mobile Equipment Hydraulics;* and *Heating, Ventilation, Air Conditioning & Refrigeration,* with the latter including a detailed examination of trailer reefer technology, subject matter that falls outside of the learning objectives of a general textbook. While technicians specializing in all three areas are in demand in most areas of the country, there are as yet no national certification standards in place.

In addition, the series now includes two books that are ideal for students beginning their study of commercial vehicle technology. MDT's titles *Preventive Maintenance and Inspection* and *Diesel Engines* are written so that they can be used in high school programs. Each uses simple language and a no-nonsense approach suited for either classroom or self-directed study. That some high schools now offer programs specializing in commercial vehicle technology is an enormous progression from the more general secondary school "shop class," which tended to lack focus. It is also a testament to the job potential of careers in the commercial vehicle technology field in a general employment climate that has stagnated for several years. Some forward-thinking high schools have developed transitional programs partnering with both colleges and industry to introduce motive power technology as early as grade 10, an age at which many students make crucial career decisions. When a high school student graduates with credits in "Diesel Technology" or "Preventive Maintenance Practice," it can accelerate progression through college programs as well as make those responsible for hiring future technicians for commercial fleets and dealerships take notice.

Because each textbook in the MDT series focuses exclusively on the competencies identified by its title, each book can be used as a review and study guide for technicians prepping for specific certification examinations. Common to all of the titles in the MDT series, the objective is to develop hands-on competency without omitting any of the conceptual building blocks that enable an expert understanding of the subject matter from the technician's perspective. The second editions of these titles not only integrate the changes in technology that have taken place over the past five years but also blend in a wide range of instructor feedback based on actual classroom proofing. Both should combine to make these second editions more pedagogically effective.

Sean Bennett 2012

Preface

The reason for writing this textbook is to give truck technicians a solid foundation in the area of current HVAC systems. The book starts with an introduction to the system as well as to environmental and safety practices. The chapter on thermodynamics is a key building block for students to comprehend. All other chapters of this book build on the principles that are learned in that chapter. My belief is that if technicians understand how something is supposed to function, they will have a greater ability to diagnose and make the necessary repairs to the system than technicians who arbitrarily change parts until the system operates correctly and/or the complaint goes away. The text is written in a step-by-step format for the entry-level technician, in appropriate language so as to not leave new technicians behind. Once the fundamentals of air conditioning have been discussed, the text continues on to the air-conditioning components, types of systems, service procedures, air-conditioning protection units (ACPU), and troubleshooting.

The second part of the text deals with truck-trailer refrigeration equipment. Skilled technicians in this area of the trucking industry are in great demand. Again, this section of the text builds on the earlier chapter on thermodynamics and goes forward from there to an introduction of the mobile refrigeration unit (reefer), then takes the technician through the components, refrigerant flow, electrical components, and system preventive maintenance. A secondary objective of this book is to cover some of the ASE T7 and NATEF task objectives. This section is included in the instructor's manual. The learning outcome objectives are designed to meet or exceed ASE T7 and NATEF task objectives. Included in learning objectives are HVAC system service and repair; A/C system and component diagnosis, service, and repair; heating and engine cooling systems diagnosis service and repair; and refrigerant recovery, recycling, and handling.

Heating, Ventilation, Air Conditioning & Refrigeration, 2nd Edition is unique to today's market because there is currently no competitive textbook that combines truck HVAC and truck-trailer refrigeration

systems. This book should be a very usable study resource for entry-level as well as experienced technicians working on HVAC systems. In addition, mobile refrigeration technicians get an overview of refrigeration systems and maintenance tasks required in the industry.

New to this edition:

- **Chapter 12** is a completely new chapter on coach air conditioning. This chapter takes the technician through Carrier large bus system refrigerant flow schematics, system controls, performance testing, and service procedures.

I would like to thank Stuart Bottrell, corporate trainer at Freightliner Canada, LLC, for all of his help and technical expertise in the production of this textbook; Index Sensors & Controls, which provided technical information, art, and troubleshooting charts for this text; and Carrier Refrigeration Operations for its excellent training and service procedures in bus air conditioning.

John Dixon, August 2012

ACKNOWLEDGMENTS

I feel it is important to thank my apprenticeship students for their feedback over the years. While developing this text, I was able to teach from it a sort of field test run, if you will. This allowed my students to be my greatest critics, and I was able to make any changes as required. My rationale is that if my students didn't understand a concept, I would try another explanation until they did. Many of my students have been working in the trade for five years or more on the front line of new technology. Their feedback was and is paramount to me.

I would also like to thank my wife, Connie, and our three daughters, Alyzza, Jaymee, and Olyvia for giving me the time to work on this text. They sacrificed much of their time spent with me, allowing me to pursue my goals.

Finally, I must thank Sean Bennett for being such a great mentor to me in the production of this book. Without his encouragement, expertise, and patience, this book would not have been possible.

INDIVIDUALS

Ken Attwood, Centennial College
Jim Bardeau, Mack and Volvo Trucks
Centennial College
Sean Bennett, Centennial College
Brad Bisaillon, Proheat, Inc.
Susan Bloom, Centennial College
Dan Bloomer, Centennial College
Stuart Bottrell, corporate trainer, Freightliner
Canada, LLC
Sean Brown, Denver Auto Diesel College
Mike Cerato, Centennial College
David R. Christen, University of Northwestern Ohio
David Chyznak, Centennial College
Alan Clark, Lane Community College
Don Coldwell, Volvo Trucks Canada, Inc.
Owen Duffy, Centennial College
Boyce H. Dwiggins, Delmar, Cengage
Learning author
Danny Esch, Southwest Mississippi Community
College
Jim Gauthier, corporate trainer, Mack/Volvo
Dennis Hibbs, West Kentucky Community College
Helmut Hryciuk, Centennial College
Ray Hyduk, Centennial College
Serge Joncas, Mack and Volvo Training
Centennial College
John Kramar, Centennial College
George Liidermann, Freightliner Training
Alan McClelland, Dean School of Transportation
Centennial College
Rock Mezzone, Centennial College
John Montgomery, Mack and Volvo Trucks Canada
David Morgan, Mack and Volvo Training
Centennial College
John Murphy, Centennial College
Josephine Park, Centennial College
Daniela Perriccioli, Centennial College
Greg Schwemler, Centennial College

Glenville Sing, Mack and Volvo Trucks Training
Centennial College
Martin Sissons, Centennial College
Darren Smith, Centennial College
Angelo Spano, Centennial College
Russ Strayline, Lincoln Technical Institute
Gino Tamburro, Centennial College
Al Thompson, Centennial College
Trevor Thompson, Centennial College
Pierre Valley, Mack and Volvo Trucks Canada
David Weatherhead, Canadian Tire Training
Centennial College
Gus Wright, Centennial College

CONTRIBUTING COMPANIES

We would like to thank the following companies that provided technical information and art for this book:

ASE
Battery Council International
Carrier Refrigeration Operations
Caterpillar, Inc.
Espar Heater Systems
Freightliner LLC
Index Sensors & Controls
Proheat, Inc.
Robinair, SPX Corporation
Snap-On Tools Company
Thermo King Corporation
Toyota Motor Sales, U.S.A.
Volvo Trucks North America, Inc.

INSTRUCTOR RESOURCES

Time-saving instructor resources are available at the Instructor Companion Website for the text or on CD. Either delivery option offers the following resources: PowerPoint chapter presentations with selected images, an ExamView test bank, an Image Gallery containing images from the book, an Instructor's Guide which includes an answer key to chapter review questions, Word documents containing the chapter review questions, a chart correlating NATEF tasks to text pages, and a set of job sheets for use in the shop.

CHAPTER 1

Heating, Ventilation, and Air Conditioning

Learning Objectives

Upon completion and review of this chapter, the student should be able to:

- Describe the evolution of the modern-day air-conditioning system.
- Explain the purpose of the compressor as used in an air-conditioning system.
- Describe the function of the condenser.
- Explain the key differences between an orifice tube and a thermostatic expansion valve.
- Explain the purpose of a drier as used in an air-conditioning system.
- Describe the function of the evaporator.
- Explain how the accumulator works and what its function is in an air-conditioning system.
- Describe the uses for the manifold gauge set.
- List the different types of leak detectors and explain the purpose of a leak detector.
- Explain the functions of a vacuum pump as used on an air-conditioning system.
- Outline the reasons for refrigerant recovery.
- Describe refrigerant recycling.
- Explain why antifreeze must be recycled.
- List the advantages of a ventilation system.
- Outline the advantages to a technician of having the use of a scan tool.
- Explain why a refrigerant identifier should be used before servicing an air-conditioning system.

Key Terms

accumulator

compressor

condenser

evacuation

evaporator

humidity

HVAC

leak detector

manifold gauge set

orifice tube

receiver-drier

recovery

recycle

refrigerant identifier

scan tool

thermometer

thermostatic expansion valve

vacuum pump

ventilation

INTRODUCTION

This is the first of many chapters intended for the technician in the **HVAC** (heating, ventilation, and air-conditioning) field. It is interesting to see just how far humanity has come in such a short time regarding the development of climate control systems in modern vehicles. A technician should understand what functions an HVAC system is intended to perform and how the system accomplishes these tasks. Next the technician will be introduced to the components that make up a modern HVAC system and the tools required to maintain these ever-evolving systems.

SYSTEM OVERVIEW

In this chapter, the technician will first be given a brief history of the modern HVAC system. The technician will then be introduced to the purpose of the heating, ventilation, and air-conditioning system and be given a brief description of the components making up modern HVAC systems. These components will be discussed in detail in later chapters. This chapter will finish with an introduction to some of the specialty tools used by technicians in the HVAC field.

HISTORY OF AIR CONDITIONING

People who lived as far back as the ancient pharaohs of Egypt were probably the first to actively try to control the temperature of their environment. Evidence shows that each night, thousands of workers were used to disassemble the inner walls of the pharaoh's palace, and the thousand-pound blocks were carried into the desert, where they were left to cool during the night. The next morning they were taken back to the pharaoh's palace and the inner walls were reassembled. This extreme amount of work allowed the palace to remain a relatively cool 80°F (27°C) when the temperatures outside the palace were as high as 120–130°F (49–54°C).

In 1884, the Englishman William Whiteley placed blocks of ice in a tray under a horse carriage and used a fan attached to a wheel to force air inside. Later, a bucket of ice in front of a floor vent became the automotive equivalent.

Railway passenger cars also used to have large blocks of ice loaded into containers built underneath the passenger compartment; a fan was used to blow air over the ice and circulate cool air through the rail car.

Automobiles were not very comfortable in the early years because the cabs were open. Passengers had to

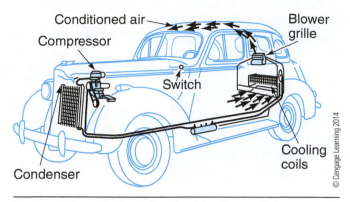

Figure 1-1 A 1939 Packard with air conditioner.

wear many layers of clothing in the winter, and in the summer the only ventilation was what could be brought in through the windows or open top of a vehicle that could cruise at a speed of 15 mph. Car companies then began closing up the cabins on cars; this required a change in temperature control systems. First, vents were put in the floors of cars, but this brought in more dirt and dust than it did cool air. In early attempts to cool the air, drivers placed buckets of water on the floor of their cars, thinking that air flowing over the surface of the water would cool the occupant compartment of the vehicle.

Evaporative cooling systems soon followed. In 1939, Packard produced the first passenger cars using refrigeration components. The huge evaporator was mounted in the trunk, leaving little room for luggage, and the only way to shut the evaporator off was to stop, raise the hood, and remove the drive belt from the compressor. Cadillac followed suit in 1941 with an air-conditioned car, and in 1954, Delphi Harrison Thermal Systems engineered an air-conditioning system that located all the major components of the air-conditioning system under the car's hood **(Figure 1-1).**

TODAY'S AIR-CONDITIONING SYSTEMS

Thanks to recent advances in modern technology, today's vehicles are extremely comfortable no matter what the weather is like outside the vehicle. Innovations such as computerized automatic temperature control (which allows you to set the desired temperature and have the system adjust automatically) and improvements to overall durability, have added complexity to today's air-conditioning systems. When today's truck drivers travel through regions of differing climates throughout the United States and Canada, they can enjoy the same comfort levels that they are accustomed to at home. With the simple slide of a lever or the push of a button, the climate-control system will make the transition from heating to cooling and back without the driver ever wondering how these changes occur.

For vehicles operating in northern United States, or Canada, heating systems keep the occupants warm and comfortable and also keep the windshield clear of ice and snow, improving visibility dramatically.

For vehicles operating in southern United States, or Canada, air conditioning greatly improves the comfort level of the occupants by cooling the cabin of the vehicle far below the temperatures outside the vehicle and, as an added benefit, also removes humidity (water vapor) from the circulating air.

Due to the complexity of today's air-conditioning systems, the "do it yourself" approach to air-conditioning repair is a thing of the past. To add to the complications, technicians are now faced with stringent environmental regulations that govern even the simplest of tasks. The technician is required to be certified to purchase refrigerant and to repair air-conditioning systems. The shop in which the technician works must also incur the cost of purchasing expensive dedicated equipment that is capable of removing all of the refrigerant from a vehicle, in order to prevent any of the ozone-depleting chemicals from escaping into the environment. This is required any time the air-conditioning system must be opened for repairs.

VEHICLE HEAT AND COLD SOURCES

The heat and cold that an HVAC (heating, ventilation, and air-conditioning) system is required to overcome originate from many different sources. Ambient air temperature (the outside air temperature), whether hot or cold, is one such source. Another source of heat is solar radiation. Solar radiation is the reason that the interior of a truck can be much hotter than the ambient temperature when the vehicle is parked in the sun. The tinting of windows can reduce the effects of solar radiation. Other sources of heat are those generated by the engine and cooling system. These include heat from the transmission, heat from the exhaust system, and heat that is radiated up through the floor of the vehicle from the surface of the road. Heat is also generated by the driver and, if applicable, the passenger in the vehicle. The heat that the human body constantly radiates to the air in the cab, as well as the warm moist air expelled from the human lungs, all add to the heat and moisture that must be removed from an HVAC system **(Figure 1-2)**.

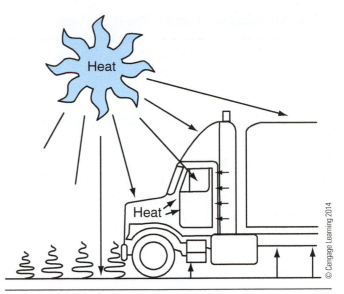

Figure 1-2 Heat enters the cab through windows, engine heat enters through the firewall, and heat radiates up through the floor from the hot pavement.

Another source of hot or cold air is the fresh air **ventilation** system. This system helps drivers stay more alert by changing or refreshing the air in the cab once or twice per minute. The air is circulated by a fan, usually referred to as a blower motor. The outside air coming into the cab must be either heated or cooled before it reaches the vehicle interior, depending upon whether the driver has selected heating or air conditioning. The ventilation system improves the performance of the air-conditioning or heating system by improving air flow within the vehicle. These air currents inside the vehicle guarantee that all areas inside the vehicle receive fresh air, whether heated or cooled.

PURPOSE OF THE HVAC SYSTEM

In today's trucks, the heating, ventilation, and air-conditioning (HVAC) systems perform three very important functions:

- Temperature control: The HVAC maintains the temperature within the passenger compartment as selected by the operator. It accomplishes this by adding or removing heat from the vehicle interior.
- Humidity control: The HVAC system reduces the **humidity** (water level in the air) within the passenger compartment, preventing condensation on the windows. Dehumidification or drying of the air helps the driver feel much more comfortable.

■ Air circulation control: The HVAC refreshes the air in the vehicle's interior by circulating and replacing stale air, while maintaining the selected interior air temperature.

AIR-CONDITIONING COMPONENTS

Today there are two different types of air-conditioning systems, which differ only slightly. The concept and design of these two types are very similar. The most common components that make up these truck air-conditioning systems are as follows:

1. Compressor
2. Condenser
3. Pressure regulating devices
   a. Orifice tube
   b. Thermostatic expansion valve
4. Thermostatic expansion valve
5. Evaporator
6. Receiver-drier
7. Accumulator

Compressor

The **compressor** can be referred to as the heart of the system. Compressors are bolted to the engine and are belt-driven by either a V-belt or a serpentine belt. The compressor is responsible for compressing and transferring refrigerant gas (**Figures 1-3** and **1-4**).

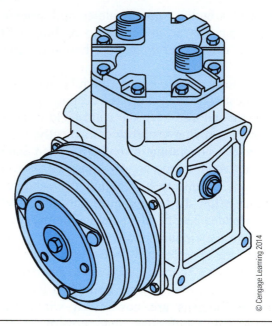

Figure 1-4 A two-piston type compressor.

The air-conditioning system may be divided into two different sides: the high-pressure side (commonly referred to as the discharge side) and the low-pressure side (also known as the suction side). The compressor is the dividing point between the suction and discharge sides of the air-conditioning system.

The suction side of the compressor draws in refrigerant gas from the outlet of the evaporator. In some cases, it does this via the accumulator. Once the refrigerant is drawn into the suction side, it is compressed. This concentrates the heat in the vapor, raising its temperature. The vapor leaving the compressor must be hotter than the atmosphere so that while it is in the condenser, it will dissipate the heat that it carries to the cooler ambient air. It is important to remember that these pumps are designed to compress only vapor. If liquid refrigerant gets into the inlet side of the compressor, it will damage the compressor by breaking valves or will cause the compressor's pistons to lock up.

Condenser

The **condenser** is the component that dissipates the heat that was once inside the cab of the truck. In most cases, the condenser has an appearance very similar to that of the radiator, because the condenser and radiator have very similar functions. The condenser is designed to radiate heat and is usually located in front of the radiator. In some retrofit applications, it may be located on the cab roof (**Figure 1-5**).

Condensers must have air flow any time the system is in operation. This is accomplished by the ram air

Figure 1-3 A swash plate compressor. Compressors are mounted in the engine compartment and are belt-driven by the truck's engine. The compressor includes an electromagnetic clutch to engage or disengage the compressor.

Condenser

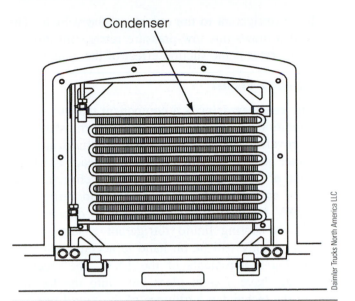

Figure 1-5 The condenser is the component in which the refrigerant surrenders heat from the cab to the ambient air.

effect of the truck as it runs down the road or by the engine cooling fan. Some manufacturers lock up the clutch fan whenever the air-conditioning system is operating.

The compressor pumps hot refrigerant gas into the top of the condenser. As the refrigerant is circulated through the condenser, the gas is cooled and condenses into high-pressure liquid refrigerant at the bottom of the condenser or condenser outlet.

Pressure Regulating Devices

As you will soon learn as you study thermo-dynamics **(Chapter 3)**, the desired temperature of an evaporator can be maintained by controlling the refrigerant pressure. Over the years, many types of pressure regulating devices have been used. Today, the most common are the orifice tube and the thermostatic expansion valve.

Orifice Tube. The **orifice tube** is a simple restriction located in the liquid line between the condenser outlet and the evaporator inlet. In a properly running air-conditioning system, this will be a transition point at which the line is hot coming from the condenser and will immediately become cool as the refrigerant passes through the orifice tube. This restriction may be identified by small indentations placed in the line that keep the orifice tube from moving within the liquid line. Most orifice tubes used in today's trucks are approximately 3 inches long and consist of a small brass tube surrounded by plastic and covered with a filter

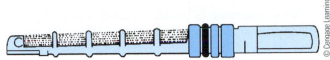

Figure 1-6 An orifice tube, used to meter the flow of refrigerant into the evaporator of an orifice tube air-conditioning system.

screen at each end. The inside diameter of the brass tube restricts the amount of liquid refrigerant that is able to pass through the valve. The orifice tube contains no moving parts. Truck manufacturers use different sized orifice tubes in order to balance the size of the air-conditioning system **(Figure 1-6)**.

Thermostatic Expansion Valve. The other common pressure regulating device is the **thermostatic expansion valve**, or TXV for short. Thermostatic expansion valves are used by many truck manufacturers **(Figure 1-7)**.

The thermostatic expansion valve, like the orifice tube, is situated between the condenser outlet and the evaporator inlet. This valve can sense both temperature and pressure, and is very efficient at controlling refrigerant flow through the evaporator. The expansion valve's job is to regulate the flow of refrigerant so that any liquid refrigerant metered through it has time to evaporate or change states from liquid to gas before leaving the evaporator. This is an important function because liquid refrigerant will destroy the compressor.

Expansion valves, although efficient, have maintenance characteristics different from those of orifice tubes. They can become clogged with debris just as orifice tubes can, but they also have small moving parts that may stick and malfunction due to corrosion; they may even freeze if enough water is able to enter the system.

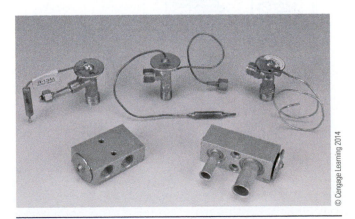

Figure 1-7 An assortment of thermostatic expansion valves.

Evaporator

The **evaporator** is usually located within the controlled space or is in some way isolated from the outside of the vehicle. The evaporator's primary function is to remove heat from within the cab of the vehicle. A secondary function is dehumidification (removing moisture from the air). A blower motor circulates air from the cabin through the evaporator coil. As the warmer air travels through the cooler fins of the evaporator, the moisture in the air condenses on their surface. Dust and pollen passing through stick to the wet surface and are expelled with the water to the ground outside the truck through the evaporator drain tube. On humid days, you may have noticed this water dripping from beneath the vehicle, especially when the air conditioner is turned off. In order to keep the evaporator from freezing, several different temperature- or pressure-regulating devices may be used; these will be discussed in a later chapter. Keeping the evaporator from freezing is extremely important because a frozen evaporator will not absorb very much heat **(Figure 1-8)**.

Refrigerant enters the evaporator as a low-pressure liquid. The temperature of the refrigerant is lower than that of the air inside the truck cab, so heat just follows its natural inclination to flow from a warm substance to a cooler one. The warm air from the cabin passes through the evaporator fins and it is this heat that causes the liquid refrigerant within the evaporator to boil (refrigerants have very low boiling points). The boiling refrigerant absorbs large quantities of heat from the cabin, and this is how the driver gets relief on a scorching summer day. This heat is then carried off with the refrigerant to the outside of the vehicle. The force that draws this low-pressure refrigerant through the evaporator is the suction effect of the compressor.

Receiver-Drier

The **receiver-drier** is a component that is used on air-conditioning systems that use a thermostatic expansion valve. The receiver-drier is a cylindrical metal container generally located on the bulkhead. The main function of the receiver-drier is to store refrigerant and separate any gas refrigerant from liquid refrigerant. The TXV requires liquid refrigerant to operate efficiently, so storing liquid refrigerant ensures that a constant supply will be on hand to accommodate the fluctuating requirements of the TXV. The receiver-drier may have a sight glass built into the top that allows the technician a glimpse of the liquid refrigerant as it passes through the receiver-drier. Under normal operating conditions, vapor bubbles should not be visible in the sight glass. There are various types of receiver-driers, and several different desiccant materials are in use. The receiver and desiccant types are chosen for the type of system and refrigerant used within the system **(Figure 1-9)**.

Accumulator

An **accumulator** is used on air-conditioning systems that employ a fixed orifice tube as the means of controlling the flow of refrigerant into the evaporator. The accumulator is plumbed into the system between the exit of the evaporator and the inlet of the compressor. The main purpose of the accumulator is to

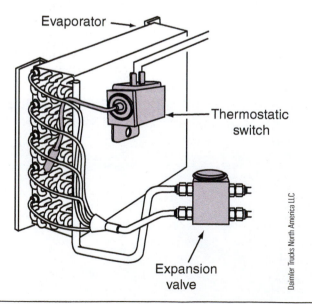

Evaporator

Thermostatic switch

Expansion valve

Daimler Trucks North America LLC

Figure 1-8 The evaporator is the component that absorbs heat from the truck's cab.

© Cengage Learning 2014

Figure 1-9 The receiver-drier provides storage filtration and moisture removal for passing refrigerant.

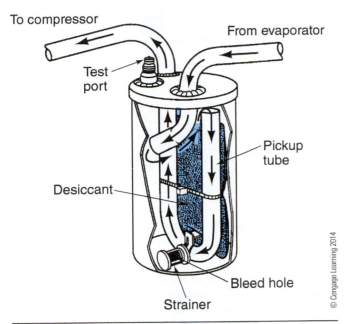

Figure 1-10 The accumulator ensures that only vaporous refrigerant may be returned to the compressor.

Figure 1-11 A manifold gauge set is probably the technician's best diagnostic tool because it provides information about what is happening inside the air-conditioning system.

prevent liquid refrigerant from reaching the compressor. The accumulator also contains a desiccant (as does a receiver-drier) that removes debris and moisture from the passing refrigerant. Moisture is enemy number one for an air-conditioning system because moisture in a system mixes with the refrigerant and forms a corrosive acid **(Figure 1-10).**

SPECIAL AIR-CONDITIONING TOOLS

In order to service air-conditioning systems, technicians must become familiar with the use of tools designed specifically for the mobile air-conditioning field. One of the tools that must be mastered by any air-conditioning or refrigeration technician is the manifold gauge set.

Manifold Gauge Set

A technician must be able to read the **manifold gauge set** and interpret the pressures of the air-conditioning system as it operates. These pressures tell the technician if the system is operating correctly or if there is a problem with the system. The manifold gauge set is usually the first tool installed on an air-conditioning system before any diagnostic work takes place. A manifold gauge set consists of a manifold block, two hand valves, three refrigerant hoses, and two pressure gauges **(Figure 1-11).**

The refrigerant hoses are usually color-coded to indicate where they should be connected. The hose on

the left is color-coded blue and is connected to the low-pressure/suction side of an air-conditioning system. Connected to the low-pressure hose through the manifold is a gauge that reads either vacuum or pressure and is also usually blue. Because the gauge reads in two different ranges of pressure, it is usually referred to as a compound gauge. On the vacuum side, the gauge will read to 30 inches of mercury. On the positive pressure side, the gauge will read accurately up to 120 pounds per square inch (psi) with a retard section of the gauge reading up to 250 psi. This means that the gauge will read accurately up to a positive pressure of 120 psi, while pressures from 120 psi to 250 psi can't be measured accurately but will not damage the gauge.

The hose on the right side of the gauge set is color-coded red and is connected to the high-pressure/discharge side of the air-conditioning system. Connected to the high-pressure hose through the manifold is a gauge that reads in pounds per square inch or kilopascals. This gauge is usually red, like the hose to which it is connected. The high side is usually calibrated from 0 psig (0 kPa) to 500 psig (3447 kPa). This gauge is usually referred to as the high-pressure gauge.

Safety Eyewear

Safety eyewear should be worn any time a person enters a shop environment. This is especially true when working with refrigerants because refrigerant that comes in contact with the eye can freeze the delicate tissue of the eye, causing blindness. Goggles may be worn over eye glasses. Safety glasses should also be equipped with side shields. In addition, full face shields are available for technicians working on air-conditioning systems. The type of safety eyewear worn by the technician should be a type that is approved for working with liquids or gases and must meet ANSIZ87.1-1989 standards **(Figure 1-12).**

Leak Detectors

The purpose of a **leak detector** is to determine the origin of a refrigerant leak. Special tools are required to find refrigerant leaks because often the gas will escape, leaving no visible trace as to where it exited the system **(Figure 1-13).**

Many different leak detectors are available to find the refrigerant leaks common in air-conditioning systems. Leak detectors can be low-tech or state-of-the-art electronic equipment. Dish soap and water can be sprayed or applied by brush to components and will bubble as the leaking refrigerant tries to pass through the soap and water. Leak detection solutions are also commercially available and generally cling to vertical surfaces better than soap and water.

Electronic leak detectors have been used for many years and are extremely sensitive in finding leaks. These units are capable of detecting leaks as small as 0.5 oz (14 ml) per year. Electronic leak detectors are called halogen leak detectors and may be used to test

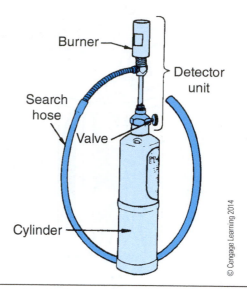

Figure 1-13 A flame-type leak detector used to identify refrigerant leaks; the flame color changes in the presence of refrigerants.

Figure 1-14 Electronic refrigerant leak detector for finding very small leaks.

Figure 1-12 Safety eyewear, glasses, goggles, or shields should be worn by everyone entering the shop, but this is especially important for air-conditioning technicians.

for refrigerant leaks with HFC-134a. Another style of leak detector uses fluorescent dye **(Figure 1-14).** The dye is injected into the system, mixes with the refrigerant and oil, and is circulated throughout the air-conditioning system by the compressor. The dye has no detrimental effects on the air-conditioning system,

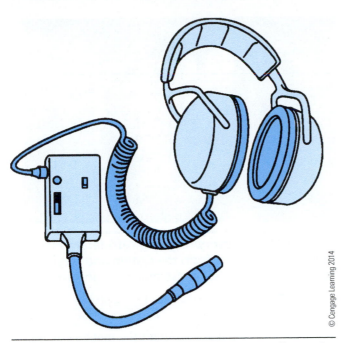

Figure 1-15 Ultrasonic leak detectors allow the technician to hear the refrigerant leak in the ultrasonic range.

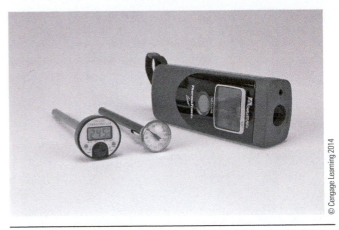

Figure 1-16 Thermometers are used by technicians to make accurate diagnoses of air-conditioning systems; shown are a few examples of temperature measuring devices.

nor does it hinder the system's performance. When refrigerant leaks, it also pushes out some compressor oil (if the leak is large enough) and some of the fluorescent dye. The technician can then sweep an ultraviolet lamp over the refrigerant lines and components. The dye will fluoresce under the ultraviolet lamp, indicating the source of the leak to the technician.

One other way of finding refrigerant leaks is to listen for them. This is accomplished with the use of an ultrasonic tester. Ultrasonic testers are able to detect sounds in the ultrasonic frequency that can't be detected by the human ear. The detector then converts and amplifies the sound so that the technician can hear it using a head set. Some detectors will also display the sound/leak rate **(Figure 1-15)**.

Thermometers

A **thermometer** is used by the technician to measure temperatures throughout the air-conditioning system. Regardless of the style of thermometer, it must be reasonably accurate for correct diagnostics of the system.

Dial-type thermometers come in analog or digital form, and are used by many air-conditioning technicians. The temperature range of the thermometer should be between 0°F and 220°F (−18°C and 104°C).

When accuracy is the main concern, an electronic thermometer may be required. They also come in analog or digital form.

Infrared temperature guns are frequently used to measure radiator temperatures and coolant lines as well as engine operating temperatures. These temperature-measuring devices do not require direct contact with the surface that they are measuring **(Figure 1-16)**.

Shop Specialty Tools

In addition to the specialty tools already mentioned, the technician will require the use of some big-ticket (expensive) tools that are generally supplied by the shop. Some of the tools to be supplied by the shop include a vacuum pump, a refrigerant recovery and recycling system, an antifreeze recovery and recycling system, an electronic scale, a refrigerant identifier, and an electronic thermometer. Also, there are scan tools used for diagnosis of the automatic temperature control system and specialty tools required for compressor service. Following is a brief description of these tools. They will be discussed in detail later in this book.

Vacuum Pump

The **vacuum pump** is used by the technician to remove moisture and air, which is able to enter the system whenever it has been opened for service procedures or when a leak has been repaired. This process of removing air and moisture that has entered the air-conditioning system is called **evacuation**. Air, if left in the system, will cause higher than normal pressures and carbonizing of the compressor oil. If moisture is left in the system, it will mix with the oil, causing acids to form in the system. This acid will destroy the components of the air-conditioning system from the inside out. The vacuum pump reduces the pressure within the system to such a low level that all the air is

Figure 1-17 A vacuum pump, used to remove air moisture and impurities from the air-conditioning system.

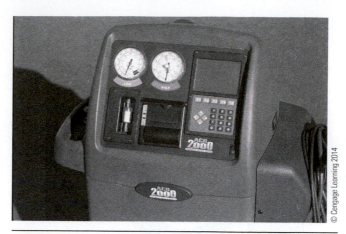

Figure 1-18 A recovery/recycling refrigerant machine removes refrigerant from the system and recycles the refrigerant for reuse; most machines incorporate a vacuum pump.

drawn out. Due to the reduced pressure, the boiling point of the water is also reduced. This causes the water to boil even at room temperature, allowing the vapor to be drawn out along with the air. These concepts will be explained further in **Chapter 3.**

A vacuum pump may be a separate unit or it may be incorporated into a refrigerant management center **(Figure 1-17).**

Refrigerant Recovery and Recycling Equipment

Any shop that services air-conditioning equipment must have **recovery** and **recycling** equipment so that refrigerant is not released into the atmosphere. This is also mandated by the EPA (Environmental Protection Agency). Refrigerant hoses from the recovery and recycling machine are connected to the suction and discharge ports of the air-conditioning system that is being serviced. The recovery part of the machine incorporates a pump to draw the entire refrigerant charge out of the air-conditioning system. The recycling part of the machine then cleans the refrigerant so that it can once again be used in an air-conditioning or refrigeration system. The machine accomplishes this by circulating the refrigerant through replaceable filter elements and drier elements that remove contaminants and moisture from the refrigerant. The recycling machine will also separate the compressor oil from the refrigerant. Some compressor oil is drawn out with the refrigerant during the recovery process. The amount of oil drawn out of the system is measured by the technician and added by the recycling machine when the system is recharged. Some equipment incorporate a weigh scale that indicates the amount of refrigerant removed from an air-conditioning/refrigeration system.

The technician can then program the machine to pump the exact refrigerant and compressor oil charge into the system without ever operating the air-conditioning system.

Recovery and recycling equipment is generally dedicated to one type of refrigerant to prevent cross contamination of refrigerant. Some machines have separate systems that can recover two different types of refrigerants, but not at the same time, because the recovery system shares some components required by both systems. The refrigerant hose hookup for R134a differs from the refrigerant hoses used by other air-conditioning systems so that cross contamination of refrigerants can't take place inside the machine.

Some other features incorporated by some recovery/recycling machines include an automatic air purge, a high-performance vacuum pump, and an automatic shutoff to prevent overfilling the refrigerant recovery tank **(Figure 1-18).**

Antifreeze Recovery and Recycling Equipment

Eventually, the mixture of antifreeze and water in a vehicle's cooling system will need to be replaced. This is because over time the corrosion-inhibiting additives are gradually used up and the coolant loses its ability to effectively protect the metal parts within the cooling system. Leaving the antifreeze in this condition will lead to radiator and heater core failure, erosion of the water pump impeller, as well as rust and scale buildup inside the engine and radiator, contributing to poor cooling system performance.

Waste antifreeze may contain heavy metals such as cadmium, lead, and chromium in high enough

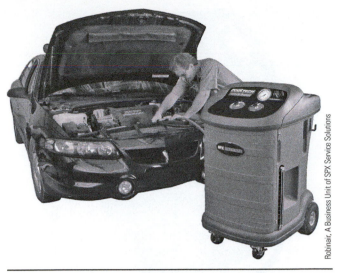

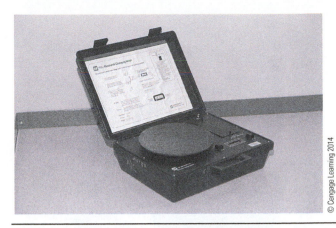

Figure 1-20 A portable electric scale, used by technicians to accurately weigh in the correct refrigerant charge.

Figure 1-19 An antifreeze recovery machine, used to recycle antifreeze so it can be reused in the engine cooling system.

concentrations to be considered hazardous waste. Dumping waste antifreeze into storm drains, waterways, or into the ground, where it can mix with groundwater, is a violation of the Clean Water Act.

The main ingredient in antifreeze (ethylene glycol) never wears out and can be recycled. Many shops use recovery/recycling equipment to avoid the high cost of disposing of their used antifreeze **(Figure 1-19)**. All recycling machines use a two-step procedure:

- Removing contaminant either by distillation, filtration, reverse osmosis, or ion exchange.
- Restoring the antifreeze to its original properties with an additive package. Additives usually contain chemicals that inhibit rust and corrosion, raise and stabilize the pH level of the antifreeze, reduce water scaling, and slow the breakdown of the ethylene glycol.

Electronic Weigh Scales

Electronic weigh scales are used to accurately dispense refrigerant by weight. Air-conditioning systems usually have a sticker placed somewhere on the vehicle that indicates the type and capacity of refrigerant required for that particular system. Accuracy is important because too much refrigerant in the system creates high compressor discharge pressures and too little refrigerant creates low compressor suction and discharge pressure, and insufficient cooling **(Figure 1-20)**.

The electronic weigh scale may be a simple portable unit that stands on its own and has a pad to mount a refrigerant cylinder of up to 50 pounds (23 kg). These scales usually incorporate a liquid crystal

display that can be switched to show either pounds or kilograms. The resolution of these scales is 0.05 pounds (0.02 kg).

The electronic weigh scale may also be incorporated into refrigerant recovery and recycling equipment. In these machines, the scale communicates to an onboard computer that displays the weight of refrigerant recovered from a system; it also accurately installs the correct charge of refrigerant as programmed by the technician.

Scan Tools/Onboard Diagnostics

Scan tools are used to improve troubleshooting capabilities, allowing the technician to accurately get to the origin of a problem. There are a variety of scan tools manufactured with different capabilities. These tools can display trouble codes for the technician, and some of the more highly sophisticated tools will allow the technician to monitor and view sensor and computer information. This allows the technician to pinpoint a heating, ventilation, or air-conditioning (HVAC) problem. Some scan tools may even take the place of a manifold gauge set by showing system pressures through the use of pressure transducers in the refrigerant lines **(Figure 1-21)**.

Refrigerant Identifier

In order to determine the contents of an air-conditioning system, a **refrigerant identifier** should be used. It is important to know what type of refrigerant is in a system so that cross contamination within a recovery machine can be prevented. The identifier should be used whenever the technician is not certain of the contents of an air-conditioning/refrigeration system. The refrigerant identifier can also be used to

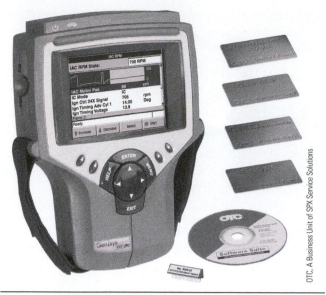

OTC, A Business Unit of SPX Service Solutions

Figure 1-21 A scan tool; pictured is a Prolink 2000, used to improve the technician's trouble-shooting capabilities. HVAC system faults can be displayed on the reader. Prolink Web site.

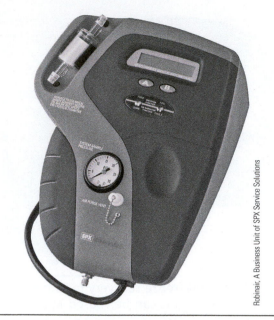

Robinair, A Business Unit of SPX Service Solutions

Figure 1-22 Refrigerant identifier, used to test the type and purity of refrigerant within the system.

determine the purity and quality of a refrigerant sample taken directly from an air-conditioning system or refrigerant storage container **(Figure 1-22)**.

CAUTION *If the sample contains a flammable hydrocarbon, it should not be serviced unless extreme care is taken to prevent serious personal injury.*

The identifier will display the following:

Green Pass LED R-12 if the refrigerant is CFC-12 and the purity of the refrigerant is better than 98% by weight.

Green Pass LED R-134a if the refrigerant is HFC-134a and the purity of the refrigerant is better than 98% by weight.

Red Fail LED Fail if either CFC-12 or HFC-134a is not identified or is not 98% pure. Alarm horn will sound 5 times.

Red Fail LED if the refrigerant sample contains hydrocarbon, a flammable substance. An audible warning will also sound 30 times.

Compressor Servicing Tools

Many different types of compressor servicing tools can accommodate the wide range of compressors manufactured. These tools are used to remove the compressor's electromagnetic clutch assembly and to remove and replace the compressor's rotary front drive shaft seal. These tools are usually quite compact so that the technician may be able to service the compressor without having to remove it from the vehicle.

Summary

- The pharaohs of ancient Egypt were probably the first people to actively try to control the temperature of their environment.

- The field of mobile mechanical air conditioning is relatively new in human existence.

- Air-conditioning systems have evolved at a very fast pace.

- The compressor is the component responsible for compressing and transferring refrigerant gas.

- The condenser is the component that dissipates the heat that was once inside the cab of the truck.

- The orifice tube is a simple restriction located in the liquid line between the condenser outlet and the evaporator inlet.

- The expansion valve functions to regulate the flow of refrigerant so that any liquid refrigerant that is metered through it has time to evaporate or change states from liquid to gas before the refrigerant leaves the evaporator.

- The evaporator's primary function is to remove heat from within the cab of the vehicle.

- The main function of the receiver-drier is to store refrigerant and separate any gas refrigerant from liquid refrigerant.

- The main purpose of the accumulator is to prevent liquid refrigerant from reaching the compressor.

- The manifold gauge set is a tool used by the technician to measure the operating pressures of an air-conditioning system.

- Safety glasses should be worn any time a person enters a shop environment, especially when the person is working with refrigerants.

- Leak detectors are used to pinpoint refrigerant leaks within the air-conditioning system.

- Thermometers are used by an air-conditioning technician to measure temperatures throughout the air-conditioning system.

- The vacuum pump is used to remove moisture and air that is able to enter the system.

- Recovery machines are used to remove the entire refrigerant charge from an air-conditioning system.

- Recycling machines are used to clean used refrigerant so that it may once again be used in an air-conditioning system.

- Electronic weigh scales are used to accurately dispense refrigerant by weight.

- Scan tools allow the technician to accurately troubleshoot an HVAC system.

- Refrigerant identifiers are used to determine the type and purity of the refrigerant in an air-conditioning system.

- Compressor servicing tools are specially designed tools used by technicians in the industry to service compressor clutches and seals.

Review Questions

1. How was the pharaoh's palace made to remain cool in the daytime by having the stones removed and placed in the desert at night?

   A. The stones were able to radiate heat during the day.

   B. The palace was able to ventilate during the night.

   C. The stones were able to cool at night; therefore, they could absorb heat from the palace during the daytime.

   D. None of the above statements is correct.

2. The interior temperature of a truck parked in direct sunlight will be:

   A. Lower than the ambient temperature

   B. Higher than the ambient temperature

   C. The same as the ambient temperature

3. Technician A says that the manifold gauge set may be used to add or remove refrigerant from an air-conditioning system. Technician B says that the manifold gauge set can be used by a skilled technician as an essential diagnostic tool. Which technician is correct?

   A. Technician A is correct.

   B. Technician B is correct.

   C. Both Technicians A and B are correct.

   D. Neither Technician A nor B is correct.

4. What standard should safety glasses and goggles meet in order to be used by a technician working on an air-conditioning system?

   A. OSHA

   B. EPA

   C. ANSI

   D. SAE

5. Technician A says that a vacuum pump may be used to remove air from a refrigeration system. Technician B says that a vacuum pump will remove moisture from an air-conditioning system. Which technician is correct?

    A. Technician A is correct.   C. Both Technicians A and B are correct.

    B. Technician B is correct.   D. Neither Technician A nor B is correct.

6. What is the state of the refrigerant as it enters the condenser?

    A. Liquid   C. Solid

    B. Hot gas   D. Cold

7. What is the state of the refrigerant as it enters the evaporator?

    A. Liquid   C. Solid

    B. Hot gas   D. Cold

8. How is air circulated through the evaporator?

    A. Convection air currents   C. Blower motor

    B. Conduction   D. Ram air

9. What will happen if moisture is present in an air-conditioning system?

    A. Moisture will help to keep   C. Moisture will mix with the compressor oil and cause harmful
       the compressor cool during   acids to form in the system.
       operation.
                                         D. Moisture will plug the filter drier, preventing the flow
    B. Moisture will perform much   of refrigerant.
       like the refrigerant in the
       system.

10. Two technicians are discussing the evacuation process. Technician A says that moisture may be removed from the air-conditioning system during the evacuation process. Technician B says that air may be removed from an air-conditioning system during the evacuation process. Which technician is correct?

    A. Technician A is correct.   C. Both Technicians A and B are correct.

    B. Technician B is correct.   D. Neither Technician A nor B is correct.

11. What is the difference between an orifice tube and a thermostatic expansion valve?

    A. Orifice tubes can't become   C. Thermostatic expansion valves have small moving parts, unlike
       plugged.   the orifice tube, which has no moving parts.

    B. Orifice tubes have small   D. Thermostatic expansion valves are always located between the
       moving parts, unlike   compressor discharge and the inlet to the condenser.
       thermostatic expansion
       valves.

12. Technician A says that antifreeze recycling machines are good for the environment because they can prevent the unlawful dumping of hazardous materials. Technician B says that the main ingredient in antifreeze (ethylene glycol) never wears out. Which technician is correct?

    A. Technician A is correct.   C. Both Technician A and B are correct.

    B. Technician B is correct.   D. Neither Technician A nor B is correct.

2 Environmental and Safety Practices

Learning Objectives

Upon completion and review of this chapter, the student should be able to:

- Describe the effects of CFCs (chlorofluorocarbons) on the world's ozone layer.
- Explain the Clean Air Act as described.
- Identify the pros and cons of CFCs, HCFCs (hydrochlorofluorocarbons), and HFCs (hydrofluorocarbons).
- Work safely with mobile air-conditioning and refrigeration equipment.
- Explain the differences between virgin, recycled, and recovered refrigerant.
- Explain what the greenhouse effect is and the impact it has on our environment.
- List the greenhouse gases.
- Explain the differences between disposable and refillable refrigeration containers.
- Discuss reasons for the manufacturing of alternate refrigerants.
- Understand the health hazards of working with refrigerants.
- Explain how refrigerants may produce poisonous gases.
- Comprehend any issues regarding the safe handling of refrigerant containers.
- List general workplace precautions for working with refrigerants.

Key Terms

alternative refrigerants

chlorofluorocarbons (CFCs)

Clean Air Act

disposable refrigerant cylinders

Environmental Protection Agency

greenhouse effect

hydrochlorofluorocarbons (HCFCs)

hydrofluorocarbons (HFCs)

ozone layer

recovered refrigerant

recycled refrigerant

refillable refrigerant cylinders

refrigerants

Section 609

stratospheric ozone depletion

ultraviolet

virgin refrigerant

INTRODUCTION

"No later than January 1, 1993, any person, repairing or servicing motor vehicle air conditioners shall certify, to the **Environmental Protection Agency (EPA)** that such person has acquired, and is properly using approved equipment, and that each individual authorized to use the equipment is properly trained and certified under **Section 609** of the **Clean Air Act**. In addition, only Section 609 Certified Motor Vehicle A/C technicians can purchase refrigerants in containers of 20 pounds or less."

SYSTEM OVERVIEW

The following chapter is written as a guide for technicians working in the air-conditioning and mobile refrigeration trades. It outlines some of the safe handling and environmental concerns in working with today's refrigerants. You will also find the EPA's penalty for undocumented refrigerant gas usage as it relates to ozone depleting substances.

> **CAUTION**   This chapter by no means exempts a technician from getting proper certification or licensing, as mandated by the EPA.

STRATOSPHERIC OZONE DEPLETION

The **ozone layer** is located in the stratosphere, high above the earth's surface, at an altitude of between 7 and 30 miles (11 and 48 kilometers) **(Figure 2-1).**

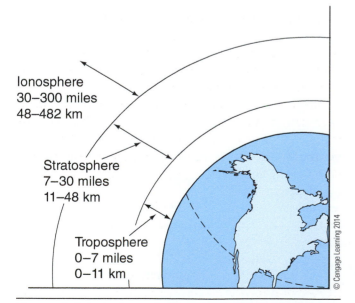

Figure 2-1 Location of the stratosphere, far above the earth's surface.

Ionosphere
30–300 miles
48–482 km

Stratosphere
7–30 miles
11–48 km

Troposphere
0–7 miles
0–11 km

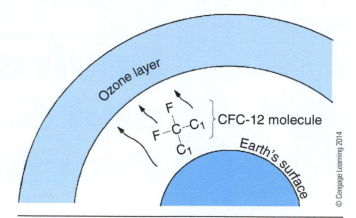

Figure 2-2 The CFC molecules gradually float up to the stratosphere, where they damage the ozone layer.

The ozone layer is formed by **ultraviolet** (UV) light from the sun acting on oxygen molecules. The ozone layer is often referred to as a protective layer because it absorbs and scatters ultraviolet light from the sun, preventing some of the harmful ultraviolet light from reaching the earth's surface **(Figure 2-2).**

Ozone is a gas with a slightly bluish color and a pungent odor. Ozone is a molecular form of oxygen that consists of three atoms of oxygen in each molecule; the oxygen we breathe contains two atoms in each molecule. Chemically, oxygen is O_2 and ozone is O_3.

Chlorine is a chemical that can deplete the ozone layer. **Chlorofluorocarbons (CFCs)** contain chlorine but not hydrogen and are so stable that they do not break down in the lower atmosphere even 100 years or more after being released. These chemicals gradually float up to the stratosphere, where the chlorine reacts with the ozone, causing it to change back into oxygen.

When the ozone layer decomposes, more UV radiation penetrates to the earth's surface **(Figure 2-3).**

The health and environmental concerns caused by the breakdown of the ozone layer include:

- Increase in skin cancers
- Suppression of the human immune response system
- Increase in cataracts
- Damage to crops
- Damage to aquatic organisms
- Increase in global warming

Stratospheric ozone depletion is a global concern. It will take the cooperation of many nations to bring this process under control, as CFCs and halons (chemicals in handheld fire extinguishers) are used by many nations. As a result, the release of these chemicals in one country could unfavorably affect the stratosphere

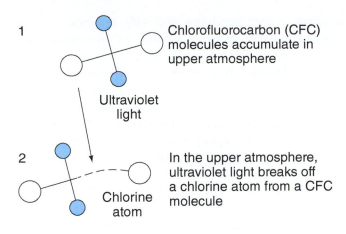

1. Chlorofluorocarbon (CFC) molecules accumulate in upper atmosphere

Ultraviolet light

2. In the upper atmosphere, ultraviolet light breaks off a chlorine atom from a CFC molecule

Chlorine atom

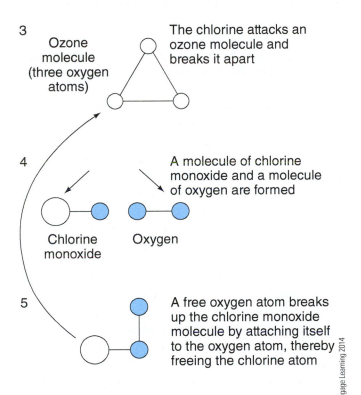

3. Ozone molecule (three oxygen atoms)

The chlorine attacks an ozone molecule and breaks it apart

4. A molecule of chlorine monoxide and a molecule of oxygen are formed

Chlorine monoxide   Oxygen

5. A free oxygen atom breaks up the chlorine monoxide molecule by attaching itself to the oxygen atom, thereby freeing the chlorine atom

6. The chlorine atom is free to repeat the process

© Cengage Learning 2014

Figure 2-3 How chlorofluorocarbons (CFCs) destroy the ozone layer.

above another country, and therefore, the health and welfare of its people.

The Montreal Protocol

In response to many nations recognizing the global nature of ozone depletion and deciding that something had to be done, the Montreal Protocol was established. On September 16, 1987, in Montreal, Canada, 24 nations and the European Economic Community (EEC) signed the Montreal Protocol on substances that deplete the ozone layer. Most of the nations that are

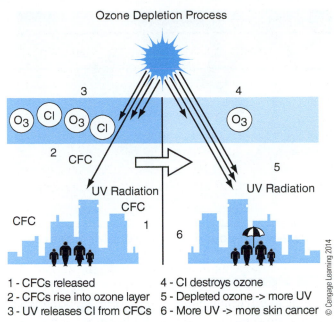

Ozone Depletion Process

1 - CFCs released
2 - CFCs rise into ozone layer
3 - UV releases Cl from CFCs
4 - Cl destroys ozone
5 - Depleted ozone -> more UV
6 - More UV -> more skin cancer

© Cengage Learning 2014

Figure 2-4 The ozone depletion process.

major producers and consumers of CFCs and halon signed the agreement. On August 1, 1988, the U.S. EPA put this agreement into regulations for the United States **(Figure 2-4).**

THE CLEAN AIR ACT

The Clean Air Act of November 15, 1990, directs the EPA to establish regulations to prevent the release of ozone-depleting substances. The Clean Air Act has many sections dealing with air quality and emissions; the sections dealing with ozone depletion are Sections 608 and 609.

Section 608 pertains to stationary air-conditioning equipment, mobile refrigeration equipment, and air-conditioning equipment that uses R-22, which is commonly used in bus air-conditioning systems as well as in truck/trailer refrigeration equipment.

Section 609 deals with the mobile motor vehicle open-driven air-conditioning industry. The sale of refrigerant containers weighing less than 20 pounds, including 1-pound cans, is **restricted to technicians certified in Section 609 (Figure 2-5).**

The purpose of Section 609 is to teach and test a technician's ability to properly handle and recover refrigerants. Technicians are also trained in the laws enacted to protect the stratospheric ozone layer.

Technicians working with HFC-134a mobile vehicle air conditioning must be trained and certified by an EPA-approved organization. Technicians already certified to handle CFC-12 are not required to recertify in order to work with HFC-134a.

Figure 2-5 One-pound disposable refrigerant can.

EPA Penalties

(The following was taken directly from the EPA.)

The EPA Enforcement Office will issue fines for undocumented refrigerant gas usage as it relates to ozone-depleting substances.

The phase-out program of refrigerant gas is now in full swing and facilities that use equipment requiring the use of ozone-depleting substances (ODS) are at risk for a substantial EPA penalty if they fail to follow the requirements outlined in the U.S. Clean Air Act related to data management and usage reporting. Equipment that must be tracked includes refrigeration and air-conditioning systems, commercial refrigeration, heating, ventilation and air conditioning systems, and fire protection systems.

To avoid an EPA penalty, companies, municipalities and property managers that utilize refrigerant equipment must monitor its usage and submit documentation outlining refrigerant management efforts. Those who fail to do so face substantial fines. As such, many facilities are relying on refrigerant tracking and reporting programs that automatically manage their use of refrigerant, identify leaks, track repairs, and guide in proper disposal. This allows them to keep current with government policies, compliance requirements, and penalties for non-compliance.

Overseeing the EPA penalty aspect of the U.S. Clean Air Act is the Office of Enforcement and Compliance Assurance. They are aggressively pursuing enforcement of the requirements to curb harmful gas emissions. Auditors and inspectors are permitted to make spot inspections to review a facility's records pertaining to regulated gases. Those unable to produce proper documentation, or those who have incomplete or missing data are subject to heavy fines.

The EPA penalty applies to facilities that improperly emit, vent, or dispose of refrigerant gas. The law requires proper servicing and safe removal of any equipment using restricted substances. Noncompliance could result in fines of up to $25,000 a day, per violation. Additional fines are added if the refrigerant gas is not properly recovered. Because of the substantial penalties involved, the government has developed a technician certification program for anyone who provides service, repair, maintenance, or disposal of equipment containing refrigerant gases.

Under the EPA penalty guidelines, refrigerant leaks not fixed within 30 days are subject to a $32,500 fine per day, per unit. Furthermore, purchasing used or imported refrigerant gas calls for fines of $300,000 per 30-pound cylinder of refrigerant gas. With so much money at stake, it is crucial for business entities, organizations, and municipalities to track every pound of gas and manage its inventory, especially those with more than one location where records management and ease of reporting becomes difficult.

Any amendments to the environmental laws usually allow for a period when comments are accepted and updates to the regulations are proposed and implemented. Currently, the EPA and other governmental agencies are taking civil and criminal actions against companies nationwide who violate the law. The total of fines collected is in the billions of dollars each year. With the added incentives related to carbon emissions management and the world's heightened awareness of climate change it is fully anticipated that more stringent and more restrictive measures will be placed on all substances that harm the environment.

Because refrigerant gas contains chlorofluorocarbons and hydrochloro-fluorocarbons, identified as the major causes of ozone depletion, its use is being reduced, and eventually eliminated, worldwide. The plan reduces the use of R-22 refrigerant gas by 75% by 2010 and eliminates it by 2015. The EPA penalty increases as the complete phase-out comes to a close.

EPA penalty policies are based on the guidelines established by the U.S. Clean Air Act, and its international counterparts, the Montreal Protocol and the Kyoto Protocol, to control the use of refrigerant gas as a means of reducing the damaging impact it has on the ozone layer and lowering the potential for global warming. Penalties are being issued to protect the environment and to encourage facilities to improve their carbon footprint.

GREENHOUSE EFFECT

The **greenhouse effect** is a naturally occurring process that helps to heat the earth's surface and atmosphere. The earth absorbs incoming solar radiation

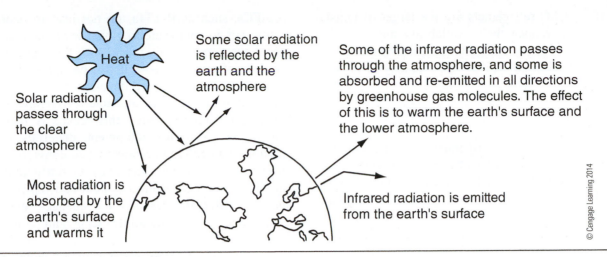

Figure 2-6 The greenhouse effect.

and then cools by emitting long-wavelength infrared radiation. This radiation is absorbed by greenhouse gases, thereby preventing the heat from escaping. The increase in greenhouse gases may increase average global temperature. Without the greenhouse effect, life on earth might not exist because the average temperature would be $-0.4°F$ ($-18°C$) instead of $59°F$ ($15°C$), as we presently know it.

The greenhouse effect causes the atmosphere to trap more heat energy at the earth's surface and within the atmosphere by absorbing and re-emitting long-wavelength energy. Of the long-wavelength energy emitted back to space, 90% is intercepted and absorbed by greenhouse gases. In the last few centuries, the activities of human beings have directly or indirectly caused the concentration of the major greenhouse gases to increase. Scientists predict that the greenhouse effect will cause the planet to become warmer. Experts estimate that the average global temperature has already increased by 0.3 to 0.6 degrees Celsius since the beginning of this century. Predictions about the climate are that by the middle of the next century, the earth's global temperature may be 1 to 3 degrees Celsius higher than it is today.

For an animated view of the greenhouse effect, visit the following Web site: http://earthguide.ucsd.edu/earthguide/diagrams/greenhouse/.

Greenhouse Gases

The following is a list of some of the major greenhouse gases and their sources:

- CO_2 burning of carbon-based fuels
- CH_4 anaerobic bacteria in rice fields, cows, sewage
- N_2O fossil fuels and fertilizer
- CFCs   refrigeration and spray cans

CFCs are extremely harmful and will dominate the greenhouse gases in our atmosphere if their global usage remains high. However, worldwide production of CFC has been significantly reduced because of concern about the ozone layer **(Figure 2-6).**

Methane (CH_4) is another concern. Methane is more directly related to food production and population growth; therefore, it could also dominate the greenhouse gases in the near future.

Frozen methane is also found in the Arctic ice cap and will be released due to global warming, thus aggravating the problem. This source of methane is a far more serious condition than most people realize or has been reported.

REFRIGERANTS

Most of the **refrigerants** in use today are compounds containing carbon, fluorine, usually chlorine, and sometimes hydrogen, bromine, or iodine. Refrigerants used in motor vehicle air-conditioning systems may be referred to as CFCs, HCFCs, or HFCs. A refrigerant referred to as a CFC contains chlorine, fluorine, and carbon. A refrigerant referred to as an HCFC contains hydrogen, chlorine, fluorine, and carbon. A refrigerant identified as an HFC contains hydrogen, fluorine, and carbon.

CFCs

Because CFCs contain no hydrogen, they are chemically very stable, even if released into the atmosphere. CFCs contain chlorine; they are very damaging to the ozone layer, a protective layer far above the earth's surface.

These two characteristics give CFC refrigerants a high ozone-depletion potential, or ODP. Due to this side effect, CFC refrigerants are the target of legislation that will reduce their availability and use. The manufacture of these refrigerants was discontinued as of January 1, 1996. R-12 is a CFC and may be referred to as CFC-12.

HCFCs

Another category of refrigerants currently available is the category of **hydrochlorofluorocarbons (HCFCs)**. As mentioned earlier, these refrigerants contain chlorine, which is damaging to the ozone layer, but they also contain hydrogen, which makes them less chemically stable once they are released into the atmosphere. These refrigerants decompose in the lower atmosphere so that very little chlorine ever reaches the ozone layer. These refrigerants have a lower ozone-depletion potential (ODP). HCFC-22, also referred to as R-22, is used extensively in commercial air conditioning, transport refrigeration equipment, home air conditioners, refrigerators, freezers, and dehumidifiers. Any new equipment today is not designed to operate with R-22. This refrigerant is being phased out over a period of time.

Additional restrictions in 2020 will end the production and importing of R-22. You will still be able to find supplies of recovered and recycled R-22, but no new additions to the stock will be available.

HFCs

Hydrofluorocarbon (HFC) refrigerants have largely replaced CFC-12 in the automotive field. These refrigerants contain no chlorine and have an ozone-depletion potential of zero. However, these refrigerants are considered greenhouse gases and probably contribute to global warming. The refrigerant used in most automotive and truck trailer applications to replace CFC-12 is HFC-134a. In addition, the truck trailer industry also uses R404A.

Alternative Refrigerants

Due to public awareness concerning the depletion of the earth's protective ozone layer, the use of CFC refrigerants (R-12) was to be phased out by January 1, 1996. This meant that **alternative refrigerants** would have to take the place of the CFC refrigerants.

Considerations for any new refrigerant are chemical stability in the system, toxicity, flammability, thermal characteristics, efficiency, ease of detection in the event of leaks, environmental effects, compatibility with system materials, compatibility with lubricants, and cost. In general, HFC-134a has replaced R-12 in all truck/automotive applications.

HFCs, such as R-134a, do not lead to ozone depletion but do contribute to global warming due to the greenhouse effect. This means that the recycling and recovery of refrigerants will still be required, regardless of the new refrigerant development.

There are no simple substitute refrigerants for any equipment category. What this means is that some changes in a system's equipment, materials, or construction are always necessary when converting to an alternative refrigerant. The existing refrigerant can't simply be removed from a system and replaced with another refrigerant. Usually the changes amount to replacement of incompatible seals and compressor lubricants.

HFC-134a still carries some concerns about compatible lubricants. Lubricants typically used with CFC-12 do not mix with HFC-134a. Polyalkylene glycols (PAGs) mix properly with R-134a at low temperatures but have upper temperature problems, as well as incompatibility with aluminum bearings and polyester hermetic motor insulation. Ester-based synthetic (POE) lubricants for HFC-134a resolve these problems, but are incompatible with existing PAG or mineral oils. POE oils are incompatible with as little as 1% residual oil (PAG or traditional mineral) in the system.

In operation, HFC-134a is very similar to CFC-12. With proper equipment redesign, efficiencies are similar. In automotive applications, capacity suffers only minor reductions.

DISPOSABLE REFRIGERANT CYLINDERS

Disposable refrigerant cylinders are commonly used in the automotive air-conditioning market. These cylinders are available in refrigerant quantities of between 1 and 50 pounds. Intended to be used only once, these cylinders are not refillable with refrigerant or any other product. Refrigerant cylinders are usually color-coded to identify the type of refrigerant they contain. Thus, if labels are torn off, the contained refrigerant can still be identified. The color code for containers filled with R-134a is light blue. Be sure to follow EPA regulations in your area for safe legal disposal of these cylinders. Disposable cylinders are equipped with a safety relief valve to prevent overpressurizing of the cylinder. This can happen if the cylinder is subject to excessive heat. When the liquid refrigerant is subjected to heat, it expands into the vapor sitting on top of the liquid. This will cause the pressure of the cylinder to rise as long as space is available for expansion. If no space is available for the liquid refrigerant to expand, as with an overfilled cylinder, and if the cylinder is not

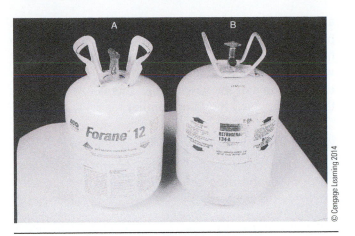

Figure 2-7 Disposable refrigerant cylinders.

equipped with a pressure relief valve, the cylinder can rupture violently, possibly causing serious bodily injury or death.

Disposable cylinders are made from steel, which will eventually rust, weakening the cylinder to the point where the walls of the container can no longer contain the compressed refrigerant. Cylinders should be stored and transported in a dry environment to prevent rusting. Never leave disposable refrigerant containers that contain residual refrigerant outdoors where they will rust. The internal pressure of a cylinder containing 1 ounce of liquid refrigerant and one that is full of refrigerant is the same. If containers are left outdoors, they will eventually deteriorate and may explode. Disposable refrigerant containers that are very rusty should have the contents removed and be properly discarded. Before a cylinder can be disposed of, it must be completely emptied. This means that the cylinder should be put on a refrigerant recovery machine and the cylinder's contents drawn into a vacuum. At this point, the cylinder may be disabled by being punctured or by having the valve disabled. This ensures that the container cannot be reused for purposes for which it was never intended. Follow EPA regulations for proper disposal of cylinders in your area **(Figure 2-7)**.

REFILLABLE CYLINDERS

Refillable refrigerant cylinders are also commonly used in the air-conditioning/refrigeration industry. These cylinders contain the same refrigerants that may also be available in disposable cylinders. Like the disposable cylinders, refillable cylinders are regulated in their design, fabrication, and testing by DOT for use in the transportation of refrigerants.

Refillable cylinders may also be marked with a color to indicate which refrigerant they contain. This type of container may be used with refrigerant recycling equipment. Cross contamination can become a problem if cylinders are interchanged, so identification distinguishing between **virgin, recycled** (removed, filtered, and oil separated), and **recovered** (dirty) refrigerant should be clearly marked on the cylinder by the technician. This is important because if a technician recovers refrigerant from a piece of equipment that has suffered a catastrophic failure, contaminants will be present in the refrigerant and also in the refrigerant oil. If new, clean refrigerant is added to this cylinder, it will also become contaminated. Due to the rising cost of refrigerants, proper recycling and storage of the refrigerant for reuse in other equipment will very quickly save large amounts of money for the shop owner.

Refillable cylinders contain a warning decal to caution the user against physical contact with or exposure to the refrigerant, as well as a warning about the working pressure of the cylinder.

Refillable refrigerant containers must be retested at 5-year intervals. This is because corrosion of the cylinder exterior or damage caused by mishandling cannot always be avoided. The valves of these tanks should be checked occasionally, especially the relief valve. Check that nothing is blocking the relief valve and that no visible deterioration or damage has occurred. If damage is visible, empty the cylinder and have the cylinder repaired **(Figure 2-8)**.

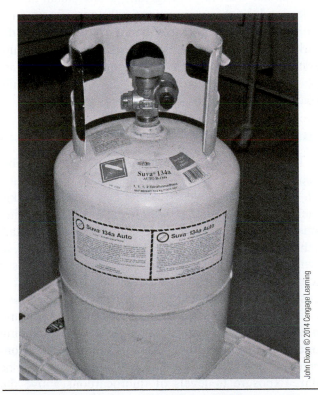

Figure 2-8 Refillable refrigerant cylinder.

> **CAUTION** *Never under any circumstance use a cylinder with a faulty pressure relief valve or a cylinder with an obvious structural problem.*

Cylinder Color Code

Refrigerant cylinders may be color-coded to indicate the type of refrigerant they contain. When a refillable refrigerant cylinder is refilled, it is critical that the refrigerant being installed is the correct type for that cylinder. Otherwise, it is very probable that another technician may install the wrong refrigerant for an intended application.

Some of the colors used for cylinder identification are:

- Refrigerant 12   White
- Refrigerant 22   Green
- Refrigerant 134a   Light Blue
- Refrigerant 502   Light Purple
- Refrigerant 401B   Yellow/Brown
- Refrigerant 402A   Light Brown
- Refrigerant 404A   Orange

REFRIGERANT SAFETY

Any technician working in the air-conditioning field should understand general safety considerations concerning fluorocarbon refrigerants. Before coming into contact with any refrigerant, a technician should understand the safety concerns for the specific product. With the new refrigerants being introduced to the marketplace, safety becomes an even bigger issue because testing is not yet complete. Long-term health effects are not yet known. Manufacturers can provide specific product information for the refrigerants they produce.

Health Hazards

Technicians working with fluorocarbon refrigerants should always do so while wearing safety goggles and non-leather work gloves. (Leather can stick to the skin when it comes in contact with refrigerant.) Skin irritations or frostbite may occur if skin comes into contact with refrigerants. Refrigerant vapor is heavier than air, so it will drop to the floor of a shop. In extreme instances, a technician working in a pit could be asphyxiated because the refrigerant will displace the oxygenated air.

Shops where work is done with refrigerants should be well ventilated to prevent technicians from inhaling large concentrations of refrigerants. Inhalation of concentrated refrigerant vapor is dangerous and can be fatal. Exposure to levels of fluorocarbons above the maximum recommended level can result in drowsiness

and loss of concentration. Cases of fatal heart attacks have been reported in people accidentally exposed to high levels of refrigerants. Technicians experienced at working with refrigerants should note that exposure time for some of the new refrigerants is lower than for those with which they are familiar.

> **CAUTION** *Refrigerant should NEVER come into contact with skin or eyes. Liquid refrigerant quickly evaporates when exposed to the air and will almost instantly freeze skin or eye tissue, so serious injury or blindness could result from contact with liquid refrigerant.*

First Aid

A victim who has inhaled refrigerant should be removed to a place where the air is fresh. If the victim is not breathing, call 911 and administer artificial respiration at once. If the victim experiences labored breathing, oxygen should be administered. The victim should avoid all stimulants/depressants, such as caffeine, tobacco, and alcohol. Get the patient to a doctor for continued treatment. If a person's eyes come in contact with liquid refrigerant, the eyes should be flushed with water for a minimum of 15 minutes, and again, it is important to get the person to a doctor immediately. If skin comes in contact with liquid refrigerant, it should be flushed with warm water to warm the skin gradually **(Figure 2-9)**.

Poisonous Gas

Most halogenated compounds decompose, producing toxic vapor when they come in contact with high temperatures such as those produced by an open flame. Hydrofluoric acid will result under these conditions. If the compound contains chlorine, hydrochloric acid will also be formed, and if a source of water (or oxygen) is present, a small amount of phosgene gas will be formed. Halogen gases are easily identified by a very sharp/sour smell. This odor can be detected at concentrations below toxic levels, and when these odors occur, the area should be evacuated and opened up to allow fresh air to dissipate the gas.

GENERAL WORKPLACE PRECAUTIONS

- Read all product labels and material safety data sheets (MSDS).
- Always work in a well-ventilated area. Inhaling refrigerant, even in small amounts, can cause

Figure 2-9 Eye wash station.

light-headedness. Refrigerants can also cause irritation to the eyes, nose, and throat.

- Never allow refrigerant vapors to come in contact with open flames, sparks, or hot surfaces.
- Do not weld or steam clean an air-conditioning system because excessive pressure could build up in the system.
- When testing for leaks, do not mix R-134a with air because the mixture could explode when under pressure.
- Do not recover or transfer refrigerant into a disposable cylinder. Always use a DOT approved cylinder.
- Do not transport refrigerant in the passenger compartment of a vehicle.
- Never expose refrigerant to open flames, high temperatures, or direct sunlight.
- Technicians who work with refrigerants should wear safety goggles, non-leather gloves, and work clothing suitable for the job or as required by the employer.
- Ensure that showers and eye wash fountains of the deluge-type are readily accessible in case of refrigerant contact with eyes or skin.

- In readily accessible areas, store breathing apparatus in case an abnormally high concentration of refrigerant vapor should develop in the storage, handling, or production areas.

HANDLING REFRIGERANT CYLINDERS

- Keep cylinders secured in an upright position.
- Store refrigerants out of direct sunlight in a clean, dry area.
- Refrigerant cylinders should not be permitted to attain temperatures above 125°F (51.6°C). Do not store refrigerant containers in a vehicle. On hot sunny days, the temperature in a parked vehicle can reach 150°F (65°C).
- Keep the outlet port of the refrigerant cylinder capped and the valve hood securely screwed onto the neck of the cylinder at all times when it is not being used.
- Never apply an external source of heat to a refrigerant cylinder.
- Never drop or strike a refrigerant cylinder with any object.
- Never lift a refrigerant cylinder by its valve or valve cover.
- Never try to repair the valve.
- Never tamper with the safety device.
- Avoid exposure to vapors through spills or leaks.
- Evacuate the area if a large refrigerant spill occurs. Return only after the area has been properly ventilated.
- Check to ensure that the cylinder label matches the color code.
- Open cylinder valves slowly, and close after every use.
- Always ventilate the work area before using open flames.
- Avoid contact with liquid refrigerant because frostbite may occur.
- Do not attempt to refill disposable cylinders.
- Do not remove or alter cylinder identification markings.
- Be careful not to cut or gouge the cylinder.
- Keep cylinders away from corrosive materials and water.
- Don't use cylinders that are extensively rusted or otherwise deteriorated. Ensure that tanks are disposed of properly.
- Never leave an empty reusable cylinder with its valve open to the atmosphere because moisture will enter the cylinder and rust will occur.

Summary

- Any person repairing or servicing motor vehicle air conditioners shall be certified by the Environmental Protection Agency.

- The ozone layer is a protective layer in the atmosphere that absorbs and scatters ultraviolet light from the sun, preventing some harmful ultraviolet light from reaching the earth's surface.

- With the ozone layer decomposing, more UV radiation penetrates to the earth's surface.

- The Clean Air Act directs the EPA to establish regulations to prevent the release of ozone-depleting substances.

- The increase in greenhouse gases may increase average global temperature.

- Refrigerants referred to as CFCs contain chlorine, fluorine, and carbon.

- Refrigerants referred to as HCFCs contain hydrogen, chlorine, fluorine, and carbon.

- Refrigerants referred to as HFCs contain hydrogen, fluorine, and carbon.

- HFCs, such as R-134a, are not ozone-depleting but contribute to global warming because HFC is considered to be a greenhouse gas.

- Disposable cylinders are intended to be used only once and are not refillable with refrigerant or any other product.

- Refillable refrigerant cylinders may be used with refrigerant recycling equipment. Cross contamination can become a problem if cylinders become interchanged.

- Refrigerant cylinders may be color-coded to indicate the type of refrigerant they contain.

- If a person has inhaled refrigerant, he or she should be moved to an area where there is fresh air. If the person is not breathing, call 911 and start artificial respiration.

- Poisonous gas may be formed when refrigerant comes in contact with a heat source.

Online Research Tasks

For more information, check the following Web site:

http://www.epa.gov/ozone/strathome.html

Review Questions

1. What safety equipment must you always wear while working with refrigerants?

   A. Safety goggles/glasses and leather gloves

   B. Insulated coveralls and gloves

   C. Safety goggles/glasses and non-leather gloves

   D. Safety equipment is not usually required in working with refrigerants.

2. Never expose refrigerant storage containers to temperatures higher than what temperature?

   A. 0°F (−18°C)

   B. 32°F (0°C)

   C. 100°F (38°C)

   D. 125°F (52°C)

3. Technician A wears safety goggles and rubber gloves while discharging the A/C system. Technician B wears safety goggles and leather gloves when charging and leak testing the system. Which technician is correct?

   A. Technician A is correct.

   B. Technician B is correct.

   C. Both Technicians A and B are correct.

   D. Neither Technician A nor B is correct.

4. Which of the following are chlorine-free HFC refrigerants?

   A. R-12 and R404A

   B. R-12 and R-134a

   C. R-134a and R404A

   D. R502 and R404A

5. Which of the following gases is responsible for the destruction of the ozone layer?

   A. Chlorine

   B. Fluorine

   C. Hydrogen

   D. Argon

6. The Clean Air Act was signed into law on:

   A. January 1, 1990

   B. November 15, 1990

   C. January 1, 1995

   D. November 15, 1995

7. Technician A states that the lack of ozone contributes to the greenhouse effect. Technician B says that thinning of the ozone layer is responsible for excessive UV radiation. Which technician is correct?

   A. Technician A is correct.

   B. Technician B is correct.

   C. Both Technicians A and B are correct.

   D. Neither Technician A nor B is correct.

8. Technician A says that refrigerant gases that come in contact with a heated surface may produce a toxic gas. Technician B says that refrigerant gases that come in contact with a heated surface will cause the refrigerant to decompose. Which technician is correct?

   A. Technician A is correct.

   B. Technician B is correct.

   C. Both Technicians A and B are correct.

   D. Neither Technician A nor B is correct.

9. Technician A and B are talking about ozone. Technician A says that ozone is pale green in color. Technician B says that ozone is a form of oxygen. Which technician is correct?

   A. Technician A is correct.

   B. Technician B is correct.

   C. Both Technicians A and B are correct.

   D. Neither Technician A nor B is correct.

10. Technician A says that the ozone layer absorbs some of the sun's ultraviolet radiation. Technician B says that ultraviolet radiation can be hazardous to human health. Which technician is correct?

    A. Technician A is correct.

    B. Technician B is correct.

    C. Both Technicians A and B are correct.

    D. Neither Technician A nor B is correct.

3 Thermodynamics

Learning Objectives

Upon completion and review of this chapter, the student should be able to:

- Describe how heat is transferred.
- Explain thermal equilibrium.
- Define a British thermal unit (BTU).
- Explain the difference between sensible and latent heat.
- Describe the three changes of state in which a substance may be formed.
- Describe the differences between latent heat of vaporization and latent heat of fusion.
- Explain the term *superheating*.
- Explain the term *subcooling*.
- Describe the effects of pressure on boiling point.
- Define atmospheric pressure.
- Describe what a vacuum is and how it is measured.
- Define the term *humidity* and explain its effects upon an air-conditioning/refrigeration system.

Key Terms

absolute zero
atmospheric pressure
boiling point
British thermal unit
change of state
condensation
conduction
convection

evaporation
freezing
heat transfer
humidity
latent heat
latent heat of condensation
latent heat of fusion
latent heat of vaporization

pressure
radiation
sensible heat
subcooling
superheat
temperature
thermal equilibrium
vacuum

INTRODUCTION

The terms *air conditioning* and *refrigeration* may be used to describe virtually the same process; these processes use the same components to perform the task of maintaining the temperature of a controlled space. The difference between the two systems is that air conditioners are not used to maintain low temperatures. If the temperatures inside a vehicle were to be kept at the temperatures of, for example, a meat locker (35°F/2°C), the occupants of the vehicle would not be comfortable. Therefore, an air-conditioning system is designed to maintain temperatures that are more suited to the average person's comfort zone (70°F to 80°F/21°C to 27°C).

When most people think of air conditioning, they think of a fan that blows cold air out of the vents on the dash of their vehicle or out of the vents in their home. This is what air conditioning feels like to the human body. Very few people understand how air conditioning really functions. However, after examining the thermodynamics of air conditioning, you will discover that there is actually something quite different going on within the system.

SYSTEM OVERVIEW

Before technicians can perform any service work or begin to diagnose problems in air-conditioning or refrigeration systems, it is essential that they have a basic understanding of thermodynamics. These principles can be confusing because we normally think of boiling temperatures as extremely hot to the touch. But when we talk about refrigerants boiling, we are talking about a substance that can boil at –20°F (–28.8°C). Whenever a person learns something new, it is very important that they build a firm foundation on the basics. In the field of air-conditioning service and repair, learning the basics includes understanding the principles behind how an air-conditioning system works. Understanding these principles will help a technician improve diagnostic skills. The more you know about the air-conditioning system, the easier it will be to identify problems within the system. The principles that will be discussed in this chapter are the basis for any air-conditioning or refrigeration system.

HEAT

The word *cold* is often used to describe the temperature of a substance or of the ambient air. Theoretically, the lowest temperature obtainable is –459°F (–273°C). This is known in scientific terms as **absolute zero** (no one has yet reached this temperature). To date, the lowest temperature achieved is –457°F (–272°C). Cold is a term used to indicate an absence of heat. Molecular action slows down as a substance becomes cooler **(Figure 3-1)**. When all molecular motion is stopped, it is considered to be at absolute zero. Anything warmer than –459°F (–273°C) contains heat. Cold, as the term is used, does not really exist because everything contains some heat (except at absolute 0). Cold can be defined in a negative way only by saying that cold is the absence of heat. When heat is removed

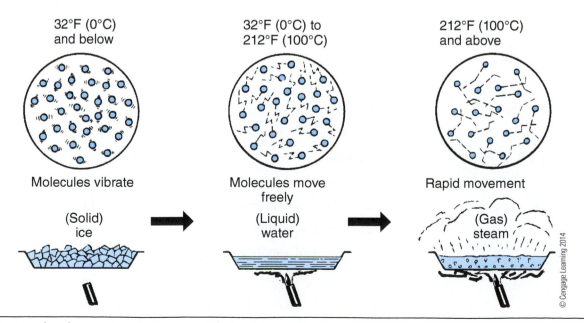

32°F (0°C) and below — Molecules vibrate — (Solid) ice

32°F (0°C) to 212°F (100°C) — Molecules move freely — (Liquid) water

212°F (100°C) and above — Rapid movement — (Gas) steam

© Cengage Learning 2014

Figure 3-1 Molecular motion increases with heat intensity.

from a substance, that substance becomes cold as a result. Heat is often defined as molecules in motion. If the temperature of a substance increases, so does the molecular motion within the substance. Likewise, as the temperature decreases, molecular motion is decreased.

Heat Transfer

The basis of all air-conditioning or refrigeration equipment is that heat flows from a warmer object to a colder object. When an object is being cooled, it is because the heat of the object is being transferred to another object. Just as water always runs downhill, heat always flows from a warm object to a colder object **(Figure 3-2)**.

Heat transfer happens in one of three ways:

Conduction is heat flow through a solid object. Conduction transfers heat from molecule to molecule through a substance by chain collision. An example of this type of heat transfer is what happens when the end of a metal bar is heated and the heat travels through the bar from one end to the other.

Convection is the transfer of heat by a flowing substance. Because solids don't flow, convection heat transfer occurs only in liquids and gases. For example, when a gas or liquid is heated it warms, expands, and will rise because it is less dense. As the gas or liquid cools, it will again become more dense and fall. As the gas or liquid warms and rises, or cools and falls, it creates a convection current. Convection is the primary method by which heat moves through gases and liquids.

Radiation is the transfer of heat from a source to an absorbent surface by passing through a medium (air) that is not heated. The efficiency with which an object radiates or absorbs radiant heat depends on the color of the surface. Lighter colors reflect more heat. Darker colors absorb more heat and, when warm, they are excellent radiators of energy. An example of this type of heat transfer is the sun heating up the interior of a car. The temperature of the space between the sun and the car can be very much cooler than the interior of the car because the car is absorbing the sun's heat.

Thermal Equilibrium

If two objects are placed together, and neither one of them undergoes any temperature change, then they are said to be at the same temperature. If the objects are not at the same temperature, the hot one gets cooler and the cold one gets warmer until both of them stop changing temperature. When the heat intensity is equal and no further change takes place, the objects are said to be in a state of **thermal equilibrium** (heat balance). For example, if we place a cup of hot coffee in a room, the heat from the coffee will transfer to the cooler ambient air of the room. In time, the temperature of the coffee reaches the temperature of the room and no further change in temperature takes place. The coffee and the air in the room are now in a state of thermal equilibrium. The same thing happens to a cold can of soda if it is forgotten on the kitchen table. The heat energy of the warmer air in the room transfers to the cooler can and its contents until both are the same temperature. When there is no more heat transfer between the room temperature and the can of soda they too are said to be in a state of equilibrium.

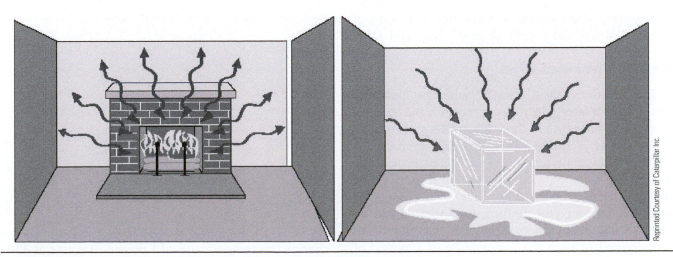

Reprinted Courtesy of Caterpillar Inc.

Figure 3-2 The direction in which heat flows is always from warm to cold.

Rate of Heat Transfer

The speed of heat transfer from one object to another is affected by the temperature difference between the two objects. Heat energy travels fast when there is a great temperature difference between two objects. For example, if you take a piece of meat out of the freezer, it will thaw out fairly quickly when immersed in a sink of hot water. If the meat were allowed to thaw at room temperature, it would take longer. And if the meat were allowed to thaw in the refrigerator, it would take even longer.

TEMPERATURE

Temperature is a term that we use almost every day of our lives to describe the weather or any of the objects around us. Many people cannot actually define the word temperature, even though they use it often.

Temperature indicates the average velocity (movement) of the molecules of a substance. When the heat energy in a substance increases, its molecules vibrate more intensely. A thermometer measures the intensity of this vibration. Temperature is not a measurement of heat energy. Temperature is a measure of the intensity of heat energy in a substance.

Temperature Scales

The Fahrenheit scale, in the customary system, divides the difference from the freezing point of water to the boiling point of water into 180 equal divisions, each division being one degree. The Fahrenheit scale sets the freezing point of water at 32°F and the boiling point of water at 212°F.

The Celsius or centigrade scale in the metric system divides the difference from the freezing point of water to the boiling point of water into 100 equal divisions. The Celsius scale sets the freezing point of water at 0°C and the boiling point of water at 100°C.

British Thermal Unit

The quantity of heat transfer cannot be measured with a thermometer. The measurement for the quantity of heat in a substance is the British thermal unit, or BTU for short. You may have noticed the abbreviation BTU used on various appliances such as barbeques and small window air conditioners. Most people would tend to believe that the higher the BTU rating, the better the product will perform, and they would be correct in this assumption. A BTU is the quantity of heat required to raise the temperature of 1 pound of water by 1°F. The metric equivalent of the BTU is the

kilocalorie (kcal). One kcal is the amount of energy required to raise the temperature of 1 kilogram of water by 1°C.

Example 1

You have 1 pound of water at 32°F. How many BTUs would be required to bring the water to the boiling point?

$$212°F - 32°F = 180°F$$
$$BTU = 1 \text{ pound} \times 180°$$
$$= 1 \times 180$$
$$= 180 \text{ BTUs}$$

Answer: It would take 180 BTUs to bring 1 pound of water at 32°F to the boiling point at sea level **(Figure 3-3)**.

Example 2

You have 15 pounds of water at 70°F and you want to cool the water to 35°F. How many BTUs of energy would be required?

$$70°F - 35°F = 35°F$$
$$BTU = 15 \text{ pounds} \times 35°$$
$$= 15 \times 35$$
$$= 525 \text{ BTUs}$$

Answer: It would take 525 BTUs of energy to cool 15 pounds of water from 70°F to 35°F.

These formulas show how to convert between English and metric units of heat:

$$\underline{\hspace{1cm}}\textbf{Watts} \times 3.409 = \underline{\hspace{1cm}}\textbf{BTUs per hour}$$
$$\underline{\hspace{1cm}}\textbf{BTUs} \times 0.252 = \underline{\hspace{1cm}}\textbf{kilocalories (kcal)}$$
$$\underline{\hspace{1cm}}\textbf{BTUs} \div 3.412 = \underline{\hspace{1cm}}\textbf{watts}$$

In reference to an air-conditioning system capacity, smaller systems are rated in BTUs and larger systems are rated in tons of refrigeration capacity. This refers to

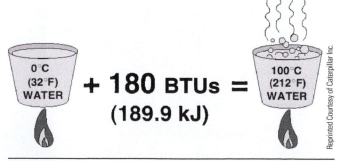

Figure 3-3 It requires 180 BTUs of heat energy to change the temperature of 1 pound of water at 32°F to 212°F.

the amount of cooling required to change 1 ton (2000 pounds) of water to 1 ton of ice in 24 hours.

This is calculated by using the following method:

1 ton = 2000 pounds

1 pound of water to 1 pound of ice requires 144 BTUs

$$\begin{aligned}\textbf{1 ton of refrigeration} &= \textbf{2000} \times \textbf{144}\\ &= \textbf{288,000 BTUs}\end{aligned}$$

This example would be based on a 24-hour period. Normally we speak of a 1-hour duration and this then would be 12,000 BTUs.

Most truck air-conditioning systems are in the range of one and three-quarters to three and a quarter tons of capacity. This is more than most two-story home central air-conditioning units, which usually range from one and one-half to two tons of capacity.

Types of Heat

Sensible Heat. This type of heat causes a change of temperature in a substance. This is a kind of heat that can be felt, such as the temperature of the ambient air. **Sensible heat** can be measured with a thermometer.

Latent Heat. This is the heat required to change the state of matter from either a liquid to a vapor or a liquid to a solid. **Latent heat** is hidden heat. It cannot be felt nor be measured with a thermometer. A way of demonstrating latent heat simply would be to put a thermometer into a pot of water. As heat is applied to the pot, the water gets warmer and the change in

temperature can be seen on the thermometer. When the water reaches its boiling point of 212°F (100°C), the water will start to change states. The whole time the water is boiling (changing states), the temperature of the thermometer will not rise above 212°F (100°C). This is latent heat, the change of state of a substance without a change in temperature.

Latent and sensible heats are the two most important types of heat with respect to refrigeration.

Superheat

Whenever the steam in a closed vessel is raised above the temperature at which it changes states, it is considered to be **superheated**. As mentioned earlier in this chapter, the water must be under pressure. For example, if the water in a radiator is under pressure and the temperature of the water is 214°F (101°C), then the water is considered to be 2°F (−17°C) above its boiling point. This water would be considered superheated.

Subcooling

Subcooling is a term used to describe the temperature of a liquid. A liquid that is at a temperature below its boiling point is considered to be subcooled. For example, if water in a pot is allowed to cool after it has boiled, it would be subcooled. If the water was 210°F (99°C), it would be considered to be 2°F (−17°C) subcooled.

The graph in **Figure 3-4** illustrates how heat energy is used to convert 1 pound of ice at 0°F (−17.7°C)

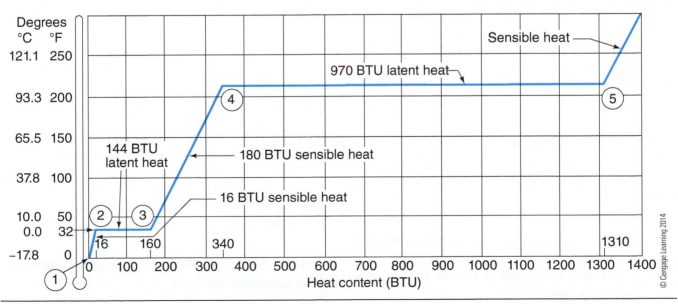

Figure 3-4 This graph shows how heat energy is used to convert 1 pound of ice at 0°F (−17.8°C) to steam at 212°F (100°C).

to steam at 212°F (100°C). The graph starts with a 1-pound (0.45 kilograms) piece of ice at 0°F (−17.7°C). Sixteen BTUs of sensible heat are added to the ice, which will cause the ice to change temperature.

Ice has a specific heat of "0.5"; therefore, it will take 16 BTUs of energy to raise the temperature from 0°F to 32°F (−17.7°C to 0°C).

$$0.5 \times 32 = 16 \text{ BTUs}$$

Once the ice reaches 32°F (0°C), it will begin to melt. It will require the addition of 144 BTUs of heat to convert all the ice into water. Note that the entire time the ice is melting, you can see that the temperature of the ice and water mixture does not change from 32°F (0°C). The 144 BTUs of added heat could not be measured with the thermometer, and therefore the mixture is absorbing latent heat.

If heat is added after all of the ice has changed into water, it will change the temperature of the water. To increase the temperature of the water to 212°F (100°C), 180 BTUs of heat energy must be absorbed. Water has a specific heat of 1.0, and in this case, the change in temperature is 180°F (212°F − 32°F).

$$1.0 \times 180 = 180 \text{ BTUs}$$

During the time that the water absorbs the 180 BTUs, the temperature of the thermometer increases. Therefore, the water is absorbing sensible heat.

When the water temperature reaches 212°F (100°C), it will begin to boil. The water will absorb an additional 970 BTUs of energy to convert all the water into steam. The entire time the water is turning to steam, the temperature of the thermometer will not rise above 212°F (100°C). The fact that the thermometer does not change means that the water absorbs latent heat energy while it is changing states.

If heat is still added to the steam, it will become superheated. Once the temperature of the steam rises above 212°F (100°C), it is considered to be super-heated **(Figure 3-4)**.

CHANGE OF STATE

Matter can commonly be found in three different states: solid, liquid, and gas. The greatest amount of heat movement (heat transfer) occurs during a change of state. Three processes describe a **change of state**:

1. Evaporation: the change in state from a liquid to a gas (vapor).
2. Condensation: the change in state from a vapor (gas) to a liquid.
3. Freezing: the change in state from a liquid to a solid.

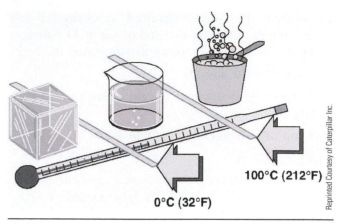

Reprinted Courtesy of Caterpillar Inc.

Figure 3-5 Water can be a solid, liquid, or gas, depending upon the temperature of the water.

In an air-conditioning/refrigeration system, the refrigerant will undergo two of these changes of state **(Figure 3-5)**.

Boiling point occurs when a pot of water is placed on a hot stove element; everyone knows that eventually the water in the pot will boil, producing steam. For this to happen, the water must reach 212°F (100°C) at sea level. The boiling point of the water is the temperature at which a liquid changes state to a vapor, in this case steam.

Evaporation occurs when enough heat is added to the water to change it from a liquid to a gas. Large amounts of heat energy are absorbed as the liquid changes state from liquid to vapor.

Condensation is the exact opposite of evaporation. If you take a vapor and remove enough heat from it, a change of state will occur that will cause the vapor to return to a liquid state. The change of vapor to a liquid is called condensation. Large amounts of heat energy are transferred from the vapor to another substance as the vapor condenses back into a liquid.

Freezing is the result of continuously removing heat from a liquid to the point that it becomes a solid. Heat is released as liquid freezes into a solid. Every known substance has a freezing point. The freezing point of refrigerant is very low, so it does not freeze within the air-conditioning/refrigeration system.

The changing of states is a very important principle to understand because refrigerant is continuously changing states, moving between a liquid and a vapor, within the system.

Latent Heat of Fusion, Vaporization, and Condensation

Ice can change into water or water into ice at 32°F (0°C) sensible heat. The energy required to change ice into water or water into ice is called the **latent heat of fusion**. Whenever ice and water are in contact with one another and thermal equilibrium is reached, the temperature will remain at 32°F (0°C).

Note: Technicians have used this ice water benchmark to accurately calibrate thermometers and thermostats.

For 1 pound of ice to change into 1 pound of water, it must absorb 144 BTUs of latent heat. Therefore, to change 1 pound of water into 1 pound of ice, the water must give up 144 BTUs of latent heat energy to a colder absorbent surface.

Water can change into steam or steam into water at 212°F (100°C). The energy required to change water into steam or steam into water is called **latent heat of vaporization**. Once water is heated to 212°F (100°C), an additional 970 BTUs of latent heat energy must be absorbed by the water to change the 1 pound of water into steam. At this temperature, steam can condense back into liquid droplets; the energy used is called **latent heat of condensation** (Figure 3-6).

Solids must absorb large amounts of heat when changing to a liquid state, and liquids, in turn, must also absorb large amounts of heat when changing into steam.

If we put a thermometer into a pot with 1 pound of water at 32°F (0°C) and place the pot on the stove element, the water would need to absorb 180 BTUs of sensible heat to increase to a temperature of 212°F (100°C). As the water heats up, you can track the progression of its temperature on the thermometer. At sea level, the water will begin to boil at 212°F (100°C). If heat is maintained on the pot, eventually all the water will boil off into steam. During the entire time the water boils, the temperature of the thermometer

does not rise above 212°F (100°C). If we increase the temperature of the element, we can make the water boil faster. Regardless of the intensity of the heat being applied, it will take an additional 970 BTUs of latent heat to boil all the water into steam.

Any time a substance changes states, large amounts of heat energy are absorbed or released. All air-conditioning/refrigeration units use these two principles to move heat from one place to another. Latent heat of vaporization is used to absorb large amounts of heat from the controlled space (the vehicle's interior). Latent heat of condensation is used to release large amounts of heat outside the controlled space to the ambient air.

PRESSURE

Pressure is defined as force per unit area. The effect that a force has on a surface depends on how the force is applied. Pressure can be applied in all directions, in just one direction, or several directions. For example, if a person walks in deep snow wearing only boots, he or she will sink down deep into the snow; yet if the person is wearing snowshoes, he or she will sink far less. The force the person applies to the snow is the same in both instances, but the force *per unit area* is different. The person wearing the snowshoes spreads the force out over a larger area, reducing the pressure.

In an air-conditioning/refrigeration system, the compressor raises the pressure of the refrigerant vapor. When the refrigerant is compressed, the temperature of the refrigerant rises, because the heat energy of a large mass of refrigerant is concentrated together. No heat is being applied by the compressor. It is crucial that the refrigerant be hotter than the temperature of the surrounding air (ambient air); otherwise, no heat transfer would take place.

ATMOSPHERIC PRESSURE

Atmospheric pressure is the weight of the atmosphere exerted on an object. If you took a 1-square-inch column of air and measured the weight of that air from the outer atmosphere to the earth's surface at sea level, it would weigh 14.7 pounds. Pressure, no matter how it is produced, is measured in pounds per square inch or ''psi,'' or it may be measured in kilopascals (Figure 3-7).

Atmospheric pressure is often measured with a mercury barometer. If an open tube is positioned vertically in a dish of liquid mercury, the mercury will not rise up the tube. This is because there is no pressure differential between the mercury inside the tube and the mercury outside the tube (Figure 3-8).

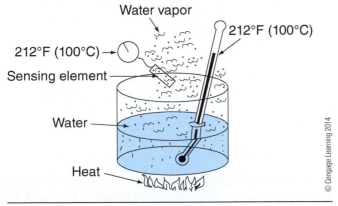

Figure 3-6 Water boils at a temperature of 212°F (100°C) at sea level.

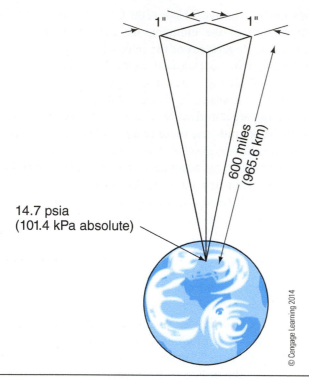

Figure 3-7 Atmospheric pressure is the weight a 1-square-inch column of air exerts on the earth's surface at sea level.

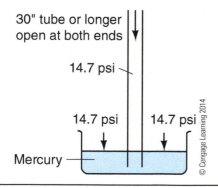

Figure 3-8 Mercury is not pushed up the tube as atmospheric pressure is applied both inside and outside the tube.

If the tube is closed at the top and all the air inside the tube is removed with a vacuum pump, it is said to be in a complete vacuum (no air, no air pressure to push down on the mercury inside the tube) **(Figure 3-9).** Atmospheric pressure at sea level acting on the surface of the mercury in the dish will push the mercury up the tube a distance of 29.92 inches (760 mm). At 29.92 inches, the weight of the mercury in the tube is equal to atmospheric pressure (14.7 psi/122 kPa). Thus, a complete vacuum at sea level is equal to 29.92 inches (760 mm) of mercury (Hg).

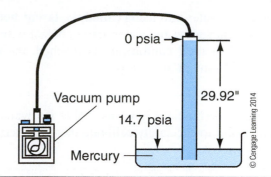

Figure 3-9 With all the air inside the tube removed, atmospheric pressure pushes the mercury up the tube a distance of 29.92 inches (760 mm).

Note: Hg is the scientific shorthand for mercury; very accurate vacuum gauges actually use a column of liquid mercury to indicate vacuum intensity.

PRESSURE GAUGES

In refrigeration work, pressure above normal atmospheric pressure is read in pounds per square inch or kilopascals. Pressure below atmospheric is read in inches or millimeters of vacuum.

Two types of gauges are used within the refrigeration industry. Technicians in the engineering field will use a gauge that reads pounds per square inch absolute (psia). Air-conditioning and truck trailer refrigeration technicians will more commonly use a gauge that reads pounds per square inch gauge or psi (psig).

The difference between these two gauges is that the gauge that reads in psia will read 14.7 pounds per square inch at sea level when it is not connected to a pressure source **(Figure 3-10).**

The gauge that reads psig will read 0 pounds per square inch at sea level when it is not connected to a pressure source **(Figure 3-11).**

Note: To convert psia to psig, subtract 14.7. And, likewise, to convert psig to psia, add 14.7.

Compound Gauge

A compound gauge will measure both positive and negative pressures. The compound gauge in **Figure 3-12** measures pressure for 0–250 psig and vacuum from 0–30 inches Hg.

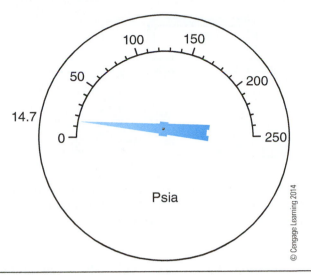

Figure 3-10 A gauge that reads in psia will read 14.7 pounds per square inch at sea level.

Figure 3-12 Manifold compound gauge reads both positive pressure and vacuum.

Figure 3-11 A gauge that reads in psig will read 0 pounds per square inch at sea level.

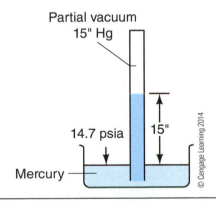

Figure 3-13 A 15-inch vacuum (381 mm) causes the mercury to rise 15 inches.

VACUUM

A perfect **vacuum** is considered to be 30 inches of mercury and still has yet to be obtained.

A partial vacuum is a vacuum that could be considered any pressure less than atmospheric. For example, a 15-inch vacuum (381 mm) will cause the mercury to rise 15 inches (381 mm) inside the tube. The amount of vacuum in the tube is expressed as 15 inches or 15 Hg or 381 millimeters of Hg **(Figure 3-13).**

PRESSURE/TEMPERATURE RELATIONSHIP

If the pressure on a given material is changed, so does the point at which the substance will boil or vaporize. Water, a common substance we all know, has a boiling (vaporization) point of 212°F (100°C) at sea level. If you put an open pot of water on a hot stove element, the water temperature will rise to 212°F (100°C). Any increase in the intensity of the heat will cause the water to boil harder, but the water temperature will never rise above 212°F (100°C). This is because the boiling water releases heat with the rising vapor. However, if the vessel is closed, like the radiator of a car or truck, the temperature of the water can rise above 212°F (100°C). Trapped vapor builds pressure on the water's surface. Increased surface pressure raises the boiling point of liquid; therefore, the water must get hotter than 212°F (100°C) before it can boil and release excess heat energy.

For each 1 pound of pressure increase, the boiling point will rise approximately 3°F (1.6°C), and correspondingly for each 1 pound of pressure removed from the surface of the water, the boiling point will be reduced.

Raising the Boiling Point

Raising the boiling point of liquid by pressurizing the container is a common practice. Truck radiators control pressure with a radiator cap designed to maintain a certain pressure but also to release excess pressure if it becomes too high. Raising the system pressure (boiling point) is necessary because truck cooling systems often operate at 230°F (110°C) or higher. Nonpressurized water would violently boil out of the radiator and the engine would overheat. Simply maintaining a cooling system pressure of 10 psi raises the boiling point to 242°F (115°C). For every pound of pressure, the boiling point of water increases approximately 3°F **(Figure 3-14)**. (With the addition of antifreeze, the boiling point will rise even further.)

> **CAUTION** *Automotive and truck radiators are designed to operate under pressure and the pressure within them is controlled by a pressure cap that will release excess pressure. If liquid is heated in a closed container, there is the possibility that the container may explode, causing serious personal injury or death.*

Lowering the Boiling Point

Because we have now seen that raising the pressure of water increases the boiling point, it stands to reason

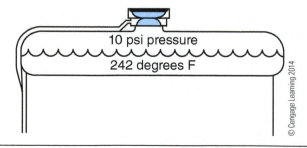

Figure 3-14 Pressure exerted upon the surface of water increases the boiling point of water by approximately 3°F per 1 pound of pressure.

Figure 3-15 In this example, the water in the flask will boil at room temperature as the vacuum lowers the pressure within the container.

that lowering the pressure will reduce the boiling point. At sea level, the weight of the earth's atmosphere bearing down on the surface of the earth is equivalent to 14.7 pounds per square inch (101.3 kPa). At this pressure, the boiling point of water is 212°F (100°C). As you move higher up in the earth's atmosphere (for example, going up a mountain), the weight of the atmosphere is reduced, which lowers the pressure exerted on the earth's surface. At 5000 feet (1524 meters), atmospheric pressure is only 12.3 pounds per square inch and the boiling point is reduced to 203°F (95.1°C). For every 550 feet (167 meters) above sea level, the boiling point of water is reduced by *approximately* 1°F. For example, if you take a sealed flask of water and reduce the pressure within the flask to 28.2 in Hg with the use of a vacuum pump, the water within the flask will boil at room temperature (70°F/21°C) **(Figure 3-15)**.

Compressing a Vapor

When a vapor is compressed, the heat contained by the vapor is concentrated. The vapor exerts pressure on all sides because it completely fills the space that it occupies. When a vapor's heat is concentrated by compression, the temperature of the vapor increases without the addition of any external heat source. Just compressing or squeezing the vapor makes it hotter **(Figure 3-16)**.

HUMIDITY

Humidity is the water vapor or moisture in the air. The amount of humidity depends on the temperature. When the weather is hot, the air can hold more moisture than when it is cold. Humidity is measured in terms of the amount of water the air can hold, and is expressed as a percentage. A relative humidity of

VAPOR COMPRESSION

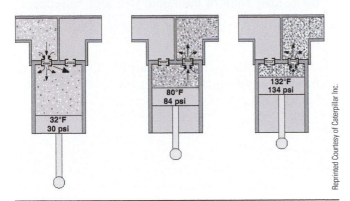

32°F
30 psi

80°F
84 psi

132°F
134 psi

Reprinted Courtesy of Caterpillar Inc.

Figure 3-16 The temperature of the refrigerant vapor increases as the pressure applied to it increases.

75% means the air can hold an additional 25% more water vapor before it is saturated (100% relative humidity).

Effects of Humidity

Before an air-conditioning system can be fully effective in transferring heat from the inside of the cab, moisture in the air must be reduced. This is the reason that an air-conditioning system must work so much harder and is less effective on a humid day. As a liquid evaporates, there is a temperature reduction. For example, if you get out of a swimming pool on a hot day, you will feel cooler than you did before you entered the pool. This is because the water is evaporating from your body, which causes a cooling effect. If you swim on an extremely humid day, it is not as easy for the water to evaporate from your body because the air already contains a large amount of water vapor, so the cooling effect will not be nearly as extreme. From this example, we can see that there is less of a cooling effect on a humid day.

Wet Bulb Temperature

Relative humidity is measured with a wet bulb or psychrometer. A wet cloth sock is placed over the bulb of a thermometer. As the psychrometer is spun on the end of a string, the wet bulb is cooled by evaporation. When the humidity is low, it registers a lower temperature than a dry bulb thermometer. When humidity is high, there is less evaporation, and, therefore, less of a cooling effect. The temperature difference is smaller between the wet and dry bulb. When there is no difference or the readings are the same, then humidity is 100% (the air is saturated with water vapor).

Summary

The principles discussed in this chapter are the basis for all air-conditioning/refrigeration systems. It cannot be overemphasized how important it is for the technician to know and understand these principles. The technician should understand that:

- Heat can be transferred by three different methods: conduction, convection, and radiation.

- Heat always travels from a high temperature to a lower temperature.

- Everything above the temperature of $-459°F$ ($-273°C$) contains some heat energy.

- The temperature at which all molecular motion stops is at absolute zero.

- When two objects reach the same temperature and no more heat transfer takes place, thermal equilibrium is reached.

- Temperature indicates the average velocity of the molecules of a substance.

- *Cold* is a term used to indicate an absence of heat energy.

- Temperature is not a measure of heat energy. Temperature is a measure of the intensity of heat energy in a substance.

- A BTU is the quantity of heat required to raise the temperature of 1 pound of water by $1°F$.

- Sensible heat causes a change in the temperature of a substance and can be measured with a thermometer.

- Latent heat is hidden heat that causes a change of state in a substance. This form of heat can't be measured with a thermometer.

- Latent heat of fusion is the energy required to change ice into water or water into ice.

- Latent heat of vaporization is the energy required to change water into steam.

- Latent heat of condensation is the energy required to change vapor into liquid.

- Superheat is the condition of a vapor when it is raised above its boiling point in a closed vessel.

- Subcooling occurs when the temperature of a substance is below its boiling point.

- Pressure is defined as force per unit area.

- As pressure is applied to the surface of a liquid, the boiling point of the liquid will increase.

- As the pressure on the surface of a liquid is reduced, the boiling point of the liquid will also be reduced.

- Humidity is the water vapor or moisture in the air.

- Hot, humid days require an air-conditioning or refrigeration system to work much harder than on a hot, dry day.

Review Questions

1. The total absence of heat is "absolute zero." This temperature is:

   A. 0°F

   B. 0°C

   C. −360°F

   D. −459°F

2. *True or False:* All substances, as we know them, contain heat, even ice.

   A. True

   B. False

3. Whenever there is a temperature difference between two substances, what will happen?

   A. Heat will move from the colder to the warmer substance.

   B. Heat will move from the warmer to the colder substance.

   C. The heat will not move.

   D. Heat will move from the west to the east.

4. When two objects of different temperatures are placed together, thermal equilibrium (heat balance) takes place when:

   A. The hot object gets hotter.

   B. The cold object gets colder.

   C. The objects touch each other.

   D. No more changes in temperature take place.

5. Heat intensity is measured in:

   A. Degrees

   B. BTUs

   C. Calories

   D. CFM

6. Heat quantity is measured in:

   A. Fahrenheit

   B. Celsius

   C. Kelvin

   D. BTUs

7. Which method of heat transfer occurs when heat moves through a solid object?

   A. Conduction

   B. Radiation

   C. Convection

   D. None of the above

8. Heat that transfers through gas or liquid but not through solids is known as:

   A. Conduction

   B. Convection

   C. Radiation

   D. None of the above

9. Heat is transmitted through empty space by:

   A. Conduction

   B. Convection

   C. Radiation

   D. None of the above

10. "The quantity of heat required to raise the temperature of 1 pound of water 1°F" is the definition of:

    A. Specific heat

    B. Heat of respiration

    C. Ambient temperature

    D. A British thermal unit

11. When a substance changes from one physical state to another (such as a solid to a liquid), this is called:

    A. Absolute zero

    B. Thermal equilibrium

    C. Change of state

    D. Respiration

12. The process of a liquid substance changing to a vapor is called:

    A. Evaporation

    B. Condensation

    C. Freezing

    D. Thermal equilibrium

13. The change of state from a liquid substance to a solid is called:

    A. Evaporation

    B. Condensation

    C. Freezing

    D. Thermal equilibrium

14. The change of state from a vapor to a liquid is called:

    A. Evaporation

    B. Condensation

    C. Freezing

    D. Thermal equilibrium

15. The process of a substance changing states from a liquid to a vapor with no increase in temperature is called:

    A. Latent heat of vaporization

    B. Latent heat of condensation

    C. Latent heat of fusion

    D. None of the above

16. The process of a substance changing states from a vapor to a liquid with no decrease in temperature is called:

    A. Latent heat of vaporization

    B. Latent heat of condensation

    C. Latent heat of fusion

    D. None of the above

17. *True or False:* Any pressure below normal atmospheric pressure may be referred to as a vacuum or partial vacuum.

    A. True

    B. False

18. *True or False:* If you want to raise the boiling point of a liquid, you need to increase the pressure on its surface.

    A. True

    B. False

19. *True or False:* If you want to lower the boiling point of a liquid, you need to decrease the pressure on its surface.

    A. True   B. False

20. When a vapor is compressed, its temperature:

    A. Increases   C. Doesn't change

    B. Decreases   D. None of the above, because a vapor cannot be compressed

CHAPTER 4

Air-Conditioning Components: Compressor, Condenser, and Receiver-Drier

Learning Objectives

Upon completion and review of this chapter, the student should be able to:

- Explain the purpose of the air-conditioning compressor.
- List five different styles of compressors used in the truck industry.
- Describe how each of the five different styles of compressors differs from the others in its operation.
- Explain the construction of the condenser.
- Describe the operating principles of the condenser.
- Explain what happens to the refrigerant as it passes through the condenser.
- Describe the purpose of the receiver-drier.
- List the reasons for using a desiccant in the receiver.
- Explain the purpose of the pickup tube in the receiver-drier.
- Describe the term *hygroscopic*.
- List the uses of the sight glass.
- Explain the precautions for installing a receiver-drier.

Key Terms

compressor	ram air effect	strainer
condenser	receiver-drier	swash plate compressor
desiccant	rotary vane compressor	two-piston-type compressor
hydrochloric acid	scroll compressor	variable displacement compressor
hygroscopic	sight glass	

INTRODUCTION

In this chapter, we look at some of the basic components involved in every mechanical refrigeration/air-conditioning system. In **Chapter 1,** we discussed the basic components in general. It is now time to take a closer look at these components **(Figure 4-1).**

A typical air-conditioning system is made up of five major components:

- Compressor
- Condenser

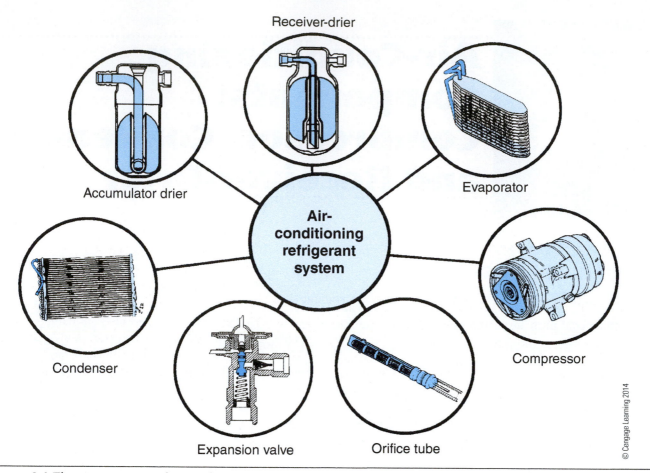

Receiver-drier

Accumulator drier

Air-conditioning refrigerant system

Evaporator

Compressor

Condenser

Expansion valve

Orifice tube

© Cengage Learning 2014

Figure 4-1 The components that make up an air-conditioning system.

- Receiver-drier or accumulator
- Thermostatic expansion valve or orifice tube
- Evaporator

This chapter discusses the compressor, condenser, and receiver-drier. There are also a few other important components that are necessary to allow the five components to do their job:

- Refrigerant
- Oil
- Hoses, fittings, and seals
- Electrical controls

SYSTEM OVERVIEW

It is extremely important for the purpose of diagnostics that the technician understand the functions of all the air-conditioning components. In addition, the technician should be able to identify the state of the refrigerant as it flows through the system.

Once all the parts of the air-conditioning system are identified, the system will be plumbed together, so to speak, and the path and state of the refrigerant will be discussed in **Chapter 6**.

COMPRESSOR

The **compressor** is the heart of the air-conditioning system because it circulates or pumps refrigerant and oil through the system. The compressor is responsible for two main functions required by air-conditioning systems. One function of the compressor is to raise the pressure of the refrigerant. You have learned from **Chapter 3** that when the refrigerant is compressed, the temperature of the refrigerant rises. No heat is added to the refrigerant by the compressor. The heat has been absorbed in the evaporator section and is concentrated by the compression process. It is critical that the refrigerant be hotter than the temperature of the ambient (surrounding) air; otherwise, no heat transfer would take place. The second function of the compressor is to create low pressure in the evaporator. The lower pressure condition in the evaporator allows the refrigerant to vaporize (boil), enabling the refrigerant to absorb large quantities of heat energy from the cab of the vehicle. The compressor also circulates the refrigerant and compressor oil, which are mixed together throughout the air-conditioning system **(Figure 4-2)**.

VAPOR COMPRESSION

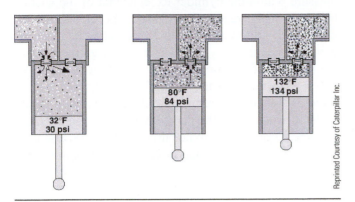

32°F
30 psi

80°F
84 psi

132°F
134 psi

Reprinted Courtesy of Caterpillar Inc.

Figure 4-2 The heat of compression.

Failure of either of the compressor's two main functions would result in a loss or reduction of refrigerant circulation within the air-conditioning system. Without proper refrigerant circulation, the cooling process of the air-conditioning system will be reduced or will stop working altogether.

Air-conditioning compressors come in many different styles, but all perform the same function. In most cases, the compressors are mounted near the front of the engine. The compressor is belt-driven from the engine crankshaft. The compressor's drive pulley is incorporated into an electromagnetic clutch. This clutch provides a means of turning the compressor on or off, depending upon the temperature requirements of the system.

Two refrigerant lines are attached to the compressor, one being the discharge line and the other the suction line. The refrigerant lines can always be identified by their physical size. The suction line is always larger in diameter than the discharge line. The suction (inlet) side draws in low-pressure, low-temperature refrigerant gas from the evaporator. The compressor then pumps out high-pressure, high-temperature refrigerant gas to the condenser.

There are many different manufacturers and styles of air-conditioning compressors, but they all operate in the same basic way. Some of the compressor variations are piston-type, vane-type, scotch yoke, and scroll compressors. These compressors may again be broken down by:

- Piston and cylinder arrangement
- Compressor mounting
- Type and number of drive belts
- Compressor displacement
- Fixed or variable displacement

TWO-PISTON-TYPE COMPRESSORS

The two-piston-type compressor was used by many truck manufacturers and can still be found on many older vehicles still on the road. This style of compressor can be mounted vertically or horizontally. These compressors may be constructed of either steel or aluminum. This compressor requires approximately 14 horsepower from the engine when it is operating **(Figure 4-3)**. You will still find these compressors on older vehicles, but because of their weight and the drag they put on the engine, they are not used much today. Today's truck manufacturers use much more efficient compressors to increase vehicle mileage. As the name implies, the compressor uses two pistons, which are driven by a crankshaft and connection rods. The rotation of the crankshaft moves the piston up and down to draw in and force out refrigerant gas. This style of compressor uses a reed valve assembly located above the pistons to control the flow of refrigerant into and out of the compression chamber. The crankshaft and housing must be sealed to prevent refrigerant and oil leaks to the atmosphere.

Two-Piston Compressor Operation

Intake Stroke.   We will start by explaining the intake stroke. Whenever the piston is moving in a downward

John Dixon © 2014 Cengage Learning

Figure 4-3 Two-cylinder compressor.

direction, it can be considered to be on an intake stroke. As the piston moves down in the cylinder, the volume becomes greater and the pressure is reduced. The refrigerant gas is then drawn into the cylinder from the suction (low pressure) side. To accomplish this, the refrigerant must first enter the intake port and push the intake reed valve down (open) so that it may be drawn into the cylinder **(Figure 4-4).**

Compression Stroke. When the piston then starts its upward travel, the volume within the cylinder is

reduced, causing a rise in pressure. The instant the pressure within the cylinder exceeds that of the suction line, the intake reed valve closes. The closing of the inlet valve allows the piston to continue to raise the pressure within the cylinder. The discharge reed valves are held closed by springs so that they act like a check valve. Once the pressure within the cylinder exceeds the force of the closing springs and the refrigerant pressure in the discharge line, the valve opens, allowing the high-pressure refrigerant gas to be pushed out into the discharge port and eventually into the discharge line **(Figure 4-5).**

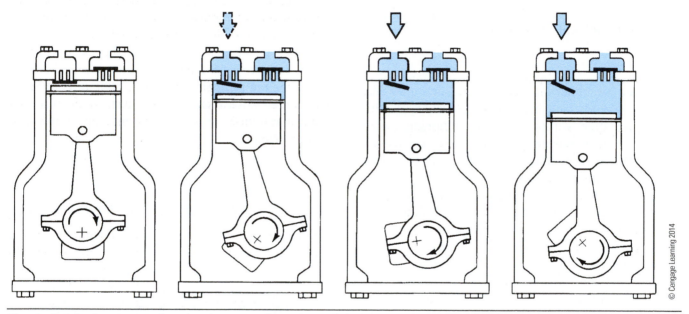

Figure 4-4 An intake stroke; the downward motion of the piston draws refrigerant vapor past the intake reed valve.

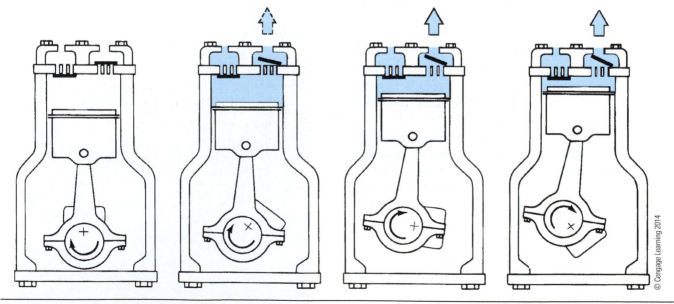

Figure 4-5 A compression stroke; the upward motion of the piston squeezes the refrigerant vapor, increasing the pressure as it forces it out the discharge valve.

Two-Piston Compressor Maintenance

This compressor has a few basic areas of maintenance to be addressed by the technician.

- The oil level of this compressor should be checked, but this can only be performed when the air-conditioning system is discharged (refrigerant removed). The oil level of a new replacement compressor must also be checked before installation.
- Compressor crankshaft seals are consumable items that will eventually leak and require replacement. For this procedure, the refrigerant must be removed from the system, unless the compressor is equipped with stem-type service valves (explained later).
- Replacing the valve plates or gaskets is another maintenance task. Again, for this procedure the refrigerant must be removed from the system, unless the compressor is equipped with stem-type service valves.
- The compressor's electromagnetic clutch may also be replaced. This is one of the few repair procedures that may be performed while the unit is fully charged.
- Belt tension should be checked on a regular basis. Because the compressor is belt driven, proper belt tension is important for proper air-conditioning operation.

SWASH PLATE COMPRESSORS

Swash plate compressors get their name from the means by which the pistons are driven, a swash (slanted) plate mounted on a revolving shaft. As the swash plate (sometimes called a wobble plate) turns, it pushes and pulls the pistons back and forth to draw in, compress, and discharge the refrigerant gas. These compressors are available in different configurations and numbers of cylinders. Each piston has a rod with ball-and-socket-type bearings that fit into a slipper (foot of the piston). The rotation of the swash plate causes reciprocating (up and down) movement of each piston inside its cylinder. The valve plate contains a reed valve assembly. These reed valves allow refrigerant to flow in only one direction, not unlike a check valve. Many manufacturers use swash plate compressors because they put less drag on the engine, approximately 7 horsepower, compared to the 14 horsepower consumed by the two-piston-type compressor **(Figure 4-6)**.

Swash Plate Compressor Operation

Intake Stroke. As the swash plate pulls the piston down in the cylinder, it creates a drop in pressure. This pressure is lower than that of the refrigerant in the suction line. This imbalance in pressure forces the higher pressure of the refrigerant within the suction line to open the suction reed valve, allowing the cylinder to be filled with refrigerant gas.

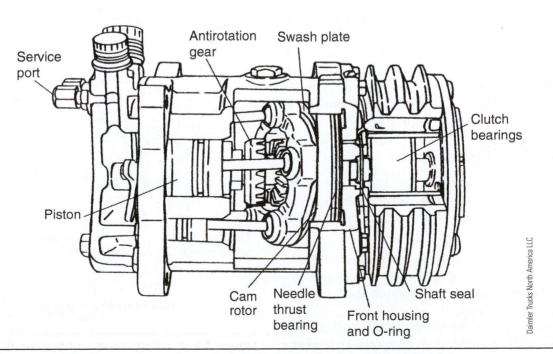

Service port · Antirotation gear · Swash plate · Clutch bearings · Piston · Cam rotor · Needle thrust bearing · Front housing and O-ring · Shaft seal

Figure 4-6 Swash plate compressors are extremely popular in the heavy truck market.

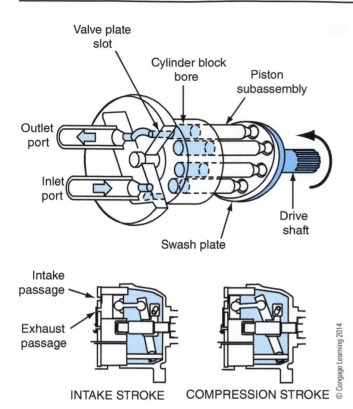

Figure 4-7 A swash plate. As the swash or wobble plate is turned, the piston is moved back and forth in its bore to draw in and compress refrigerant vapor.

Compression Stroke. Further rotation of the swash plate changes the direction of the piston within the cylinder. As the refrigerant is compressed slightly, it creates a pressure higher than that contained within the suction line. This pressure imbalance causes the suction reed valve to close, preventing refrigerant from exiting the cylinder through the suction port. The pressure within the cylinder rises to a point where the discharge reed valve opens, allowing the superheated refrigerant to be pushed out of the cylinder into the discharge port and on out to the other components of the air-conditioning system **(Figure 4-7)**.

Swash Plate Compressor Maintenance

Just as with the two-piston-type compressors, swash plate compressors have some service/maintenance issues.

- The compressor oil must be checked whenever the air-conditioning system is discharged.
- Replacement of the compressor shaft seal is a maintenance issue.
- Replacement of the valve plate or gasket (for a nonoperating or leaking compressor) is also a maintenance issue.

- Belt tension is also part of the required routine maintenance for this system. Because the compressor is belt driven, proper belt tension is important for proper air-conditioning system operation.

The two compressors described so far operate very differently, but despite these differences, they perform the same functions within the A/C system.

ROTARY VANE COMPRESSORS

The **rotary vane compressor** does not use a piston to raise the pressure of the refrigerant. These compressors were used for many years in the light truck market. This style of compressor also does not use a suction valve. However, it does incorporate a discharge valve. The discharge valve serves as a check valve, preventing high-pressure refrigerant vapor from migrating back into the compressor discharge port during the compressor off cycle or when the air-conditioning system is not operating. This compressor performs exactly the same function in the air-conditioning system as the piston-type or swash-plate-type compressor, but the operation of the rotary vane compressor is entirely different. The rotary vane compressor operates much like a vane-style power steering pump **(Figure 4-8)**.

Rotary Vane Compressor Operation

The compressor drive pulley is belt-driven from the engine's crankshaft pulley. Turning the rotor assembly causes the vanes to extend by centrifugal force, and causes them to seal against the cylinder wall. As the

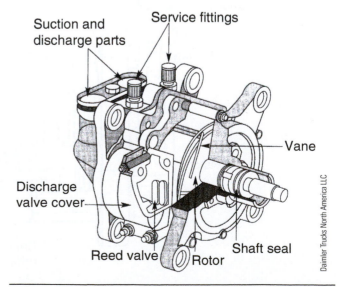

Figure 4-8 A rotary vane compressor does not use pistons to compress refrigerant.

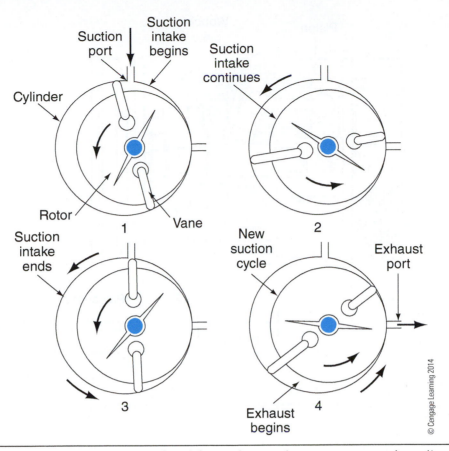

Figure 4-9 A rotary vane compressor; centrifugal force throws the vanes outward, sealing them against the cylinder wall.

vane passes the suction inlet, the volume of the chamber increases, causing a lower pressure that draws refrigerant into the cylinder. Once the vane passes the largest portion of the eccentric, the vane will be pushed back into its bore as it follows the eccentric shape. This causes the volume of the chamber to decrease, squeezing the refrigerant between the vane and the cylinder wall, and increasing the pressure and temperature of the refrigerant. The high-pressure, high-temperature refrigerant is then pushed out the discharge port and through the discharge valve out into the discharge line **(Figure 4-9).**

VARIABLE DISPLACEMENT COMPRESSORS

Variable displacement compressors automatically adjust their displacement to match the vehicle's air-conditioning demands. These compressors are used in both the light- and medium-duty truck market. A control valve senses the evaporator load and automatically changes the displacement of the compressor. Unlike cycling clutch systems, these variable displacement compressors run continuously without any clutch cycling. Temperature is maintained by changing

the capacity of the compressor, not by engaging or disengaging the clutch. This feature allows the system to cool more softly and uniformly. It also eliminates the noise problems associated with cycling clutch systems. Dehumidification and fuel economy are also improved **(Figures 4-10, 4-11,** and **4-12).**

Figure 4-10 A variable displacement compressor found in the light- and medium-duty markets.

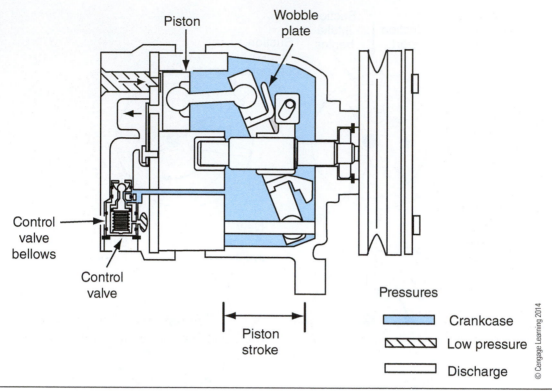

Figure 4-11 A variable displacement compressor at maximum displacement, with the wobble plate at maximum angle.

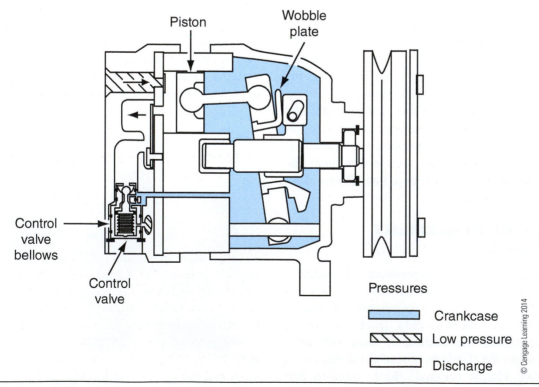

Figure 4-12 A variable displacement compressor at minimum displacement, with the wobble plate at minimum angle.

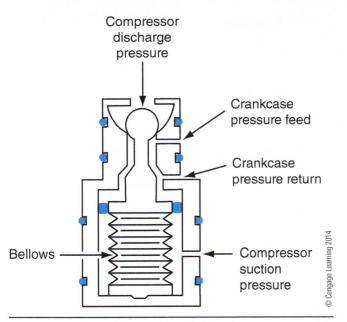

Figure 4-13 A control valve assembly for a variable displacement compressor.

Figure 4-14 A typical scotch yoke, sometimes referred to as an R4.

Variable Displacement Compressor Operation

The pistons of the compressor are driven by a variable angle swash plate. The angle of the swash plate is changed by a bellows-activated control valve located in the rear head of the compressor. The control valve senses the suction pressure and controls the swash plate angle based on crankcase suction pressure differential. Operation of the control valve is dependent on differential pressure (Figure 4-13).

SCOTCH YOKE COMPRESSORS

Scotch yoke compressors have been used for many years in the air-conditioning industry. On this compressor, opposed pistons are pressed onto opposite ends of a yoke. The yoke rides on a slider block located on the shaft eccentric of the compressor. Rotating the shaft also turns the yoke with its attached pistons. This causes the pistons to reciprocate, following the eccentric contour. Each of the four pistons contains a suction reed valve, and a reed valve plate is located on top of each cylinder (Figure 4-14).

Scotch Yoke Compressor Operation

The operation of this compressor is similar to that of all the other piston compressors. During the intake stroke, the piston moves down in its bore and refrigerant vapor is drawn into the cylinder through the suction reed valve. The piston changes its direction of travel on the exhaust stroke and compresses the refrigerant vapor

to a high-pressure gas, pushing it through the discharge valve plate on the exhaust stroke (Figure 4-15).

SCROLL COMPRESSOR

The scroll compressor is not a new design; it was first patented in 1909, but was not really used until 1988 when the Copeland Corporation used it in residential air-conditioning units and heat pumps. The transport refrigeration equipment industry also climbed aboard with a scroll compressor of its own design. The first company to use the scroll compressor in the automotive industry was Sanden in 1993. The beauty of the scroll compressor is that it has only one moving part. The scroll compressor has not yet been used in the heavy truck market, but because it puts less drag on the engine, it is probably only a matter of time before it gains popularity (Figures 4-16 and 4-17).

Note: Less engine drag equals improved fuel economy.

Scroll Compressor Operation

The scroll compressor operates by rotating one scroll within a stationary scroll. The ends of the rotating scroll scoop up the refrigerant vapor at the suction port of the compressor. As the scroll continues to rotate, the inlet passage is sealed off and the volume of the passage becomes smaller, increasing the pressure of the refrigerant vapor. The refrigerant vapor is squeezed through to the discharge passage at the center of the scrolls. The refrigerant vapor is at a higher pressure and temperature as it leaves the discharge port.

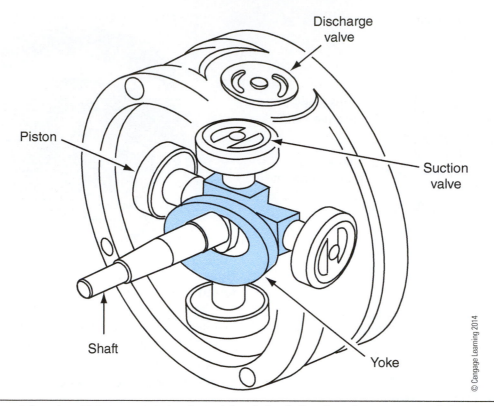

Figure 4-15 A scotch yoke. Pistons pump back and forth by following the contour of the eccentric.

Figure 4-16 A scroll compressor.

The scroll's vapor passages are continuously at various stages of compression at the same time. This way the scroll compressor is able to supply a smooth, stable suction and discharge pressure **(Figure 4-18).**

Note: To retain engine power, most manufacturers cut power to the compressor's clutch, which turns the compressor off when the engine is under load.

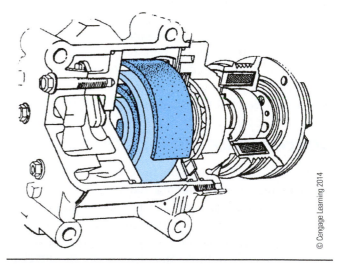

Figure 4-17 A scroll compressor, showing the inner scroll section of the compressor.

LUBRICATION

Lubrication is the life's blood of the compressor. Without oil, the compressor would destroy itself. Compressor oil lubricates the compressor piston, cylinders, and other moving parts, preventing metal-to-metal contact, which in turn reduces friction. A thin film of oil also coats the surfaces of the shaft seals to help prevent refrigerant leaks.

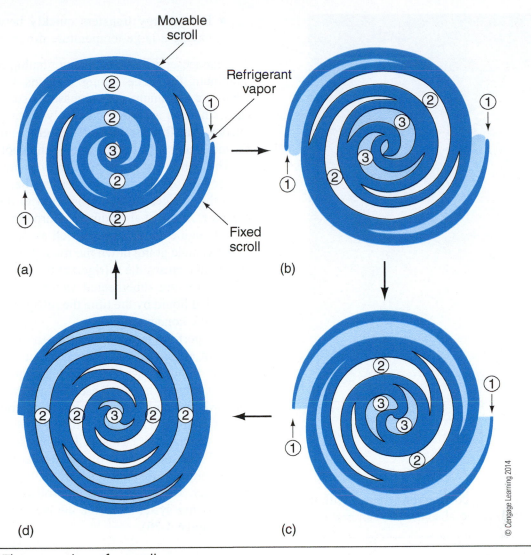

Figure 4-18 The operation of a scroll compressor.

....

Note: The number one cause of all compressor failure is lack of lubrication. Lubrication is a fundamental requirement of all compressors because of their moving parts. Lubrication reduces friction; without it, the wear to moving parts increases dramatically.

....

There are a number of ways that oil can be removed from a closed air-conditioning system:

- Any time there is an external leak (refrigerant leak to the atmosphere), the refrigerant carries out some of the compressor oil with it.
- Whenever the air-conditioning system is discharged into a refrigerant recovery machine, compressor oil may also be drawn out with the refrigerant.

- Any time an air-conditioning component is replaced, oil may be trapped within the component.

No matter what causes a compressor to lose oil, it must be replaced before the unit can be put back into service. This protects the compressor and allows the system to operate with the required amount of lubrication.

When topping up the oil level of a system, it is important to add the correct type of compressor oil, because the compressor oil must be chemically stable. This oil must come from a sealed container, or one that has been kept tightly closed. Refrigerant oils are hygroscopic (drawing moisture into themselves). Also, the technician must use a clean dry tube, funnel, or oil pump to install the oil. Some refrigerant management centers are equipped to inject compressor oil directly into the system. In that case, all the technician needs to do is select the desired amount to be installed.

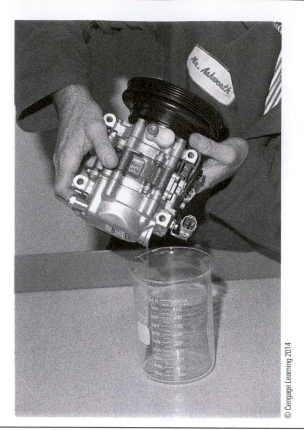

Figure 4-19 Checking compressor oil in a new compressor.

....

Note: Anything coming into contact with the compressor must be absolutely clean and dry.

....

Figure 4-19 illustrates checking the oil level in a typical swash-plate-style compressor. Because there is no means of checking the oil level when the compressor is mounted to the engine, the compressor must be removed and the oil drained out and measured. It is much easier to check compressor oil level in two-piston-type compressors. Once the compressor has been relieved of refrigerant, a side plug on the compressor block can be removed and a curve-shaped dipstick can be inserted and routed through to the bottom of the sump. The oil level can then be read.

CONDENSER

The **condenser** is typically mounted in front of the radiator on most trucks. Its function is to release the heat from the vehicle's interior to the outside air.

The condenser relies on two key principles of heat transfer:

- Heat energy always moves from hotter to colder.

- Heat energy transfers quickly between objects having a large temperature difference.

The condenser consists of tubing running through sharp aluminum fins. The tubing contains the refrigerant, while the aluminum fins increase its surface area. Because the fins are attached firmly to the tubing, conduction transfers the heat of the refrigerant to the fins for maximum exposure of heat to the cool ambient air.

The compressor pumps high-pressure superheated refrigerant through the condenser; heat is removed from the refrigerant and transferred to the cooler ambient air. Ambient air is directed through the condenser by the engine cooling fan as well as the ram air effect of the vehicle going down the road. This cooling of the refrigerant causes the refrigerant to change states from a high-pressure superheated vapor to a high-pressure subcooled liquid by the time the refrigerant reaches the end of the condenser.

The performance of the condenser relies on two forms of air circulation. The more air that can be circulated through the condenser, the greater the heat transfer rate of the condenser. When the truck moves forward, the ambient air is driven through the condenser, helping it to transfer the heat of the refrigerant to the ambient air. This is known as the **"ram air effect."** The other form of air circulation is produced by the engine-mounted fan that pulls air through the condenser. The fan clutch is controlled by both the air-conditioning system and the engine coolant temperature **(Figure 4-20).**

Condenser Service

The condenser is usually a maintenance-free component, as long as it is kept clean and the fins are not bent or damaged. Due to the condenser's typical mounting position, it is susceptible to damage from

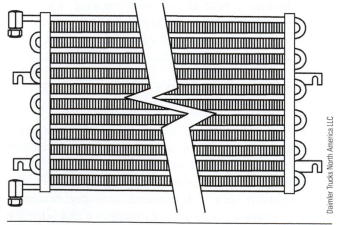

Figure 4-20 A typical condenser assembly.

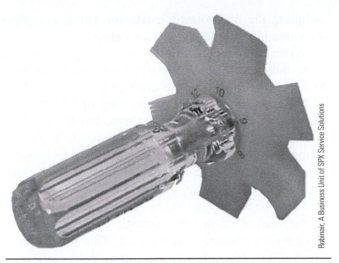

Figure 4-21 A fin comb used to straighten damaged condenser fins.

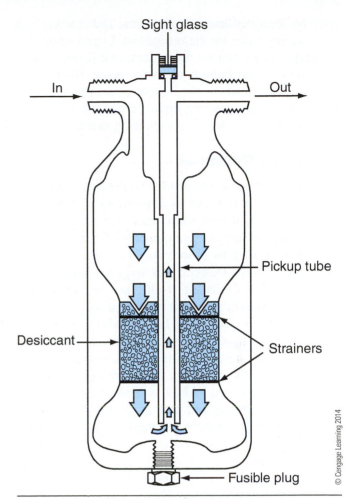

Figure 4-22 The internal components of a receiver-drier.

debris that can puncture the delicate tubing and fins, causing a refrigerant leak. In addition, bent fins and debris clogging the fins will restrict air circulation through the condenser, reducing heat transfer and increasing system pressures. To keep the performance of the air-conditioning system at an acceptable level, the condenser should be cleaned regularly with a mild detergent and rinsed with a garden hose. High-pressure washers should be avoided because they can damage the fins and cause them to flatten out. Damaged fins can be straightened with a tool known as a fin comb. The technician simply starts at a good section of the fins and gently pulls the comb through the damaged section, straightening the damaged fins **(Figure 4-21)**.

From the condenser, the high-pressure liquid refrigerant enters a component called the receiver-drier.

RECEIVER-DRIER

Liquid refrigerant from the condenser enters the receiver-drier, which is a metal cylinder generally located on the firewall **(Figure 4-22)**. The **receiver-drier** performs three important functions that are vital to the air-conditioning system:

- It filters the refrigerant.
- It removes moisture from the refrigerant.
- It stores excess refrigerant.

Filter

The liquid refrigerant enters the receiver-drier and passes through a filter. Contaminants can cause internal damage to other components within the air-conditioning system. Contaminants may include dirt,

dust, sludge (from compressor oil breakdown), or even **desiccant** material that may be carried with the refrigerant throughout the system. Some receiver-driers have filters on their inlet and outlet. Some driers use a **strainer** instead of a filter to catch all foreign particles that would otherwise be able to flow throughout the system. The strainer consists of fine wire mesh that removes impurities as the refrigerant passes through the receiver-drier or accumulator drier. (The accumulator is explained in **Chapter 5.**) The receiver-drier may contain two strainers, one on either side of the desiccant. It is important to note that the refrigerant must pass through a filter or a strainer before leaving the receiver in order to prolong the life of the air-conditioning system.

The pickup tube used in a receiver-drier ensures that only liquid refrigerant can exit the drier to feed the thermostatic expansion valve. (The thermostatic expansion valve is explained in **Chapter 5.**) Because under certain conditions both liquid and gaseous refrigerant may enter the receiver-drier, the refrigerant must be separated into its two forms. The receiver tank acts as a separator for the refrigerant. Liquid refrigerant is heavier than vaporous refrigerant, so it flows by force of gravity to the bottom of the receiver. Because the pickup tube extends to the bottom of the receiver tank, this ensures only gas-free liquid refrigerant can be fed through to the thermostatic expansion valve.

Moisture Removal

Moisture must be removed from the air-conditioning system because if water is allowed to mix with the compressor oil, it will form acids that are extremely harmful to the internal components of the system. A desiccant is used to absorb moisture that is able to enter the system. Desiccant is **hygroscopic** (absorbs moisture). The desiccant itself may be found in a couple of different forms. It may be a solid, granular, or beaded form. Desiccant has the ability to remove moisture from liquid or vapor. Desiccants known as XH-7 or XH-9 are used with R-134a, and can also be used in an R-12 system. The desiccant may be held in place by two screens, known as strainers, or it may be placed in a mesh bag and held in place by a spring, or the bag of desiccant may be simply placed inside the receiver, free to move around.

The amount of water that a drier can absorb depends upon the volume and type of desiccant used, as well as the temperature. The warmer the ambient temperature, the less moisture the desiccant can retain.

Refrigerant Storage

One of the functions of the receiver-drier is to store excess refrigerant. The thermostatic expansion valve controls the flow of refrigerant into the evaporator. When the TXV orifice closes, the liquid refrigerant requires a place to be stored so it won't back up into the condenser causing discharge pressure to skyrocket. Refrigerant is stored in the receiver until it is required by the TXV.

Receiver-Drier Location

The positioning of the receiver-drier has an effect on the amount of moisture the drier can retain. As indicated earlier, temperature affects the amount of moisture the desiccant of the drier can retain. As the ambient temperature increases, the desiccant loses its ability to retain moisture. The removal of moisture within an air-conditioning system is very important because moisture reacts with refrigerant to form **hydrochloric acid** (HCl). If moisture remains in the system, the acid will affect all the metal components,

including the thermostatic expansion valve, compressor valve plates, and the service valves.

When the drier gets to its saturation point, it will allow moisture to be circulated throughout the system. At the saturation point, the drier may be able to hold all the moisture in the early morning or late evening when ambient temperatures are low, but in the afternoon, when the ambient temperatures are high, the temperature of the desiccant will also rise to its saturation point, and some of the moisture will be released into the air-conditioning system.

Moisture that is not held by the desiccant can freeze in the orifice of the thermostatic expansion valve, blocking the passage of refrigerant. This restriction in the flow of refrigerant will stop the cooling effect of the air-conditioning system.

Moisture in an air-conditioning system is not always obvious because it takes time for the water droplets to gather together and freeze into a blockage. An indication to the technician that there is water within the system is a complaint from a customer that the air-conditioning system works great for a little while, and then stops working. If the customer turns the air-conditioning unit off for a spell and then turns it back on again, it will again work for a while and then gradually stop working again. This is because the water in the system that had frozen the first time the unit was turned on was able to thaw out during the off cycle, which is why the air-conditioning system could cool normally when cycled on again.

Repairing this condition requires that the refrigerant from the system be removed and recycled and the receiver-drier must be replaced. In addition, the system should be given a long, deep evacuation in order to boil and remove any moisture that is present in the system.

There are many ways in which moisture can enter an air-conditioning system. Any time there is a refrigerant leak or the system has been opened for service, moisture from the humidity in the air is able to enter the system. This is why it is critical that the receiver-drier/accumulator drier be changed any time the system has been opened to the outside air. On a humid day, if the protective caps on a new receiver-drier or accumulator are removed, the desiccant will absorb all the moisture it can up to its saturation point in about five minutes.

The filter/strainer within the drier can also become blocked by debris, causing a restriction. Restrictions within an air-conditioning system will cause a pressure drop and can usually be detected by a change in temperature. The human hand is one of the quickest and best tools that a technician can use to detect these changes in temperature. The inlet and outlet lines to the drier should be the same temperature (slightly warm in a properly operating system). The drier should cause no restriction to the air-conditioning system. If the inlet temperature of the receiver-drier is warmer than the outlet line temperature, there is a restriction in the drier. Under the right ambient conditions, frost may be observed on the outlet line of a restricted receiver-drier. Any time a receiver-drier becomes restricted, it must be replaced immediately because the air-conditioning system will not function in this condition.

Receiver-Drier Service and Installation

The receiver-drier is usually located under the hood of the vehicle and should be positioned where it can stay as cool as possible. Some manufacturers position the receiver-drier on the outside of the main frame rail where it can be cooled by fresh air and can also be as far away from the hot engine as possible.

The receiver-drier may be equipped with a number of different fitting styles to connect it to the refrigerant lines of the air-conditioning system. Fittings may use an SAE flare style, an O-ring style, or may even have a barbed-style fitting that is pushed inside the refrigerant line and then clamped into position. It is important to replace a receiver-drier with exactly the same part as per the manufacturer's instructions because the receiver must be able to store the correct amount of liquid refrigerant and the drier desiccant must be able to absorb the amount of moisture for the size of the air-conditioning system in which it is being installed.

The drier must also be mounted in a vertical position for proper operation because the pickup tube must be able to draw liquid refrigerant from the bottom of the drier. Usually, a drier should not be mounted more than 15 degrees from vertical.

The refrigerant lines from the air-conditioning system must be correctly connected to the drier or the system will not cool properly. The refrigerant line coming from the condenser must be installed to the fitting marked **(IN)** on the drier, and the refrigerant line leading to the thermostatic expansion valve must be attached to the drier fitting marked **(OUT).** If the drier is not stamped with an IN or OUT marking, then it will have a directional arrow indicating the direction of refrigerant flow through the drier.

Note: This is just one of the reasons it is important that the technician understand the flow of refrigerant through the air-conditioning system.

Rough handling of the drier can displace components within the drier, possibly causing internal restrictions in a new component. As stated previously, this condition can be indicated by a difference in temperature between the inlet and outlet sides of the drier while the unit is in operation. If the pickup tube gets broken, it will not supply liquid refrigerant to the thermostatic expansion valve, and the system will not cool properly. Unlike the serviceable screens at the inlet to the compressor and the inlet to the thermostatic expansion valve, the inlet screen to the receiver-drier is not serviceable, and when it becomes restricted, the entire drier unit must be replaced.

WARNING *NEVER remove the protective caps installed on a new receiver-drier or accumulator until it is to be installed into the air-conditioning system. Remember that the desiccant is hygroscopic. The receiver-drier or accumulator should also be the last component installed before the vacuum pump is used to remove the air and moisture that entered the air-conditioning system during servicing.*

SIGHT GLASS

Many receiver-driers contain a **sight glass** that allows the technician to observe the flow of refrigerant in an operating air-conditioning system. The sight glass may be located in the top of the receiver-drier, on the outlet side of the receiver, or anywhere in the liquid line up to the thermostatic expansion valve. From the sight glass, the technician can observe the state of the refrigerant. When a check of the sight glass is performed, the engine must be running, the windows should be open, and the air-conditioning controls should be set to maximum cool with the fan at its highest speed. Allow the vehicle to operate for approximately 5 minutes to allow the air-conditioning system to stabilize itself. If the system is operating properly, you should see a steady stream of liquid with no bubbles.

- Bubbles or foaming of the refrigerant indicates a problem with the system, usually a loss of refrigerant.
- Oil streaking also indicates that the system is empty.
- A clear sight glass with no bubbles indicates that the refrigerant level is correct with no contamination. Or it might indicate that the system is completely empty.

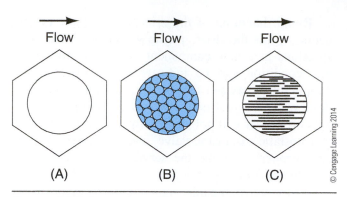

Figure 4-23 A receiver-drier sight glass.

Figure 4-24 A receiver-drier containing a moisture indicator.

> **Note:** Bubbles may be present until the air-conditioning system has time to stabilize. If the air-conditioning system is a retrofit system to R-134a, bubbles may be present in a properly operating and charged system.

■ Cloudy refrigerant indicates that the desiccant has broken down and has probably contaminated the entire air-conditioning system.

The sight glass can be used to check the refrigerant level, but only on an R-12 system and only when the ambient temperature exceeds 70°F (21°C). It is normal to see bubbles on a cool day, and this does not indicate a shortage of refrigerant. Bubbles may also be observed in the sight glass whenever the compressor clutch cycles (on/off) or when the heat load demands of the air-conditioning system change **(Figure 4-23).**

> **Note:** Not all R-12 air-conditioning systems use a sight glass, and most R-134a systems do not use them because bubbles are usually present when this refrigerant is used. On units without sight glasses, the manifold gauge set is used to check the operating conditions within the air-conditioning system. Some receiver-driers used on R-134a systems contain a moisture indicator. The one in **Figure 4-24** is blue, indicating a dry system. Pink would indicate the presence of moisture in the system.

The most accurate way of charging an air-conditioning system is to know the capacity of the system and charge the system by weight **(Figure 4-25).** (This is explained in a later chapter.)

Figure 4-25 A warning decal on the firewall of a tractor that also lists refrigerant type and charge weight.

MAINTENANCE PROCEDURES

For the most part, air-conditioning systems do not require a large amount of regular routine maintenance. The technician should be able to performance test the system to ensure that the system is performing satisfactorily. The condenser should be cleaned regularly to ensure good heat transfer. The compressor belt should be checked to make sure it is in good condition and that the belt tension is correct. Receiver-driers should be changed at intervals indicated by the individual OEM or whenever the system has been opened for service. The compressor oil level should be checked whenever the compressor has been replaced or when the oil level has been depleted by a refrigerant leak.

Summary

- The compressor raises the pressure of the refrigerant.

- The compressor creates the low pressure in the evaporator that is required for the liquid refrigerant to change states.

- The two-piston compressor operates along the same principle as a small internal combustion engine. A spinning crankshaft causes the pistons to move up and down in their cylinders.

- In a swash plate compressor, the crankshaft turns a swash or wobble plate, which causes the pistons to move in their cylinders.

- The rotary compressor uses no pistons. The crankshaft spins vanes that seal against the wall of the eccentric shaped cylinder, very much like a power steering pump.

- Variable displacement compressors automatically change the angle of the swash plate, which increases or decreases the piston travel (stroke) within its cylinder.

- The scroll compressor uses one stationary and one revolving scroll to compress the refrigerant vapor.

- Compressor oil lubricates the compressor piston, the cylinders, and other moving parts, preventing metal-to-metal contact, and this in turn reduces friction. Without oil, the compressor would destroy itself.

- The number one cause of all compressor failure is a lack of lubrication.

- In the condenser, the refrigerant changes state from a high-pressure superheated vapor to a high-pressure subcooled liquid by the time it reaches the end of the condenser.

- The filter in the receiver-drier removes contaminants from the liquid refrigerant before they are able to block refrigerant flow.

- The desiccant inside the receiver-drier is used to absorb moisture that is able to enter the air-conditioning system.

- The receiver-drier separates liquid refrigerant from vaporous refrigerant.

- The pickup tube in the receiver-drier ensures that the thermostatic expansion valve is supplied with liquid refrigerant only.

- The receiver-drier should be positioned where it can stay as cool as possible.

- The sight glass allows the technician to see the flow of refrigerant in an operating air-conditioning system.

- The sight glass can be used to check refrigerant level, but only on some R-12 systems.

Review Questions

1. Two technicians are discussing the purpose of the compressor. Technician A states that the compressor turns low-pressure vapor into high-pressure liquid. Technician B says that the compressor is responsible for pumping refrigerant through the components of the air-conditioning system. Which technician is correct?

   A. Technician A is correct.   C. Both Technicians A and B are correct.

   B. Technician B is correct.   D. Neither Technician A nor B is correct.

2. The compressors used for air conditioning in the trucking industry are driven directly or indirectly off the:

   A. Accessory pulley   C. Crankshaft pulley

   B. Alternator pulley   D. Water pump pulley

3. The compressor will draw in refrigerant _____ while the compressor is on its _____ stroke.

   A. vapor/intake

   B. vapor/exhaust

   C. liquid/intake

   D. liquid/exhaust

4. On a swash-plate-style compressor, what prevents the refrigerant that is being compressed from being pushed back into the suction side of the compressor?

   A. The piston action

   B. The check valve

   C. The intake reed valve

   D. The discharge reed valve

5. Two technicians are discussing the discharge stroke of the compressor. Technician A states that the discharge stroke may be referred to as the compression stroke. Technician B states that during the compression stroke, the discharge valve in the compressor must seal tightly. Which technician is correct?

   A. Technician A is correct.

   B. Technician B is correct.

   C. Both Technicians A and B are correct.

   D. Neither Technician A nor B is correct.

6. What change happens to a variable displacement compressor to change its performance?

   A. Piston stroke

   B. Piston bore

   C. Control valve

   D. Suction pressure

7. Which of the following compressors has the fewest moving parts?

   A. Two-piston compressor

   B. Swash plate compressor

   C. Rotary compressor

   D. Scroll compressor

8. What is the most common cause of compressor failure?

   A. Broken suction reed valves

   B. Defective bearings

   C. Lack of lubrication

   D. Improper clutch engagement

9. A refrigerant reacts chemically with moisture to:

   A. Create sludge in the crankcase of the compressor

   B. Block the flow of refrigerant in the evaporator

   C. Form harmful deposits on the valve plates

   D. Form hydrochloric acid in the air-conditioning system

10. Two technicians are discussing the drier. Technician A states that the drying agent is called the absorber. Technician B states the drying agent is called a desiccant. Which technician is correct?

    A. Technician A is correct.

    B. Technician B is correct.

    C. Both Technicians A and B are correct.

    D. Neither Technician A nor B is correct.

11. The receiver-drier is used in an air-conditioning system that also uses a _____ as a metering device.

    A. fixed orifice tube (FOT)

    B. thermostatic expansion valve (TXV)

    C. Both A and B

    D. Neither A nor B

12. Technician A states that the receiver-drier should be replaced any time major repairs are made to the air-conditioning system. Technician B says that the desiccant must be suitable for the type of refrigerant used within the air-conditioning system. Which technician is correct?

    A. Technician A is correct.   C. Both technicians A and B are correct.

    B. Technician B is correct.   D. Neither technician A nor B is correct.

13. The main function of the compressor is to:

    A. Release heat energy   C. Compress refrigerant vapor

    B. Absorb heat energy   D. Store liquid refrigerant

5 Air-Conditioning Components: Metering Devices, Evaporator, Accumulator

Learning Objectives

Upon completion and review of this chapter, the student should be able to:

- List the two different types of metering devices used by air-conditioning systems.
- Explain the operation of a thermostatic expansion valve (TXV).
- Explain the purpose of the remote bulb and capillary tube.
- Describe what conditions are present if an evaporator is said to be operating in a flooded state.
- Describe what conditions are present if an evaporator is said to be in a starved condition.
- Explain what effect a starved evaporator has on the performance of an air-conditioning system.
- List the differences between internally and externally equalized thermostatic expansion valves.
- Explain the term *superheat*.
- Perform superheat calculations.
- Describe the operation of the fixed orifice tube.
- Explain the function of the evaporator.
- Describe the term *flash gas*.
- Explain the purpose of the accumulator and describe how it functions.

Key Terms

accumulator	fixed orifice tube	starved
capillary tube	flash gas	superheat
equalizing	flooded	thermostatic expansion valve
evaporator	internally equalized	
externally equalized	remote bulb	

INTRODUCTION

In **Chapter 1,** we discussed the basic components in general. This chapter deals with the remaining components involved in every mechanical refrigeration/air-conditioning system.

A typical air-conditioning system is made up of the following five major components:

- Compressor
- Condenser
- Receiver-drier or accumulator (depending upon the type of system)

- Thermostatic expansion valve or orifice tube (depending upon the type of system)
- Evaporator

This chapter will deal with the thermostatic expansion valve, the orifice tube, the evaporator, and the accumulator.

METERING DEVICES

Regardless of the type of air-conditioning system that a truck manufacturer uses, there are generally only two ways of controlling the amount of refrigerant that is allowed to enter the evaporator, these being the **thermostatic expansion valve** and the **fixed orifice tube**. The amount of refrigerant that the evaporator can handle constantly changes because heat loads are continuously fluctuating. If the metering device does not allow enough refrigerant into the evaporator, cooling performance will be poor. If too much refrigerant is allowed to enter the evaporator, there is a probability that some of the liquid refrigerant will not have time to change states from a liquid to a gas before it reaches the inlet of the compressor. If liquid refrigerant is able to enter the compressor, it produces untold damage because the compressor is not capable of compressing liquid. The thermostatic expansion valve is used with a receiver-drier, and the fixed orifice tube incorporates an accumulator/drier. The accumulator/drier will be discussed later in this chapter. The receiver-drier was discussed in **Chapter 4,** so the next component to be discussed is the thermostatic expansion valve.

THERMOSTATIC EXPANSION VALVE

Thermostatic expansion valves can be divided into two different styles, internally equalized and externally equalized. The externally equalized valve will be discussed later in this chapter.

INTERNALLY EQUALIZED THERMOSTATIC EXPANSION VALVE

The **internally equalized** thermostatic expansion valve, as you will discover, has three forces at work that are constantly trying to open the valve (allowing more refrigerant to pass into the evaporator) or trying to close the valve (restricting or entirely blocking the flow of refrigerant into the evaporator) **(Figure 5-1).**

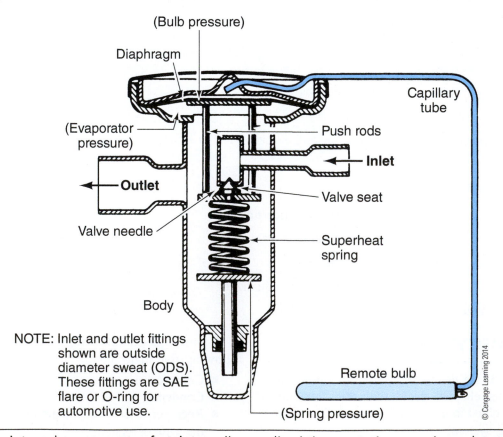

Figure 5-1 The internal components of an internally equalized thermostatic expansion valve.

Remote Bulb

The **remote bulb** or sensing bulb of the thermostatic expansion valve may contain one of several types of inert gas/liquid. Quite frequently, the inert gas/liquid is the same refrigerant used within the air-conditioning system. The sensing bulb is attached to the tailpipe (outlet side) of the evaporator so that it will monitor the temperature of the vapor as it leaves the evaporator. The bulb must be clamped tightly to the tailpipe and wrapped with a sticky cork tape to insulate it from the surrounding air. This ensures that the bulb will accurately transmit the temperature of the evaporator outlet to the top half of the diaphragm.

Capillary Tube

The **capillary tube** resembles a length of copper wire, but the inside of the wire is hollow, allowing the pressure from the remote bulb to be transmitted through the capillary tube and to be applied to the top portion of the diaphragm assembly.

Technicians working with thermostatic expansion valves must be extremely careful when handling the capillary tube. If the tube is too long for the intended application, the excess capillary tube must be coiled without kinking the tube. If the tube becomes kinked, the thermostatic expansion will cease to function because the increases or decreases of pressure sensed by the remote bulb will not be transmitted to the diaphragm.

One way to prevent kinking of the capillary tube is to wrap the excess tube around a cylindrical object with a diameter of about three or four inches, then remove the object and tie wrap the excess to prevent vibrations from causing the capillary tube to rub against itself.

THERMOSTATIC EXPANSION VALVE OPERATION

The thermostatic expansion valve controls the flow of refrigerant into the evaporator to maintain a constant superheat. The operation of the valve is determined by three fundamental pressures:

1. The sensing bulb pressure applied to one side of the diaphragm tries to open the valve against spring pressure. The bulb is usually charged with the same refrigerant used by the system in which it is installed.

Note: Sensing bulb pressure is the only force that tries to open the valve.

2. The evaporator inlet pressure applied to the opposite side of the diaphragm helps to make the valve responsive to compressor suction pressure.

Note: This pressure tries to close the valve.

3. The spring pressure, which is applied to the needle assembly and diaphragm on the evaporator side, constantly applies pressure to close the valve.

If the sensing bulb pressure were removed, the expansion valve would be closed by the spring pressure and equalizer line pressure. It is very important that the sensing bulb accurately senses evaporator outlet temperature.

As the temperature of the vapor in the evaporator rises, the sensing element temperature also rises, increasing the pressure of the liquid/vapor in the sensing bulb. This pressure is transmitted through the capillary tube and is applied to the diaphragm, which in turn opens the needle valve and allows more refrigerant to enter the evaporator. As all this occurs, evaporator pressure rises, in turn causing the valve to close slightly to maintain equilibrium.

If at this time the valve does not feed enough refrigerant, the evaporator pressure will fall. The sensing bulb temperature will now increase due to the warmer vapor leaving the evaporator, and the valve will again open further, allowing more refrigerant to enter the evaporator until the three pressures are again in balance.

As the load on the evaporator increases, the liquid refrigerant evaporates at a faster rate, thereby increasing the evaporator pressure. The increase in pressure causes an increase in temperature in the evaporator so that the increasing sensing bulb pressure on the top side of the diaphragm is equal to the pressure of the evaporator on the lower side of the diaphragm. The two pressures tend to cancel each other out as the valve adjusts to changes in load **(Figure 5-2)**.

Once the air-conditioning system is running and has had a chance to stabilize, the valve performs three functions within the system:

- It throttles the refrigerant.
- It modulates the refrigerant.
- It controls the refrigerant.

Throttling

The thermostatic expansion valve is the division between the high-pressure and low-pressure sides of

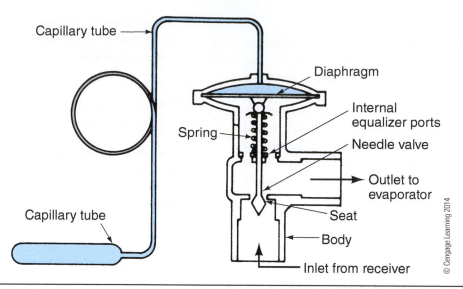

Figure 5-2 Refrigerant flow through a typical internally equalized expansion valve.

the air-conditioning system. Because there is a pressure differential between the inlet of the valve and the outlet of the valve, it can be said that the valve restricts or throttles the flow of refrigerant throughout the system. Refrigerant enters the thermostatic expansion valve as a high-pressure liquid and passes through the metering orifice. The metering orifice lowers the pressure of the refrigerant and preserves the refrigerant in its liquid state. This low-pressure, low-temperature liquid can now flow through the evaporator.

Note: There is always a small percentage of liquid refrigerant that changes to gas as it passes through the orifice. The drop in refrigerant pressure and heat contained within the refrigerant is the energy source that will cause some of the liquid refrigerant to flash into a gas, which is thus called **flash gas**. Flash gas is not desirable within an air-conditioning system. The more flash gas, the more inefficiently the air-conditioning system performs. This is because refrigerant that changes states from a liquid to gas before it reaches the evaporator will not remove much heat from the controlled space and this is considered a loss of air-conditioning capacity. The more the refrigerant is subcooled before it enters the TXV, the less flash gas will be produced.

Modulation

The job of the TXV is to meter the correct amount of refrigerant into the evaporator so that the

air-conditioning system can operate efficiently under the ever-changing heat load conditions. The TXV is constantly opening and closing as the heat loads change.

Controlling Action

The TXV has the ability to change the amount of liquid refrigerant that enters the evaporator as heat loads change. When heat loads increase, more liquid refrigerant must be metered into the evaporator, and when heat loads decrease, the TXV must close down, metering less liquid refrigerant into the evaporator. If the TXV allows too much refrigerant into the evaporator, it may not all evaporate by the time the refrigerant reaches the end of the evaporator. When liquid refrigerant exits the outlet of the evaporator, the evaporator is said to be operating in a **flooded** state.

For optimum evaporator performance, it is desirable to have the last bit of liquid refrigerant evaporate (change state) just as it leaves the evaporator. Due to sudden changes of heat load on the evaporator, it might be possible for liquid to return to the compressor, thereby causing possible compressor damage. The TXV should allow enough liquid refrigerant into the evaporator so that it can absorb enough heat from the controlled space and evaporate completely before it reaches the last couple of passes through the evaporator coil. If the TXV does not allow enough liquid refrigerant to enter the evaporator, it will all evaporate long before it reaches the end of the evaporator. This condition is referred to as a **starved** evaporator. Once the liquid refrigerant has all changed states, the cooling process slows down drastically. Therefore, if the TXV does not allow enough liquid refrigerant to enter the

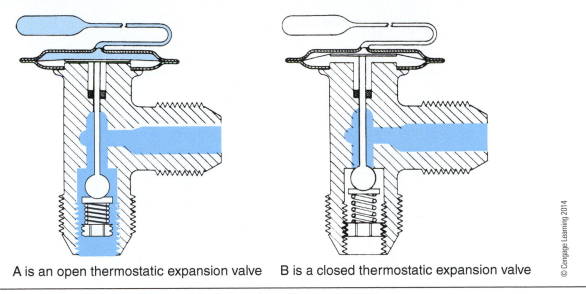

A is an open thermostatic expansion valve   B is a closed thermostatic expansion valve

Figure 5-3 The valve in both the open and closed positions.

evaporator, the air-conditioning system will not cool sufficiently **(Figure 5-3)**.

EXTERNALLY EQUALIZED THERMOSTATIC EXPANSION VALVES

As mentioned earlier, the thermostatic expansion valve may be one of two types, internally equalized or externally equalized. The term *equalized* refers to the way evaporator pressure is exerted on the underside of the diaphragm. On an internally equalized thermostatic expansion valve, the pressure to the underside of the diaphragm comes from the inlet side of the evaporator and reaches the underside of the diaphragm through a drilled passage. On an **externally equalized** TXV, the pressure to the lower side of the diaphragm comes from the outlet or tailpipe of the evaporator.

The operation of the externally equalized expansion valve is identical to that of the internally equalized expansion valve. The externally equalized valve overcomes the effect of refrigerant pressure drop on systems with large evaporators. The externally equalized valve uses an equalizer tube that is attached to the evaporator tailpipe and runs to the underside of the diaphragm. This arrangement allows the TXV to accurately adjust to the outlet pressure of the evaporator, eliminating the pressure drop of the evaporator. The superheat setting of the valve is dependent on the tension of the superheat spring **(Figures 5-4 and 5-5)**.

INTERNALLY EQUALIZED   **EXTERNALLY EQUALIZED**

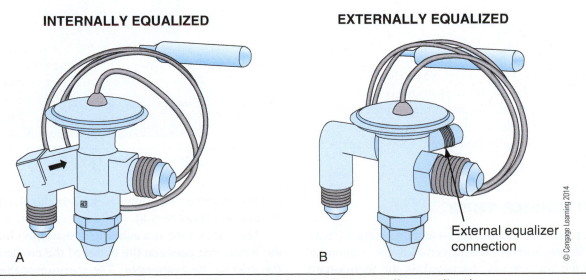

External equalizer connection

A   B

Figure 5-4 Two typical TXVs; A is internally equalized and B is externally equalized.

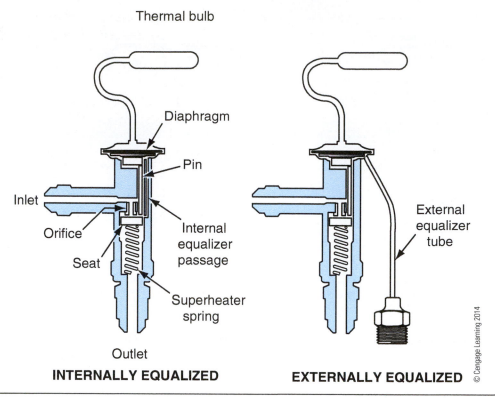

Thermal bulb

INTERNALLY EQUALIZED

Diaphragm

Pin

Inlet

Orifice

Seat

Internal equalizer passage

Superheater spring

Outlet

EXTERNALLY EQUALIZED

External equalizer tube

© Cengage Learning 2014

Figure 5-5 An internally and externally equalized TXV.

THE H VALVE

The H valve, sometimes referred to as the H-block, is another type of TXV used extensively in the heavy truck industry. It gets its name from the H shape of the internal passages. This valve contains two refrigerant passages, the first one of which is from the condenser to the evaporator and contains the ball and spring metering orifice that controls the amount of refrigerant allowed to pass into the evaporator. The other passage is the refrigerant line from the outlet of the evaporator to the inlet of the compressor. This section of the valve contains the valve's temperature sensing element. This section is like the capillary tube and diaphragm section of a normal TXV and operates the same way. Because of this second refrigerant line, there is no need for a sensing bulb or capillary tube. The diaphragm senses pressure from the outlet side of the evaporator, essentially making it an externally equalized TXV **(Figures 5-6 and 5-7).**

John Dixon © 2014 Cengage Learning

Figure 5-6 A typical H-block TXV (external view).

FIXED ORIFICE TUBE

There are two ways that air-conditioning systems control the pressure of the refrigerant in the evaporator. One, which was already discussed, is the thermostatic

expansion valve. The other method is the fixed orifice tube (FOT) **(Figure 5-8).**

The orifice tube is a calibrated restriction located in the liquid line between the outlet of the condenser and the inlet of the evaporator. The purpose of the orifice

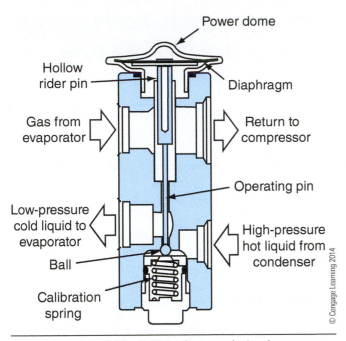

Figure 5-7 An H-block TXV (internal view).

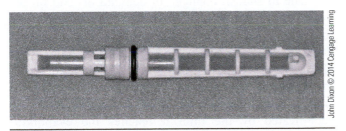

Figure 5-8 A fixed orifice tube.

tube is to meter refrigerant into the evaporator. It is also the division between the high- and low-pressure sides of the air-conditioning system. The restriction of the orifice tube causes a change in the pressure of the refrigerant. Refrigerant enters the orifice tube as a high-pressure, high-temperature liquid and leaves the orifice tube as a low-pressure liquid. (A small amount of the refrigerant is in a vaporous state due to flash gas.)

The amount of refrigerant the orifice tube allows to pass into the evaporator depends upon three factors:

- The size of the orifice
- The amount of subcooling that the refrigerant has been allowed to achieve
- The pressure differential between the inlet and the outlet of the orifice tube

The orifice tube is commonly referred to as a fixed orifice tube (FOT), because it is tubular in shape and the diameter of the orifice is fixed (does not change in size).

The passage through which the refrigerant flows is available in sizes ranging from 0.047 inch (1.19 mm) to 0.072 inch (1.83 mm), depending upon the size of the air-conditioning system in which the orifice tube will be used.

Orifice tubes may be color-coded to indicate the size of the orifice.

Color	Orifice Size (Diameter)
Blue or Black	0.067 inch (1.70 mm)
Red	0.062 inch (1.57 mm)
Orange	0.057 inch (1.45 mm)
Brown	0.053 inch (1.35 mm)
Green	0.047 inch (1.19 mm)

The fixed orifice tube usually has a filter screen on the inlet and the outlet of the tube **(Figure 5-9)**. If foreign material collects on the inlet or outlet screen, it seriously hinders the performance of the air-conditioning system. If the blockage is severe, the air-conditioning system will stop functioning altogether. The orifice tube can be removed and cleaned, but in severe instances it should be replaced. Most manufacturers recommend that the accumulator be replaced whenever the screen of the orifice tube is found to be clogged. Quite frequently, the foreign material that collects on the screen is small particles of desiccant. Because the desiccant material is inside the accumulator and it is not serviceable, the accumulator must be replaced. (The accumulator will be discussed later in this chapter.)

On the body of the fixed orifice tube is an O-ring. This O-ring creates a seal between the orifice tube body and the refrigeration line in which it is installed. This seal prevents refrigerant from bypassing the orifice. If the O-ring appears to be imperfect, the entire orifice tube assembly should be replaced.

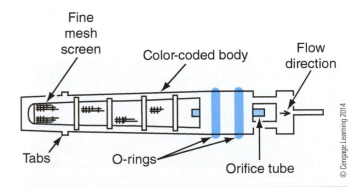

Figure 5-9 The parts of the orifice tube.

The orifice tube is generally pushed into a cavity in the liquid line between the outlet of the condenser and the inlet of the evaporator. The orifice tube is usually fairly easy to access by opening the refrigerant line and simply pulling out the old one and replacing it with a new one. Note all refrigerant must be removed to perform this task.

Note: Some orifice tubes have an arrow embossed in the plastic body. This arrow indicates the direction of refrigerant flow through the orifice so that the valve can be installed in the correct direction **(Figure 5-10).**

The orifice tube contains no moving parts, so it can never change the amount of refrigerant that can flow through it. Because the orifice tube will never close, the refrigerant inside the system will equalize when the engine or compressor is shut off. The term **equalizing** means that the high-pressure refrigerant within the system will flow through to the low-pressure side until both sides are at the same pressure. If the engine is shut off, the flow of the equalizing refrigerant can be detected by a soft gurgling sound.

Air-conditioning systems that use an orifice tube as a means of metering refrigerant into the evaporator also use a component called an **accumulator**.

Variable Orifice Valve

The variable orifice valve (VOV) is an aftermarket orifice tube that can be used with many manufacturers' air-conditioning systems. These valves use system pressure and refrigerant flow to move a metering piston relative to a fixed opening in the sleeve. The piston movement is resisted by an attached spring **(Figure 5-11).** Some of the advantages of VOV claimed by manufacturers include:

- Air that is 5°F (−5°C) to 12°F (−11.1°C) colder (at hot idle)

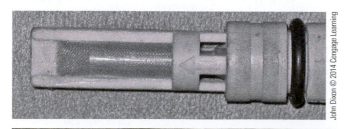

Figure 5-10 A close-up view of a fixed orifice tube; note that the arrow indicates the direction of refrigerant flow.

Figure 5-11 A variable orifice valve (VOV).

- Reduced compressor load and extended compressor life
- Improved performance when converting from R-12 to R-134a
- Improved city fuel economy and emissions
- Improved performance in factory R-134a systems

SUPERHEAT

For optimum operation of the evaporator, it would be desirable to have the last bit of liquid refrigerant evaporate (change state) just as it leaves the evaporator. Due to sudden changes in demand or load on the evaporator, it might be possible for liquid to return to the compressor, thereby causing possible compressor damage. To prevent this, the TXV is designed to allow all of the refrigerant to evaporate far enough back in the evaporator so that the refrigerant vapor can absorb enough heat to be 8°F (−13.3°C) to 12°F (−11.1°C) superheated by the time it reaches the evaporator outlet. This helps prevent liquid refrigerant from entering the compressor.

As mentioned in **Chapter 3,** the term **superheat** refers to the temperature of the refrigerant vapor above its boiling point. You might also recall from **Chapter 3** that pressure has an effect on the boiling point of a substance. (The higher the pressure, the higher the boiling point.)

Determining Superheat

To determine the superheat setting of a valve, it is necessary to measure the evaporator outlet temperature and the evaporator pressure, preferably at or near the outlet of the evaporator. Use a pressure-temperature chart **(Figures 5-12** and **5-13)** to look up the pressure of the refrigerant and then determine the temperature.

The temperature found on the pressure-temperature chart is the saturation point of the refrigerant at its current pressure. This means that this is the temperature at which the refrigerant has started to change states from a liquid to a vapor (the boiling point). Any temperature above the saturation point is considered to be superheat. Now take that temperature and subtract it

Temp. °F	Press. psig	Temp. °F	Press. psig	Temp. °F	Press. psig	Temp. °F	Press. psig	Temp. °F	Press. psig
0	9.1	35	32.5	60	57.7	85	91.7	110	136.0
2	10.1	36	33.4	61	58.9	86	93.2	111	138.0
4	11.2	37	34.3	62	60.0	87	94.8	112	140.1
6	12.3	38	35.1	63	61.3	88	96.4	113	142.1
8	13.4	39	36.0	64	62.5	89	98.0	114	144.2
10	14.6	40	36.9	65	63.7	90	99.6	115	146.3
12	15.8	41	37.9	66	64.9	91	101.3	116	148.4
14	17.1	42	38.8	67	66.2	92	103.0	117	151.2
16	18.3	43	39.7	68	67.5	93	104.6	118	152.7
18	19.7	44	40.7	69	68.8	94	106.3	119	154.9
20	21.0	45	41.7	70	70.1	95	108.1	120	157.1
21	21.7	46	42.6	71	71.4	96	109.8	121	159.3
22	22.4	47	43.6	72	72.8	97	111.5	122	161.5
23	23.1	48	44.6	73	74.2	98	113.3	123	163.8
24	23.8	49	45.6	74	75.5	99	115.1	124	166.1
25	24.6	50	46.6	75	76.9	100	116.9	125	168.4
26	25.3	51	47.8	76	78.3	101	118.8	126	170.7
27	26.1	52	48.7	77	79.2	102	120.6	127	173.1
28	26.8	53	49.8	78	81.1	103	122.4	128	175.4
29	27.6	54	50.9	79	82.5	104	124.3	129	177.8
30	28.4	55	52.0	80	84.0	105	126.2	130	182.2
31	29.2	56	53.1	81	85.5	106	128.1	131	182.6
32	30.0	57	55.4	82	87.0	107	130.0	132	185.1
33	30.9	58	56.6	83	88.5	108	132.1	133	187.6
34	31.7	59	57.1	84	90.1	109	135.1	134	190.1

© Cengage Learning 2014

Figure 5-12 A pressure-temperature chart for R-12.

Temperature °F	Pressure psig	Temperature °F	Pressure psig
–5	4.1	39.0	34.1
0	6.5	40.0	35.0
5.0	9.1	45.0	40.0
10.0	12.0	50.0	45.4
15.0	15.1	55.0	51.2
20.0	18.4	60.0	57.4
21.0	19.1	65.0	64.0
22.0	19.9	70.0	71.1
23.0	20.6	75.0	78.6
24.0	21.4	80.0	86.7
25.0	22.1	85.0	95.2
26.0	22.9	90.0	104.3
27.0	23.7	95.0	113.9
28.0	24.5	100.0	124.1
29.0	25.3	105.0	134.9
30.0	25.3	110.0	146.3
31.0	27.0	115.0	158.4
32.0	27.8	120.0	171.1
33.0	28.7	125.0	184.5
34.0	29.5	130.0	198.7
35.0	30.4	135.0	213.5
36.0	31.3	140.0	229.2
37.0	32.2	145.0	245.6
38.0	33.2	150.0	262.8

© Cengage Learning 2014

Figure 5-13 A pressure-temperature chart for R-134a.

from the temperature measured at the evaporator outlet. Here again are the four easy steps:

- Step 1. Determine suction pressure with an accurate gauge at the compressor suction service valve.
- Step 2. Using the refrigeration pressure-temperature charts, determine saturation temperature at the observed suction pressure.
- Step 3. Measure the temperature of the suction gas at the evaporator outlet.
- Step 4. Subtract the saturation temperature read from tables in Step 2 from the temperature measured in Step 3. The difference between the two is the superheat of the suction gas returning to the compressor.

Most manufacturers recommend a superheat setting of between 8°F and 12°F to provide a balance of safety and efficiency.

Example. A technician working on an R-134a system measures the suction pressure of the evaporator and finds it to be 24.5 psig (168.9 kPa). The technician then goes to the pressure-temperature chart and finds that the

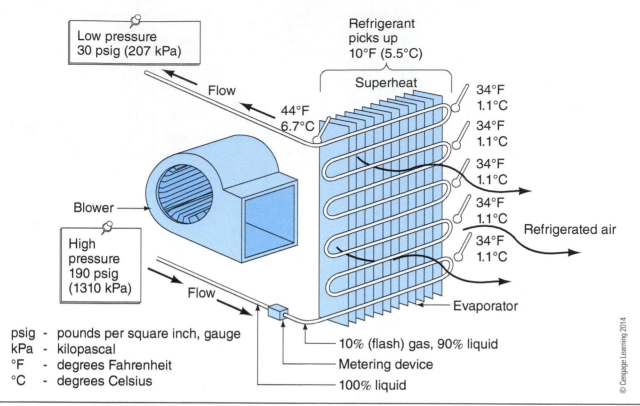

Figure 5-14 A tube- and fin-type evaporator; the refrigerant changes temperature only after the change of state from liquid to vapor is complete. The rise in temperature indicates the refrigerant's super heat value.

saturation temperature for the refrigerant at this pressure is 28°F (−2.2°C). The technician then accurately measures the temperature of the evaporator tailpipe and finds it to be 36°F (2.2°C). The difference between the measured temperature, 36°F (2.2°C), and the saturation point of the refrigerant vapor, 28°F (−2.2°C), is 8 Fahrenheit degrees of superheat or 4.4 degrees Celsius of superheat **(Figure 5-14).**

EVAPORATOR

The **evaporator** is typically mounted in the blower housing beside the heater core. This assembly is housed in the cab of the truck beneath the dashboard.

The evaporator relies on two key principles of heat transfer:

- Heat energy always moves from hotter to colder.
- Heat energy transfers quickly between objects having a large temperature difference.

The evaporator is the component of the air-conditioning system in which liquid refrigerant vaporizes (changes states).

Construction of the evaporator is similar to that of the condenser. The evaporator consists of tubing running through sharp aluminum fins. The tubing contains

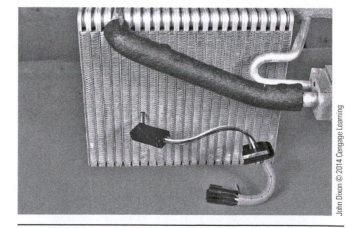

Figure 5-15 A plate- and fin-type evaporator.

the refrigerant, while the aluminum fins increase the surface area of the evaporator. Because the fins are attached firmly to the tubing, conduction transfers the heat of the circulating air through to the refrigerant within the coils of the evaporator **(Figure 5-15).**

While the air-conditioning system is operating, a blower motor circulates the warm air inside the truck cab through the evaporator. The heat from the cab is absorbed by the liquid refrigerant. When enough heat is absorbed from the cab, the liquid refrigerant vaporizes or boils. The heat that causes the refrigerant to boil is then expelled to the ambient air in the condenser section. In other words, the evaporator cools the cab air by removing the heat from it.

Another benefit of an air-conditioning system is dehumidification. The warm, moist air from the cab of the truck is circulated through the evaporator by the blower fan. The moisture from the air condenses into water droplets on the cool fins of the evaporator. The water falls into a drip pan beneath the evaporator and is expelled to the ground through a drain tube. When a truck air-conditioning system is shut down, it will sometimes form a puddle of water where the water drains from the evaporator.

Important considerations in the design of an evaporator are size, the length of the tubing, the number and size of the fins, the number of times the tubing passes through the fins, and the volume of air that passes through the fins. One other consideration is the heat load. The heat load is the amount of heat in BTUs that the air-conditioning system is required to remove.

The construction of the evaporator may contain two, three, or more cores, depending on the BTU requirements and the space available within the evaporator housing. (A core is one row of refrigerant tubing attached to the fins. Each row of tubing behind the first row becomes another core. This is the same concept as is used in the construction of truck radiators.) The evaporator must be large enough to allow the refrigerant to vaporize and become slightly superheated before it leaves the evaporator on its way back to the compressor.

When too much refrigerant is metered into the evaporator for the amount of heat load, the evaporator may become flooded. The term *flooded* indicates that liquid refrigerant is exiting the evaporator on its way to the compressor.

The refrigeration process can be affected by a flooded evaporator because the refrigerant pressure is too high and it cannot boil away as quickly. (Recall from **Chapter 3** that a rise in pressure increases boiling point.) Also remember that if the refrigerant is not changing states, the amount of heat being absorbed by the refrigerant will be much less. The greatest heat transfer occurs when the refrigerant is in a latent heat state (refrigerant is changing from liquid to vapor).

If an air-conditioning system using a TXV is allowed to operate in a flooded state, eventually serious damage will be caused to the compressor. In an air-conditioning system that uses a fixed orifice tube, the evaporator may operate in a flooded state, depending upon heat load conditions. This design is not a problem because the accumulator prevents any liquid refrigerant from entering the compressor.

When a metering device does not allow enough refrigerant to pass into the evaporator, the air-conditioning system will not cool properly, either. This is a condition known as a starved evaporator. In this condition, all the liquid refrigerant that is able to enter the evaporator vaporizes or boils off shortly after entering the evaporator. This leaves a large portion of the evaporator removing very little or no heat at all. Under this condition, the superheat of the refrigerant leaving the evaporator would be high to very high.

Evaporator Service

The evaporator, like the condenser, is generally maintenance-free as long as the fins are not damaged or blocked with dust or debris. If the evaporator develops a refrigerant leak, it usually needs to be replaced, because evaporators are generally non-repairable.

ACCUMULATOR

Air-conditioning systems that use a fixed orifice tube also use an accumulator. The accumulator acts somewhat like the receiver-drier does in a system that uses a thermostatic expansion valve. The purpose of the accumulator is to prevent liquid refrigerant from entering the compressor. It also acts as a storage vessel for excess liquid refrigerant. The accumulator contains a desiccant to absorb moisture from the refrigerant. The accumulator is located in the suction line between the outlet of the evaporator and the suction side of the compressor **(Figure 5-16).**

As discussed earlier in this chapter, the orifice tube or fixed orifice tube contains no moving parts, so when air-conditioning loads change, it is possible for the

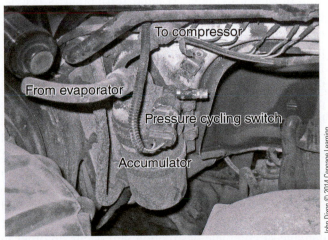

Figure 5-16 An accumulator mounted on the firewall of a (Ford) tractor.

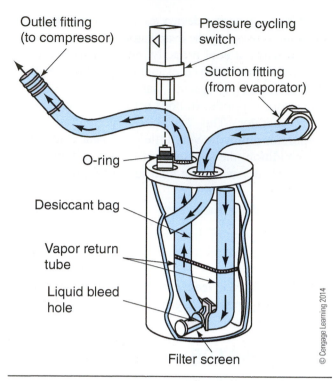

Outlet fitting
(to compressor)

Pressure cycling
switch

Suction fitting
(from evaporator)

O-ring

Desiccant bag

Vapor return
tube

Liquid bleed
hole

Filter screen

© Cengage Learning 2014

Figure 5-17 Internal piping of an accumulator.

evaporator to run in a flooded state. This means that not all the liquid refrigerant that enters the evaporator has time to evaporate; some of the refrigerant is still liquid when it exits the evaporator. The accumulator, for lack of a better term, accumulates the liquid refrigerant and compressor oil, separating it from the vaporous refrigerant.

As you look at **Figure 5-17,** follow the flow of refrigerant through the accumulator. As you can see, refrigerant enters the accumulator from the outlet of the evaporator. The inlet tube ends high up in the cylinder, allowing the liquid refrigerant to drop to the bottom of the cylinder. (Liquid refrigerant is heavier than vaporous refrigerant, so it will always flow to the bottom.) The refrigerant that returns to the compressor must pass through the pickup tube. The pickup tube is "U" shaped, with the inlet portion of the tube being the highest point. Having the pickup at the highest point ensures that only vaporous refrigerant can flow back to the suction side of the compressor. The liquid refrigerant left in the bottom of the accumulator will eventually gather enough heat to vaporize and be drawn down through the pickup tube. Compressor oil cannot be allowed to collect in the bottom of the accumulator because eventually the compressor will become starved of lubrication and severe compressor damage or complete destruction will occur. In order to prevent oil starvation by the accumulator, the pickup tube has a bleed hole drilled in its lowest point. This

small hole allows only a small amount of liquid refrigerant and compressor oil to be drawn up the pickup tube along with some vaporous refrigerant. (Because of the size of the bleed hole, not enough liquid refrigerant or oil can be drawn up the pickup tube to cause any damage to the compressor.)

As mentioned previously, the accumulator contains a desiccant. The desiccant attracts, absorbs, and holds moisture that may have entered the system due to an external leak or poor maintenance procedures. The type and makeup of the desiccant contained in the accumulator is the same as that used in a receiver-drier. The desiccant is a non-serviceable item and if replacement becomes necessary, the entire accumulator assembly should be replaced. The desiccant used in an R-12 system differs from that used in an R-134a system and the two types of desiccant are usually not compatible. To be sure of system compatibility, use only replacement parts specifically designed for the air-conditioning system that is being serviced.

The accumulator also contains a fine mesh screen to trap circulating debris, preventing it from contaminating the air-conditioning system. The mesh screen, like the desiccant, is a non-serviceable item and can only be serviced by replacing the accumulator assembly. Other filtering screens within the system are serviceable. These filtering screens are located at the compressor inlet and as part of the orifice tube itself.

Merchants of new compressors will generally not honor a compressor warranty unless the accumulator has been replaced at the same time.

CAUTION *Never remove the protective caps installed on new accumulators or receiver-driers until they are to be installed into the air-conditioning system. Remember that the desiccant is hygroscopic. The accumulator or receiver-drier should also be the last component installed before the vacuum pump is used to remove the air and moisture that entered the air-conditioning system during servicing.*

MAINTENANCE PROCEDURES

The components discussed require very little maintenance on the part of the technician. Performance testing of the system should point the technicians in the direction of any failing components. TXVs and orifice tubes are generally replaced if they fail. Accumulators, like receiver-driers, are replaced at intervals specified by the OEM, or whenever the system has been opened for repairs.

Summary

- All air-conditioning systems use some sort of metering device.

- The thermostatic expansion valve controls the flow of refrigerant into the evaporator to maintain a constant superheat.

- The pressure of the inert gas contained within the remote bulb is transmitted through the capillary tube to open the TXV.

- Evaporator inlet pressure (internally equalized) or outlet pressure (externally equalized) applied to the opposite side of the TXV diaphragm tries to close the metering portion of the TXV.

- The internal spring within the TXV tries to close the metering portion of the TXV. This spring may be referred to as the superheat spring.

- Superheat is the temperature of a refrigerant vapor above its saturation or boiling point.

- Superheated refrigerant leaving the evaporator ensures that liquid refrigerant is not able to return to the compressor, causing damage.

- To measure the superheat setting of a thermostatic expansion valve, the technician must have the following: an accurate temperature measurement of the evaporator outlet, a pressure measurement of the evaporator outlet, and a refrigerant pressure-temperature chart.

- The fixed orifice tube is a restriction in the liquid line dividing the high-pressure side of the refrigeration system from the low-pressure side. The FOT contains a precise refrigerant passage, intended for the size of the air-conditioning system in which it is to be used.

- The evaporator is the component in which refrigerant vaporizes and absorbs heat from the cab of the truck.

- The term *flooding* indicates that refrigerant is not all vaporizing in the evaporator and can damage the compressor. When liquid and vapor exist together, there is no superheat.

- The term *starved* means that the liquid refrigerant vaporizes in the evaporator far too early, causing extremely high superheat.

- The purpose of the accumulator is to prevent liquid refrigerant from entering the compressor.

- The accumulator must be used in an air-conditioning system that uses a fixed orifice tube as a metering device.

Review Questions

1. When refrigerant is metered into a properly working evaporator, the refrigerant:

    A. Becomes superheated

    B. Absorbs heat

    C. Changes states into a vapor

    D. All of the above

2. Technician A states that the condenser gives up heat to the ambient air. Technician B states that the boiling refrigerant in the evaporator absorbs heat. Which technician is correct?

    A. Technician A is correct.

    B. Technician B is correct.

    C. Both Technicians A and B are correct.

    D. Neither Technician A nor B is correct.

3. Two technicians are discussing the state of refrigerant as it enters the evaporator. Technician A says it is in a liquid state with a trace of flash gas. Technician B says that flash gas is caused by the decrease in refrigerant pressure as it passes through the metering device. Which technician is correct?

    A. Technician A is correct.

    B. Technician B is correct.

    C. Both Technicians A and B are correct.

    D. Neither Technician A nor B is correct.

4. What situation would be considered most harmful to a TXV system?

   A. Flooding of the evaporator, because the refrigerant will not vaporize due to the lack of space for expansion.

   B. Flooding of the evaporator, allowing liquid refrigerant to return and damage the compressor.

   C. Starving of the evaporator, which will drastically affect cooling efficiency.

   D. Starving of the evaporator, creating too much superheat, which could damage the compressor.

5. Which of the following statements is not true?

   A. The accumulator is located in the suction line between the evaporator outlet and the compressor inlet.

   B. The receiver-drier is located in the liquid line between the condenser outlet and the inlet of the TXV.

   C. The evaporator is considered to be in the low-pressure side of the air-conditioning system.

   D. All fixed orifice tubes are identical.

6. In order for refrigerant to change states, there must be a transfer of heat. Technician A states that if heat is removed from refrigerant, it will change states from a liquid to a vapor. Technician B states that if heat is added to refrigerant, it will change from a liquid to a vapor. Which technician is correct?

   A. Technician A is correct.

   B. Technician B is correct.

   C. Both Technicians A and B are correct.

   D. Neither Technician A nor B is correct.

7. Technician A states that the purpose of the pickup tube in a receiver-drier is to ensure liquid refrigerant is available to the TXV. Technician B says that the pickup tube in the accumulator is to prevent liquid refrigerant from returning to the compressor. Which technician is correct?

   A. Technician A is correct.

   B. Technician B is correct.

   C. Both Technicians A and B are correct.

   D. Neither Technician A nor B is correct.

8. Technician A states that the accumulator assembly should be replaced any time major repairs are made to the air-conditioning system. Technician B says that the desiccant must be suitable for the type of refrigerant used within the air-conditioning system. Which technician is correct?

   A. Technician A is correct.

   B. Technician B is correct.

   C. Both Technicians A and B are correct.

   D. Neither Technician A nor B is correct.

9. To what is the TXV remote bulb fastened?

   A. The condenser outlet pipe

   B. The condenser inlet pipe

   C. The evaporator inlet pipe

   D. The evaporator outlet pipe

10. Two technicians are discussing the purpose of the orifice tube and the thermostatic expansion valve. Technician A says that the FOT meters a fixed quantity of refrigerant into the evaporator. Technician B says that the TXV meters a fixed quantity of refrigerant into the evaporator. Which technician is correct?

    A. Technician A is correct.

    B. Technician B is correct.

    C. Both Technicians A and B are correct.

    D. Neither Technician A nor B is correct.

CHAPTER

6 The Refrigeration System

Learning Objectives

Upon completion and review of this chapter, the student should be able to:

- Differentiate between thermostatic expansion valve and orifice tube air-conditioning systems.
- Identify the high side and low side of a truck air-conditioning system.
- Describe the purpose of an air-conditioning system.
- Identify the type of refrigerant used by an air-conditioning system.
- Explain the term *ton of refrigeration capacity*.
- Identify the parts that make up the electromagnetic clutch.
- Describe what happens within the clutch when it is energized and de-energized.
- Explain how the thermostatic switch prevents the evaporator from freezing.
- Describe how the low-pressure switch prevents the evaporator from freezing.
- Explain why it is important that the evaporator not freeze on an operating air-conditioning system.
- Describe the function of a low pressure cut-off switch and how it controls compressor operation.
- Explain the purpose of the high pressure cut-off switch and how it controls compressor operation.
- Explain the purpose and function of the binary switch.
- Describe the functions of a trinary switch.

Key Terms

ambient

binary switch

de-energize

electromagnetic clutch

energize

fan cycling switch

high pressure cut-off switch

high-pressure relief valve

liquid line

low pressure cut-off switch

orifice tube air-conditioning system

saturated liquid

thermostatic expansion valve air-conditioning system

thermostatic switch

ton

trinary switch

INTRODUCTION

Once the technician/student understands the components and their purposes as explained in **Chapters 4 and 5,** it is time to put all the components together in sequential order. When the sequential order of an air-conditioning system is discussed, it is in reference to the direction or path the refrigerant takes as it flows through the system. The terms *air conditioning* and *refrigeration* can be used interchangeably because they are essentially the same thing. An air-conditioning system is really a refrigeration system.

SYSTEM OVERVIEW

The modern truck air-conditioning system uses a fluid or vapor known as refrigerant. The refrigerant is pumped to the evaporator, where it absorbs the heat from the truck cab, and then flows to the condenser to expel the heat from the truck cab to the **ambient** (outside) air.

Today's truck air-conditioning systems are made up of the components that were discussed in **Chapters 4 and 5.** There are two different styles of air-conditioning systems that can be constructed from the components. The style of air-conditioning system is usually defined by the way the refrigerant is metered into the evaporator. The two different systems are the fixed orifice tube (FOT) system and the thermostatic expansion valve (TXV) system. This chapter discusses the similarities and differences between these two styles of air-conditioning systems.

In addition to the major components already discussed, there must be some type of method to control the operation of the air-conditioning system. Here are a few reasons to control the system.

1. To control the comfort level for operator/passenger
2. To improve the cooling efficiency of the system
3. To protect the major components from damage

THE THERMOSTATIC EXPANSION VALVE SYSTEM

As the **thermostatic expansion valve (TXV) system** is discussed, refer to **Figure 6-1** and follow the flow of refrigerant through the various components.

Start at the compressor. The compressor circulates the refrigerant through the system. It raises the pressure of the refrigerant vapor. Recall from **Chapter 3** that when there is a rise in pressure, there is also a rise in temperature. The superheated refrigerant vapor returning to the compressor from the evaporator is full of heat from the cab of the truck. The compressor increases the pressure and temperature of the refrigerant and pumps the high-pressure superheated refrigerant vapor into the discharge line and on through to the condenser. The condenser is positioned in front of the truck's radiator. In this position, the condenser is subjected to the ram air effect as well as to the air circulation created by the engine's cooling fan. The more air that is circulated through the fins of the condenser, the more heat transfer can take place **(Figure 6-2)**.

Also, the difference in temperature between the condenser and the ambient air has a great bearing on how fast heat transfer can take place. As the refrigerant in the condenser gives up its heat to the ambient air (recall from **Chapter 3** that heat always flows from high temperature to low temperature), it changes from a high-pressure superheated vapor to a high-pressure **saturated liquid** (contains both liquid and vapor). This occurs somewhere near the middle of the condenser. The refrigerant continues to flow through the condenser and gives up more of its heat to the ambient air, and finally condenses into a high-pressure subcooled liquid by the time it reaches the end of the condenser. From the condenser, the refrigerant enters the **liquid line**. This line is appropriately called the liquid line, indicating the state of the refrigerant. If you look at a condenser, you will notice that the discharge line at the inlet of the condenser has a larger diameter than the outlet or liquid line. This is because liquid takes up less space than vaporous refrigerant.

The high-pressure subcooled liquid refrigerant leaving the condenser through the liquid line then enters the receiver-drier. The receiver-drier stores, dries, and filters the liquid refrigerant. The pickup tube within the receiver-drier ensures that only liquid refrigerant can be fed through to the thermostatic expansion valve.

Upon exiting the receiver-drier pickup tube, the high-pressure subcooled liquid refrigerant flows through to the thermostatic expansion valve (TXV). The TXV controls the flow of refrigerant into the evaporator by monitoring the temperature and pressure of the evaporator. The TXV also controls the degree to which the refrigerant is superheated as it leaves the evaporator. The subcooled liquid refrigerant passes through the orifice of the TXV. As this happens, the high-pressure liquid refrigerant changes to a low-pressure liquid. A very small percentage of the liquid refrigerant will have vaporized as the pressure was reduced through the TXV. This is known as flash gas. The more subcooled the

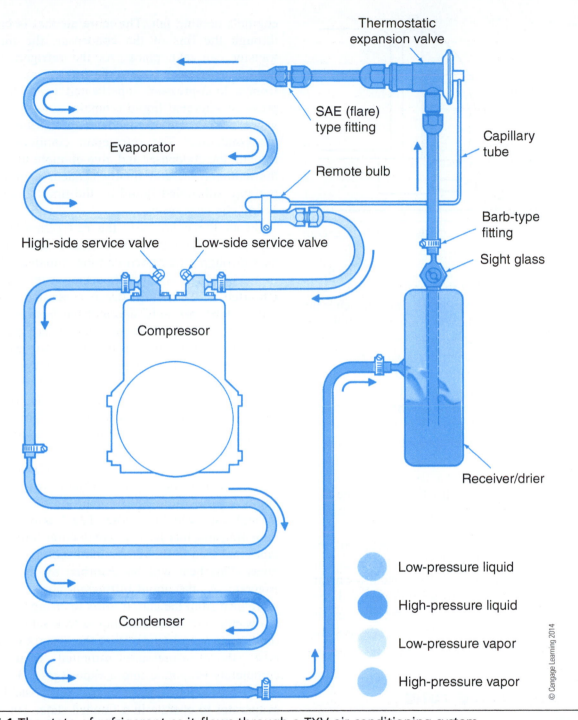

Figure 6-1 The state of refrigerant as it flows through a TXV air-conditioning system.

refrigerant when it enters the TXV, the less flash gas there will be.

The subcooled liquid refrigerant then flows through the evaporator. A fan circulates the warm air of the truck's cab, through the evaporator core. The fins of the evaporator transmit this heat through to the tubes and then to the refrigerant inside the evaporator tubes. This heat will then be absorbed by the liquid refrigerant. As the liquid refrigerant absorbs more and more heat from the cab, it starts to change states from a low-pressure subcooled liquid to a saturated liquid (both vapor and liquid). With the absorption of more heat, the low-pressure saturated liquid will completely evaporate into a vapor (gas).

Note: The heat from the cab is the heat source that causes the liquid refrigerant to boil (change states).

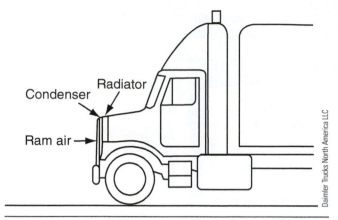

Daimler Trucks North America LLC

Figure 6-2 The ram air effect as a truck is driven forward.

The vapor can still absorb more heat, and it then becomes a superheated vapor by the time it reaches the end of the evaporator. The air from the cab that is circulated through the evaporator is cooled as it passes through the fins of the evaporator, or so it seems. You should now be able to see that the air is not actually cooled, but what really happens is that heat is removed from the air as it circulates through the evaporator, so it feels like it is being cooled. The air should drop about 10°F (6°C) as it passes through the evaporator. The next time the air passes through the evaporator, it will contain less heat than on its first pass, so the temperature in the cab of the truck will continue to be reduced until the thermostat stops the refrigerating process.

The low-pressure superheated refrigerant then leaves the evaporator and enters the suction line. From the suction line, the compressor draws the low-pressure superheated refrigerant into its suction side and compresses the refrigerant, which raises the pressure and temperature, and then the whole process starts over again.

THE FIXED ORIFICE TUBE SYSTEM

The fixed **orifice tube air-conditioning system** is another style used by many truck manufacturers. For the most part, it works very much like the TXV system, with some component changes **(Figure 6-3)**. Again, let us start from the compressor. Just as in the TXV system, the compressor pumps high-pressure superheated refrigerant vapor into the discharge line and on through to the condenser. The condenser is positioned in front of the truck's radiator. In this position, the condenser is subjected to the ram air effect as well as to the air circulation created by the

engine's cooling fan. The more air that is circulated through the fins of the condenser, the more heat transfer can take place. As the refrigerant in the condenser gives up heat to the ambient air, it changes from a high-pressure superheated vapor to a high-pressure saturated liquid (containing both liquid and vapor). This occurs somewhere around the middle of the condenser. The refrigerant continues to flow through the condenser and give up more of its heat to the ambient air, and finally condenses into a high-pressure subcooled liquid by the time it reaches the end of the condenser.

From the condenser, the refrigerant enters the liquid line. The subcooled liquid refrigerant will pass through a fixed orifice tube, situated within the liquid line. The fixed orifice tube was discussed in **Chapter 5**. The filter screen contained within the orifice tube prevents debris from being circulated back to the inlet of the compressor, where it can do the most physical damage. The liquid refrigerant is metered through a tube with a fixed diameter into the evaporator. A very small percentage of the liquid refrigerant will vaporize due to the reduction of pressure as it passes through the orifice tube. This vaporous refrigerant is known as flash gas. The more subcooled the refrigerant is when it enters the orifice tube, the less flash gas will be created. The subcooled liquid refrigerant next flows through the evaporator. A fan circulates the warm air from inside the truck's cab through the evaporator core. The fins of the evaporator transmit this heat by conduction through to the tubes and then to the refrigerant inside the evaporator tubes. This heat will be absorbed by the liquid refrigerant. As the liquid refrigerant absorbs more and more heat from the cab, it starts to change states from a low-pressure subcooled liquid to a saturated liquid (both vapor and liquid). With the absorption of more heat, the low-pressure saturated liquid should completely evaporate into a vapor (gas). Because the orifice tube cannot change in size, it is possible for the evaporator to operate in a flooded state. This is not a problem for the orifice tube system because of the accumulator.

Superheated refrigerant or saturated liquid leaves the evaporator and enters the accumulator. Due to the position of the pickup tube, only superheated refrigerant vapor is permitted to exit the accumulator. The accumulator contains a desiccant and filter to remove moisture and contaminants from the refrigerant. From the accumulator, the superheated refrigerant enters the suction line and is then drawn into the suction side of the compressor to begin the refrigeration cycle all over again.

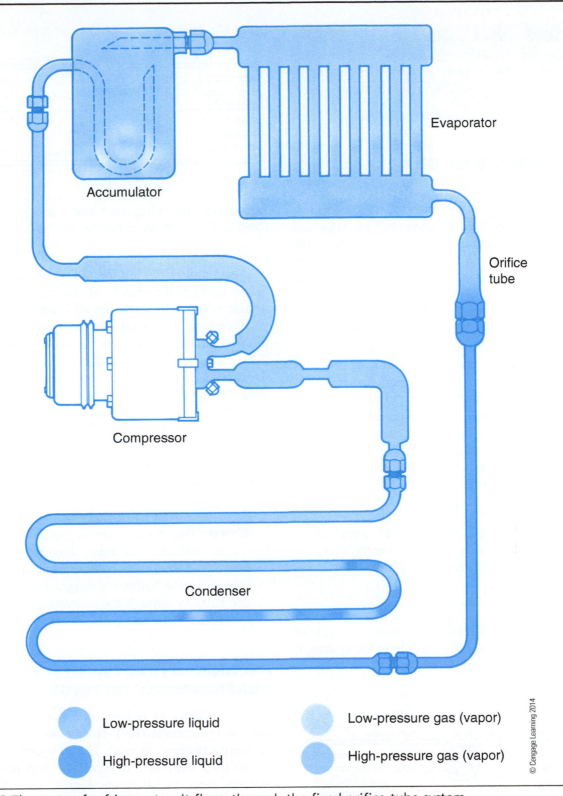

Figure 6-3 The state of refrigerant as it flows through the fixed orifice tube system.

Note: The air-conditioning system is a closed loop system. It is called this because the components of the system are connected together by hoses and fittings. During normal operation, the hoses on the high-pressure side of the system should feel warm or even hot, depending on how close you are to the compressor. The lines on the low-pressure side should feel cool or even cold. Whenever lines are disconnected from the system, they should be capped to prevent moisture and contaminants from entering the system.

REFRIGERANT PRESSURE AND STATES

From the two types of air-conditioning systems just discussed, it may be hard to remember the state and pressure the refrigerant is in at any point within the system. Here is a way in which most technicians can remember what is happening within the system. First, look at **Figure 6-4.**

In **Figure 6-4,** the compressor and the thermostatic expansion valve are the divisions between the high-pressure and low-pressure sides of the system. A vertical line is drawn through the middle of the compressor and the thermostatic expansion valve. Refrigerant should be considered to be high pressure in quadrants **A** and **B** between the compressor discharge and the inlet to the TXV while the system is operating. Refrigerant should therefore also be considered to be low pressure in quadrants **C** and **D** from the outlet of the TXV and the compressor inlet.

Note: Refrigerant pressure will equalize (high side and low side balance out each other) when the air-conditioning system is not operating.

Now that you can distinguish the high-pressure side of the system from the low-pressure side of the system, it is time to look at the refrigerant states.

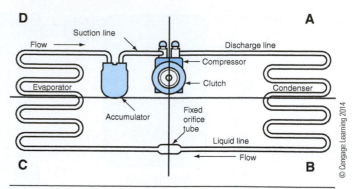

Figure 6-5 The refrigerant state and pressure in a fixed orifice tube air-conditioning system.

Again following **Figure 6-4,** a horizontal line is drawn through the middle of the condenser and the middle of the evaporator. For all intents and purposes, you should be able to assume that the refrigerant is in a vaporous state anywhere on the top half of the figure, that is, in quadrants **A** and **D**. And any refrigerant on the bottom half of the diagram, quadrants **B** and **C**, should be in a liquid state.

The above information can be applied exactly the same way for the fixed orifice tube air-conditioning system, with the only change being that the pressure division is in the middle of the orifice tube instead of in the middle of the TXV. See **Figure 6-5.**

Note: The above examples are basic rules of thumb and do not take into account the saturated liquid state, which generally occurs approximately halfway through the condenser and halfway through the evaporator.

REFRIGERATION CAPACITY— PERFORMANCE RATINGS

The capacity of refrigeration systems used to be rated in horsepower (HP). Horsepower is the amount of energy required to raise 33,000 pounds (1497 kg) 1 foot (305 mm) in 1 minute. The first refrigeration systems were rated at one-fourth, one-half, three-fourths, one HP, and so forth. The rating of horsepower was really an inaccurate method of rating the capacity of a refrigeration system because horsepower refers only to the displacement of the compressor and not the air conditioner's cooling ability.

Today, the term **ton** is used to describe the capacity of large capacity refrigeration systems. Normally, when one thinks of a ton, one relates the term to a mass

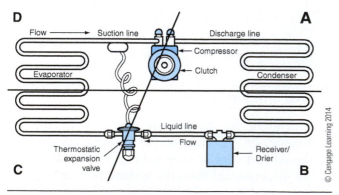

Figure 6-4 The refrigerant state and pressure in a TXV air-conditioning system.

of 2000 pounds; but a refrigeration ton is different. One ton of refrigeration capacity is a grouping of BTUs per hour and how that grouping relates to the latent heat of water, a known factor. From **Chapter 3**, recall that the amount of heat required to cause a change in state of 1 pound of ice at 32°F to 1 pound of water at 32°F is 144 BTUs. To figure out what a ton is, we need to figure out how many BTUs are required to change 2000 pounds of ice at 32°F to water at 32°F. Because 144 BTUs are required to change 1 pound, 1 ton can be found by multiplying this amount of energy by 2000 pounds.

$$\textbf{144 BTU} \times \textbf{2000 lb} = \textbf{288,000 BTUs}$$

The value of 288,000 is the amount of heat energy in BTUs required to change the state of 1 ton of ice into 1 ton of water in a 24-hour period. Now, to determine the BTU per hour rating, divide 288,000 by the 24 hours of a day.

$$\textbf{288,000 BTU} \div \textbf{24 hr} = \textbf{12,000 BTUs/hr}$$

Therefore, 1 ton of refrigeration capacity is equal to 12,000 BTUs/hr.

If an air-conditioning system has a capacity of less than 1 ton, it will usually be rated in BTUs/hr.

As an example, a half-ton air-conditioning system should have a BTU rating of 6000 BTUs. Every quarter ton of refrigeration capacity is equal to 3000 BTUs.

$$\textbf{288,000 BTUs} \times \textbf{0.25 ton} = \textbf{72,000}$$
$$\textbf{72,000} \div \textbf{24 hr} = \textbf{3000 BTUs}$$

A BTU rating is a much more accurate way of comparing the actual refrigeration capacity of an air-conditioning system.

All truck air-conditioning systems are rated at well over 1 ton (12,000 BTUs/hr) of refrigeration capacity. Anything less than 1 ton will not be powerful enough to make much of a cooling difference within the cab of the truck.

For example, factory-installed air-conditioning systems have BTU ratings in the neighborhood of 1.75 tons. By comparison, a three-bedroom two-story home (1600 square feet) can be cooled by a 1.5-ton central air-conditioning unit. Keep in mind that the home will be better insulated and not have the problem of so many heat sources all trying to infiltrate the temperature-controlled space (see **Chapter 1**). A home air-conditioning unit may take an hour or more to drop the temperature of the house 1°F when ambient temperatures are excessive. This performance level would just not be acceptable in a truck air-conditioning system.

ELECTROMAGNETIC CLUTCH

The compressors on truck air-conditioning systems are usually belt driven. The electromagnetic clutch is built into the compressor's belt drive pulley. The clutch, when energized, transmits the driving force from the belt to turn the compressor's crankshaft. This engagement occurs whenever the vehicle operator selects cooling or defrosting **(Figure 6-6)**.

When the clutch is de-energized, the drive pulley freewheels and the compressor crankshaft does not turn. When the clutch is energized, electrical power is fed to the field coil in the clutch. This magnetically pulls the clutch drive plate on the compressor crankshaft into the belt pulley drive plate, causing the compressor's crankshaft to turn.

Description

The electromagnetic clutch consists of an armature assembly, pulley assembly, and field coil. The field coil is secured to the front of the compressor housing. The armature is secured to the end of the compressor crankshaft and the belt pulley freewheels on bearings.

Operation

When current flows to the field coil, the armature assembly is pulled magnetically into the rotating belt pulley. This causes the armature to rotate the compressor crankshaft. When the field coil is de-energized, the armature and pulley assemblies separate and the compressor stops turning **(Figure 6-7)**.

Figure 6-6 A compressor's electromagnetic clutch.

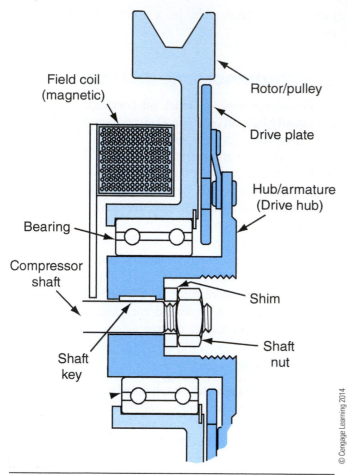

Figure 6-7 The electromagnetic clutch turns the compressor on or off.

EVAPORATOR TEMPERATURE CONTROL

Note: When the air-conditioning system is operating, moisture within the cab condenses on the fins of the evaporator. If the temperature of the evaporator is allowed to go below 32°F (0°C), this moisture freezes to the fins of the coil. This ice eventually builds up, blocking the circulation of air through the evaporator core. If the air is blocked, the cooling process will stop.

Compressor Operating Controls

Compressor operating controls are used to operate the compressor as needed. Excessive system pressure, low system pressure, and low ambient temperature can all damage the compressor, and for different reasons.

If the refrigerant level is too low and the system is undercharged, compressor oil is not circulated back through to the compressor (because oil is dissolved in the refrigerant). This lack of lubrication eventually causes the compressor to seize, causing catastrophic compressor damage.

If the compressor is operated in low ambient temperatures, the refrigerant pressure will be too low. This will essentially cause the same problems as operating with low refrigerant level. At low temperature, the compressor oil thickens to the point where it resists flow within the system.

Lastly, the compressor must be controlled to prevent damage from high system pressure. For example, if there is not enough air flow through the condenser, heat will not be removed and the pressure will increase dramatically. This can easily cause slippage of the compressor clutch, which will eventually burn the clutch out. High pressure can also burst refrigerant hoses and cause O-rings to leak.

Compressor controls include:

- Ambient temperature switch
- Thermostatic control switch
- Pressure cycling switch
- Low pressure cut-off switch
- Binary switch
- Trinary switch

Thermostatic Control Switch (Cold Switch)

One method of preventing the evaporator from freezing is with the use of a thermostatic control switch. It may be used with either a cycling clutch orifice tube system or a TXV-type air-conditioning system **(Figure 6-8)**.

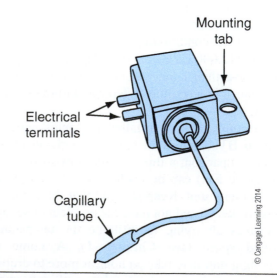

Figure 6-8 The thermostatic switch prevents the evaporator from freezing up.

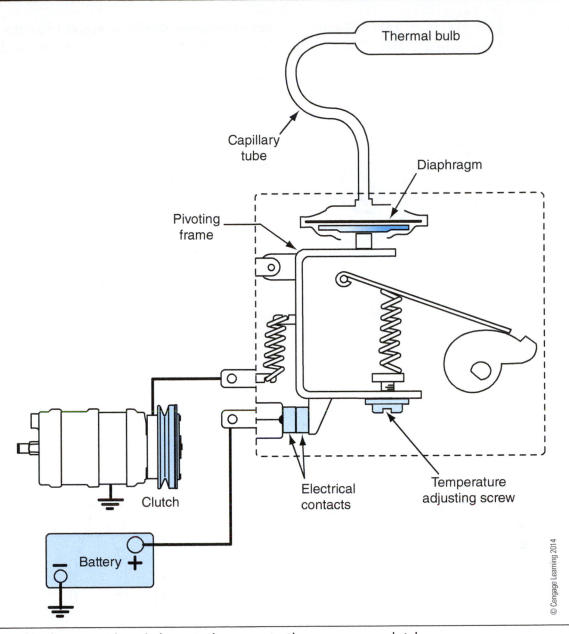

Figure 6-9 The thermostatic switch controls power to the compressor clutch.

On this system, the compressor's electromagnetic clutch is controlled by the **thermostatic switch**. The switch is usually located close to the evaporator coil in the cab of the truck. The purpose of the thermostatic switch is to keep the evaporator from freezing.

The thermostatic switch senses the temperature of the evaporator. It does this with the use of a capillary tube inserted into the middle of the evaporator. The capillary tube senses temperature exactly like a TXV sensing bulb and capillary tube. When the temperature of the evaporator drops to around 32°F, the contacts within the thermostatic switch open, breaking the electrical circuit to the compressor's electromagnetic clutch. This stops the compressor from turning, which stops the refrigeration process. The warm cab air

circulated through the evaporator quickly melts any ice that is able to freeze to the coil, and allows the water to drain to the ground through the evaporator drain tube. At a predetermined temperature (higher than the freezing point of water), the contacts within the switch close, engaging the compressor's electromagnetic clutch, and starting the cooling cycle once again **(Figures 6-9 and 6-10).**

Two varieties of thermostatic switches are available. One is designed to work within a preset temperature range. Like the switch just described, it has two temperature presets, one for off and one for on. The second type is adjustable so the on and off points can be varied according to system and operator requirements.

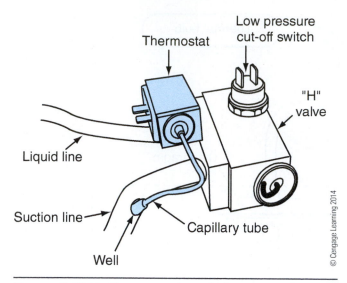

Figure 6-10 A thermostatic switch mounted to an H block TXV prevents the evaporator from freezing; the low pressure cut-off switch prevents compressor operation with low refrigerant level.

Pressure Cycling Switch

The pressure cycling switch is located on the accumulator and is used in the cycling clutch orifice tube system or CCOT. This air-conditioning system incorporates a fixed displacement compressor and pressure cycling switch. The pressure cycling switch is used to control the electrical circuit to the compressor's electromagnetic clutch. The switch senses the system pressure of either the evaporator or the accumulator **(Figure 6-11)**.

Low pressure is a sign of low evaporator temperature, just as high pressure indicates a higher temperature. If the evaporator pressure is low, it indicates that the evaporator is full of liquid refrigerant and that no

more refrigerant should be pumped into the evaporator. When the pressure reaches approximately 25 psi, the contacts within the switch open, de-energizing the clutch, and the compressor is effectively turned off. The refrigerant then has time to absorb heat, and the pressure rises. When the switch senses approximately 35 psi, the contacts within the switch close (energizing the clutch), effectively turning the compressor back on and pushing more refrigerant back into the evaporator. The pressure cycling switch serves as a freeze protection device for the air-conditioning system. Evaporator freeze-up is prevented because the switch stops the flow of refrigerant when the evaporator pressure/temperature becomes too low. During cold weather, the switch also prevents the compressor from operating as refrigerant pressure drops with temperature. This switch also detects low refrigerant levels and will not allow the compressor to operate in that condition **(Figure 6-12)**.

The pressure cycling switch is usually installed on a standard Schrader-type valve fitting. This makes it possible to remove and replace the switch without having to discharge the refrigerant from the system.

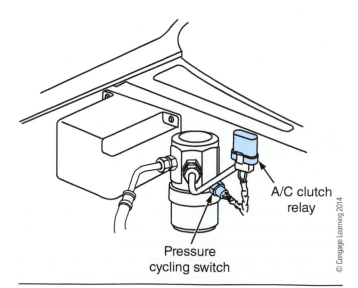

Figure 6-11 A pressure cycling switch.

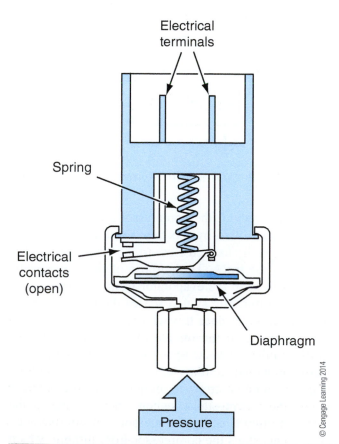

Figure 6-12 A pressure cycling switch (schematic view).

Note: Turning the compressor off stops the cooling process of the air-conditioning system.

Another system used to control the temperature of the truck cab is the variable displacement orifice tube (VDOT). This system uses a variable displacement compressor. As discussed in **Chapter 4,** whenever the air-conditioning system is turned on, the compressor's electromagnetic clutch is energized. The variable displacement compressor varies the compressor's displacement to change the amount of refrigerant it pumps through the system.

Note: Reducing the flow of refrigerant reduces the cooling effect of the air-conditioning system.

Low-Pressure Switch

Another method of preventing the evaporator from freezing is the low-pressure switch. Most orifice tube systems use a low-pressure switch instead of a thermostatic switch. The low-pressure switch is located in the low-pressure side of the air-conditioning system, usually mounted to the accumulator. The low-pressure switch opens the electrical circuit to the compressor clutch when the low-side pressure reaches 25 psig (172 kPa). This pressure corresponds to a temperature of about 32°F (0°C). When the air-conditioning system pressure rises to a predetermined level, the electrical contacts within the low-pressure switch close, restoring electrical power to the compressor clutch and the refrigeration process begins again **(Figure 6-13).**

COMPRESSOR PROTECTION DEVICES

Low Pressure Cut-Off Switch

The purpose of the **low pressure cut-off switch** is to prevent the compressor from operating if the refrigerant pressure is too low. This switch can be found in the low-pressure side of the system. It is often found on the accumulator in a fixed orifice tube system. If the refrigerant pressure becomes too low, the contacts within the switch open, de-energizing the compressor's electromagnetic clutch. This action helps protect the system from a further reduction of system pressure and

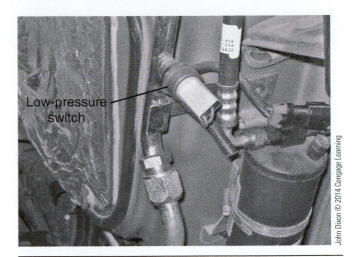

John Dixon © 2014 Cengage Learning

Figure 6-13 The low-pressure switch keeps the evaporator from freezing.

from drawing in moisture and contaminants that may enter in the event of a low-side refrigeration leak.

On some air-conditioning systems, a low pressure cut-off switch may also be found in the high side of the system and will prevent the compressor clutch from being energized if system refrigerant pressure is low. This would be the case if the refrigerant level was low due to a leak in the refrigeration system.

High Pressure Cut-Off Switch

The **high pressure cut-off switch** is used to protect the refrigeration system from high internal pressure that can damage system components or even rupture refrigerant hoses. This switch is located in the high-pressure side of the system. It is usually located on the receiver-drier on a thermostatic expansion valve system. If internal refrigeration pressure reaches a predetermined level, the contacts within the switch open and de-energize the compressor's electromagnetic clutch. When system pressures are reduced, the switch closes and the clutch is again energized **(Figure 6-14).**

Binary Pressure Switch

The **binary switch** incorporates two switching functions into one component. This switch is usually located in the receiver-drier. The binary switch protects the air-conditioning system from excessive pressure as well as from a low-pressure situation (on the high-pressure side of the system) by disengaging the electromagnetic clutch if a fault is found in either of the two functions that it controls.

If the refrigerant level is low or the ambient temperature is low, the pressure of the refrigerant within the system will also be low. (Refer to **Chapter 3** for

Figure 6-14 The high-pressure switch turns off the compressor when system pressure becomes too high.

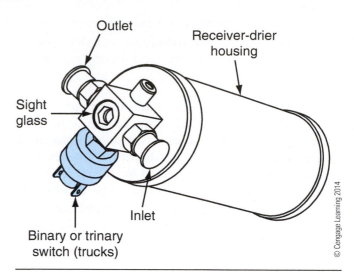

Figure 6-15 The binary switch protects the air-conditioning system from excessively high or low system pressures.

more detail.) For example, one manufacturer's binary switch opens at 28 psi, disengaging the clutch. This prevents system damage caused by low refrigerant and subsequent lubrication problems. The switch closes if system pressure reaches 30 psi, and normal compressor operation begins again.

As stated previously, the switch also protects the system from excessively high pressures. High system pressure can be caused by a couple of things, one being a loss of air flow through the condenser. This can happen if the engine cooling fan is not operating and there is no ram air effect. Other possible causes of excessive pressure include an internal blockage within the refrigeration system; dirt/debris interrupting condenser air flow; or even a refrigerant overcharge, which can cause refrigerant pressure to rise beyond the system's designed pressure. For example, one manufacturer's binary switch opens the electrical circuit to the clutch if refrigerant pressure exceeds 385 psi. When the system pressure falls back down to 300 psi, the switch closes, restoring electrical power for the clutch, and the compressor begins to pump again **(Figure 6-15)**.

Trinary Switch

The **trinary switch** controls three functions in the air-conditioning system.

First, the trinary switch prevents the compressor clutch from being energized if the system loses its refrigerant or the ambient temperature is too low. Second, if the compressor discharge pressure becomes excessively high, it stops compressor operation by de-energizing the compressor's electromagnetic clutch. The third function is a shutter/fan override switch. This helps to maintain the compressor's discharge pressure

within its designed operating range. This is accomplished by opening the radiator shutters or engaging the engine fan. This increases the air flow through the condenser, lowering the discharge pressure of the compressor. To accomplish these three functions, the switch contains three sets of contacts. One set is used to signal when system pressure drops too low (about 10–15 psi). Another set is used to signal high pressure (about 385 psi). In both of these cases, the compressor clutch is disengaged if the contacts open. The engine fan is engaged when system pressure is above 230 psi. The final set of contacts is used to engage the engine fan during normal air-conditioning operation. The trinary switch is usually located on the receiver-drier and is usually installed on a standard Schrader-type valve fitting. This makes it possible to remove and replace the trinary switch without having to discharge the refrigerant from the system **(Figure 6-16)**.

Fan Cycling Switch

The engine-mounted fan that draws air through the condenser and radiator has the ability to be turned on and off just like the compressor's electromagnetic clutch. This fan is controlled by either the engine coolant sensing switch or the air conditioning's **fan cycling switch**. The fan cycling switch is usually located on the receiver-drier. The fan cycling switch activates a solenoid-controlled air valve to turn the fan on when refrigerant pressure reaches a predetermined level. The fan draws air through the condenser, reducing the pressure of the refrigerant. When the refrigerant pressure falls to a predetermined level, the fan cycling switch turns the fan off. On newer

TRINARY SWITCH

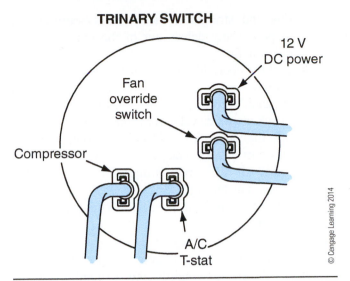

Figure 6-16 The trinary switch protects the system from high and low pressure, and also controls the engine cooling fan.

electronically controlled engines, the ECM (electronic control module) controls engine fan operation. In these cases, the ECM uses a high-side pressure switch for input to control the engine fan **(Figure 6-17)**.

Fan Timers

Too frequent cycling of the engine fan can cause premature wear of the fan clutch discs. The timer is used on many systems to lengthen the time the fan stays engaged after it has been signaled to turn on by the fan cycling switch. The engine fan typically stays engaged for a time period of between 45 and

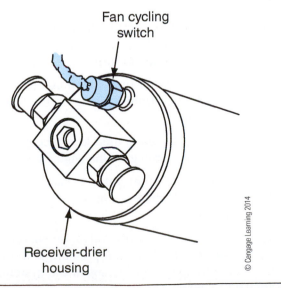

Figure 6-17 The fan cycling switch turns on the engine cooling fan when discharge pressure reaches a predetermined level.

120 seconds after the request for the fan to turn on has been received.

High-Pressure Relief Valve

The **high-pressure relief valve** can be found anywhere on the high-pressure side of the air-conditioning system. Common locations are on the receiver-drier or on the compressor body. The high-pressure relief valve works identically to the high-pressure relief valve used in the air brake system. The high-pressure relief valve exhausts refrigerant from the system when the pressure reaches a predetermined level and then closes again when the pressure is reduced **(Figure 6-18)**.

Note: Be cautious when working around the high-pressure relief valve while the air-conditioning system is operating or while charging the system, because the valve may release refrigerant without warning.

Note: More modern trucks are using pressure transducers and/or thermistors that are monitored by an electronic control unit (ECU). Simply put, a transducer is used for the pressure inputs and a thermistor for temperature inputs. The ECU determines if the compressor clutch will be engaged or disengaged.

Figure 6-18 A high-pressure relief valve.

MAINTENANCE PROCEDURES

The air-conditioning system should be checked periodically or during the truck's regular preventive maintenance (PM) service. Simple visual checks should be done to identify potential problems that could adversely affect the performance of the air-conditioning system. Following are some of the items that should be inspected.

PERFORMANCE TASKS

- Check the compressor and engine belts for condition, tension, and alignment.
- Watch the compressor clutch for proper engagement and disengagement to make sure it is operating properly.
- Examine the engine cooling system for signs of leakage around hoses and the water pump seal. Check the thermostat and radiator for condition and proper operation.
- Test the engine fan and the fan clutch operation. Check the fan shroud and shutters for damage.
- Check the condenser for anything that could block the flow of air. Check for bent condenser fins and straighten them if necessary. Clean bugs and any debris out of the condenser and radiator fins.
- Test the operation of the truck's duct system and doors to make sure the air is dispersed from the passage, according to the control lever position.
- Check the water valve for proper operation. Make sure the valve closes all the way. (Hoses should be warm or cool when the valve is closed.)
- Check all refrigerant fittings for signs of oil or refrigerant leakage.
- Inspect refrigerant hoses for kinks, punctures, and chafing.
- Check the compressor housing, gaskets, and front seal for signs of refrigerant leakage.
- Check the air-conditioning system's fuses or circuit breaker for possible opens.
- Check exposed electrical connections for corrosion, cracking, or terminal exposure.
- Check the wiring for damaged insulation and exposed wires. Focus on the more problematic areas, such as the compressor clutch and the fan solenoid switch.

Summary

- An air-conditioning system that uses a thermostatic expansion valve differs from an air-conditioning system that uses a fixed orifice tube.
- The thermostatic expansion valve system uses a receiver-drier located between the condenser outlet and the inlet to the TXV.
- The fixed orifice tube system uses an accumulator in the suction line between the outlet of the evaporator and the inlet of the compressor.
- Technicians should always wear safety goggles, non-leather gloves, and protective clothing when working with truck air-conditioning systems.
- High refrigerant pressure can be anywhere between the compressor discharge valve and the metering device.
- Low refrigerant pressure can be found anywhere between the outlet of the metering device and the compressor suction valve.
- Liquid refrigerant can be found anywhere between the midway point of the condenser and the midway point of the evaporator (in the direction of flow).
- Vaporous refrigerants can be found anywhere between the midway point of the evaporator and the midway point of the condenser (in direction of flow).
- One ton of refrigeration capacity is equivalent to 12,000 BTUs per hour.
- The electromagnetic clutch essentially turns the compressor on and off.
- When the clutch is de-energized, the drive pulley freewheels.
- When the clutch is energized, the drive pulley is locked to the compressor's drive plate, causing the compressor crankshaft to turn.
- A thermostatic switch or a low-pressure switch is used to prevent ice from forming on the fins of the evaporator coil.
- The fan cycling switch is used to reduce high refrigerant pressure by turning on the engine cooling fan.
- The high-pressure relief valve exhausts refrigerant from the system when the pressure reaches a predetermined level.
- The low pressure cut-off valve de-energizes the compressor clutch if suction pressure is too low, if there is not enough refrigerant in the system, or if ambient temperatures are too low.
- The high pressure cut-off switch de-energizes the compressor clutch if system pressure goes too high.

- The binary switch protects the air-conditioning system from excessively high refrigerant pressure as well as low refrigerant pressure.

- The trinary switch performs the two functions of the binary switch and also controls the engine fan/shutter to control compressor discharge pressure.

Review Questions

1. Technician A says that heat is dissipated to the ambient air by the condenser. Technician B says that heat is dissipated to the ambient air by the evaporator. Which technician is correct?

    A. Technician A is correct.

    B. Technician B is correct.

    C. Both Technicians A and B are correct.

    D. Neither Technician A nor B is correct.

2. When refrigerant is metered into the evaporator of a properly operating air-conditioning system, the refrigerant:

    A. Becomes superheated

    B. Changes into a vapor

    C. Absorbs heat

    D. All of the above

3. Which of the following statements is not true?

    A. The accumulator is located in the suction line.

    B. The evaporator is located in the low-pressure side of the system.

    C. Fixed orifice tubes are exchangeable in any system using this style of metering device.

    D. The receiver-drier is located between the condenser outlet and the thermostatic expansion valve inlet.

4. The high pressure cut-off switch may be found anywhere in the air-conditioning system, EXCEPT

    A. The receiver-drier

    B. The accumulator

    C. The compressor discharge line

    D. The liquid line

5. Two technicians are discussing pressure switches. Technician A says a low-pressure switch may be located in the high-pressure side of the air-conditioning system. Technician B says a high pressure cut-off switch may be located in the low-pressure side of the air-conditioning system. Which technician is correct?

    A. Technician A is correct.

    B. Technician B is correct.

    C. Both Technicians A and B are correct.

    D. Neither Technician A nor B is correct.

6. What part of the clutch assembly rotates with the compressor crankshaft?

    A. The field coil

    B. The armature

    C. Both A and B are correct.

    D. Neither A nor B is correct.

7. Which two switches can be located on the receiver-drier?

    A. The on/off switch and the binary switch

    B. The on/off switch and the thermostatic switch

    C. The thermostatic switch and the fan cycling switch

    D. The binary switch and the fan cycling switch

8. On which side of the system should you expect to find the compressor?

    A. On the high-pressure or   C. Both Technicians A and B are correct.
       discharge side of the system
                                         D. Neither Technicians A nor B is correct.
    B. On the low-pressure or
       suction side of the system

9. What safety equipment should a technician always wear when working on the refrigeration parts of the air-conditioning system?

    A. Safety goggles and leather   C. No safety equipment is really necessary.
       gloves
                                         D. Safety goggles and non-leather gloves
    B. Hat, coat, and mittens

10. For what type of compressor is belt tension an important PM service item?

    A. Piston-style compressor   C. Swash plate-style compressor

    B. Variable displacement-style   D. It is important that belt tension be correct for any one of these
       compressor   compressors.

11. When the compressor clutch is energized, power is transmitted to the compressor from the:

    A. Expansion valve   C. Condenser

    B. Evaporator   D. Engine

CHAPTER

7 Service Procedures

Learning Objectives

Upon completion and review of this chapter, the student should be able to:

- Distinguish between discharge suction and service manifold hoses.
- Explain the difference between refrigerant hoses and refrigerant lines.
- Identify the differences between R-12 and R-134a hose connectors.
- Identify the suction line, discharge line, and liquid line on any air-conditioning system.
- Explain three refrigerant hose repair techniques.
- Describe how stem-type service valves operate.
- Explain how a Schrader-type service valve works.
- Describe manifold hose installation on an R-134a service valve.
- Explain six ways of testing an air-conditioning system for refrigerant leaks.
- Explain the importance of refrigerant identifiers.
- Describe the procedures for evacuating a truck air-conditioning system.
- Explain the importance of a good evacuation.
- Describe the purpose of a thermistor vacuum gauge.
- Explain system charging procedures.
- Distinguish between recovering and recycling refrigerant.
- Explain the advantage of using a refrigerant management center.

Key Terms

charging cylinder	fluid leak detector	refrigerant hose
coupler	liquid line	refrigerant identifier
discharge line	manifold gauge set	refrigerant lines
electronic leak detector	purge	Schrader-type service valve
evacuation process	recover	service valves
flame-type leak detector	recycle	

Significant New Alternatives
    Policy (SNAP)

soapsuds solution

stem-type service valve

suction line

thermistor vacuum gauge

top up

ultrasonic leak detector

ultrasonic tester

ultraviolet leak detector

INTRODUCTION

Technicians servicing air-conditioning systems must be competent in the use of some very specialized tools used in the refrigeration industry. It takes time for technicians to learn and master the many repair techniques that are invariably used in the ever-evolving, complex field of air-conditioning repair.

SYSTEM OVERVIEW

Technicians servicing air-conditioning systems are required to master many different skill sets in order to repair the system properly. First of all, technicians need to understand where all the components are, regardless of vehicle manufacturer. Technicians must be able to install a manifold service gauge set on the vehicle, regardless of the style of service valve, and know what type of refrigerant is used in order to select the right equipment. They must be able to recover and recycle the refrigerant from the system before servicing work can be performed on internal components of the system. Technicians must know how to perform repairs to various styles of refrigerant hoses and lines. They must understand how to properly evacuate a system and know why they are performing the procedure. Technicians must know the amount of refrigerant that is to be installed in the system and how to use the required equipment to facilitate this. They must be competent in performing refrigerant leakage tests, using one or many of the different acceptable leak detection tools and methods.

MANIFOLD GAUGES

A technician must possess the ability to read the **manifold gauge set** and interpret the pressures of the air-conditioning system as it operates. These pressures tell the technician if the system is operating correctly or if there is a problem with the system. The manifold gauge set is usually the first tool installed on an air-conditioning system before any diagnostic work takes place. A manifold gauge set consists of a manifold block, two hand valves, three refrigerant hoses, and two pressure gauges **(Figure 7-1).**

The refrigeration hoses are usually color-coded to indicate where they should be connected. The hose on the left is colored blue and is connected to the low-pressure/suction side of the air-conditioning system. Connected to the low-pressure hose through the manifold is a gauge that reads either vacuum or pressure and is also usually blue. Because the gauge reads in two different ranges of pressure, it is usually called a *compound gauge.* On the vacuum side, the gauge reads to 30 inches of mercury. On the positive pressure side, the gauge will read accurately up to 120 psig, with a retard section of the gauge reading up to 250 psig. This means the gauge will read accurately up to a positive pressure of 120 psi; while pressures above 120 psig up to 250 psig can't be measured accurately, they will not damage the gauge **(Figure 7-2).**

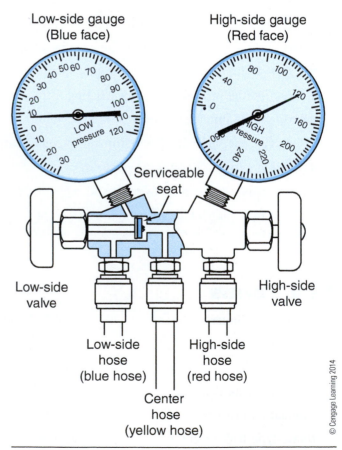

Figure 7-1 A manifold gauge set with color-coded lines and gauges.

Figure 7-2 A low-side gauge, sometimes called a compound gauge, as part of a manifold gauge set.

Some gauges incorporate an inner scale that indicates the pressure-temperature relationship of three popular refrigerant types. The gauge used for R-12 shows the corresponding evaporating temperature of R-12, R-22, and R-502. The R-134a gauge shows only the pressure-temperature relationship for R-134a **(Figure 7-3).**

Figure 7-3 The high-pressure gauge of the manifold set.

The hose on the right side of the gauge set is colored red and is connected to the high-pressure/discharge side of the air-conditioning system. Connected to the high-pressure hose through the manifold is a gauge that reads in pounds per square inch or kilopascals. The gauge is usually red in color like the hose to which it is connected. The high side is usually calibrated from 0 psig (0 kPa) to 500 psig (3,447 kPa). This gauge is usually referred to as the high-pressure gauge.

The manifold gauge set has two hand valves used to regulate the flow of refrigerant through the manifold. The flow control valves within the manifold are closed when the hand valves are turned in all the way in a clockwise direction.

When the gauge set is installed on a system and both hand valves are closed, the gauges read the pressure exerted on them.

In **Figure 7-4,** notice that both hand valves are closed, and the refrigerant pressure can still be recorded on each respective gauge.

The hand valves are opened by turning the handle counterclockwise. This opens the passage to the center hose port of the manifold set. Refrigerant will escape if there is nothing attached to the manifold center port when the hand valve is opened. Therefore, the hand valves are only opened to add or remove refrigerant from the system or to allow the vacuum pump to remove moisture and contaminants from the system.

In **Figure 7-5,** the low-side manifold hand valve is open and the passage from the low-side port to the center port is open. Note that the high-side port is blocked off from the center port. The low-side gauge will continue to record the low-side pressure only and

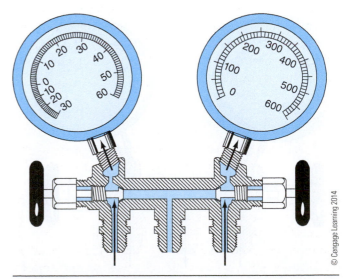

Figure 7-4 A manifold gauge set with both hand valves closed.

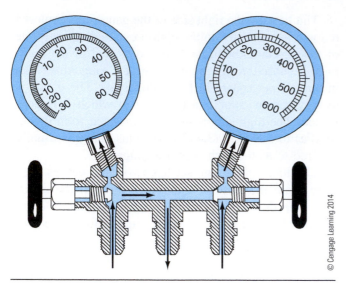

Figure 7-5 A manifold gauge set with the low-side hand valve open.

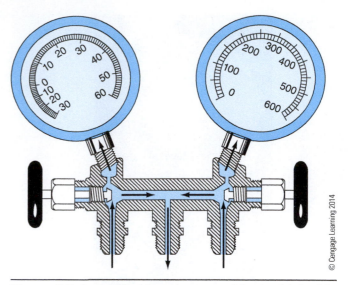

Figure 7-7 A manifold gauge set with the high-side and low-side hand valves open.

the high-side gauge will continue to record the high-side pressure only.

In **Figure 7-6,** only the high-side hand valve is open, and the passage from the high-side port to the center port is open. The low-side port is blocked off by the closed low-side hand valve. With the hand valves in this position, the low-side gauge indicates the low-side pressure, and the high-side gauge indicates the high-side pressure.

In **Figure 7-7,** both hand valves are open, allowing a common passage to the center port. With the hand valve in this position, the pressure indicated on both gauges will be the same. The high-pressure refrigerant passes through to the low side. Neither gauge in this example indicates the correct pressure, because the

low-side gauge indicates a higher-than-normal pressure and the high-side gauge indicates a pressure lower than it should. This is the position of the manifold hand valves during the evacuation process. When an air-conditioning system is being evacuated, there is no refrigerant in the system. In this case, the vacuum pump that is attached to the center hose can draw from both the high and the low side of the system simultaneously.

As stated previously, the manifold gauge set is used for almost all servicing and diagnostic procedures performed on air-conditioning systems. The manifold gauges may have the hand valve on the front or on the side of the manifold, depending upon the manufacturer. The manifold may have multiple ports as well as a sight glass that allows the technician to see the state of the refrigerant passing through the manifold. Liquid-filled gauges are also available. These gauges are more stable and more easily read by the technician because the needles don't pulsate as much, allowing a more accurate reading to be taken **(Figure 7-8).**

Manifold Gauge Calibration

Manifold gauges usually have some method of adjusting the calibration. Some use a calibration screw located on the face of the gauge, while others may be adjusted by removing and replacing the pointer. A good quality gauge will be accurate to within 2% over the range of its scale when it has been calibrated. The needles of the gauges should read zero when no pressure source is being applied. If the gauges do indicate pressure or vacuum when the hoses are open to the atmosphere, the gauge needs to be calibrated.

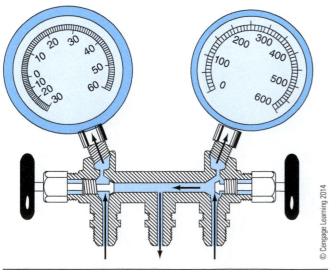

Figure 7-6 A manifold gauge set with the high-side valve open.

Figure 7-8 A typical manifold gauge set with front-mounted hand valves and manifold sight glass.

The process of calibrating the gauge requires the technician to remove the bezel or retaining ring and the gauge glass or plastic cover. The technician may then take a small slot screwdriver and insert it into the adjusting screw. The screw is then turned back and forth until the needle is lined up with the zero mark. The adjusting screw should never be forced because damage to the gauge may occur **(Figure 7-9)**.

Manifold Service Hoses

There are three color-coded service hoses on the manifold gauge set. The blue hose is for the low-pressure side, the red hose is for the high-pressure side, the yellow hose is the service hose to add or remove

refrigerant, or remove air and contaminants from the system when used in conjunction with a vacuum pump. Separate hose sets are required for R-12 and R-134a refrigerants. The hoses are high-strength nylon barrier/low permeation hoses that meet SAE J2196 standards. Hoses meeting these requirements have a burst pressure of 2100 psi (14,479 kPa) and a working pressure rating of 470 psi (3241 kPa).

REFRIGERANT LINES, HOSES, AND COUPLERS

All the components of the air-conditioning system are connected together by either refrigerant lines or hoses. These lines or hoses are then connected to the components with the use of some sort of **coupler**. The coupler may use either a flare fitting or an O-ring style fitting to make a proper seal at the component **(Figure 7-10)**.

Refrigerant Lines

Refrigerant lines are usually constructed from aluminum tubing and are used in areas where flexibility is not required. They are usually clamped to frame components and the firewall on some trucks.

Flexible **refrigerant hose** is used to connect some components where movement is required. (That is, the refrigerant hoses that connect to the compressor must be flexible to allow the compressor to move slightly due to engine torque; otherwise, the refrigerant line or hose would become damaged or broken.)

Due to the implementation of the Clean Air Act and EPA regulation of refrigerant systems and handling, there are now specified acceptable leakage rates for refrigerant. This legislation has changed the design of refrigerant lines, hoses, and fittings. New refrigerant hose is manufactured to minimize the loss of refrigerant. The different construction was necessary because the molecular structure of R-134a is much

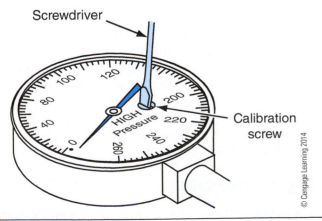

Figure 7-9 The calibration of a high-side gauge.

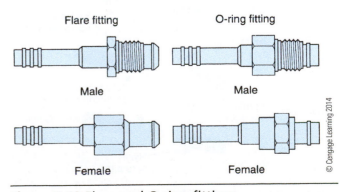

Figure 7-10 Flare and O-ring fittings.

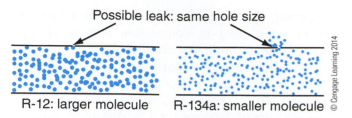

Possible leak: same hole size

R-12: larger molecule   R-134a: smaller molecule

Figure 7-11 Why air-conditioning systems using R-134 are more prone to refrigerant leaks.

smaller than that of R-12. This means that R-134a can leak from areas that would normally contain the larger R-12 molecule **(Figure 7-11)**.

The standard double braid refrigerant hose that was used for R-12 systems for many years is no longer acceptable. The new hose used for R-134a is constructed with a barrier plastic inner layer in the hose to prevent refrigerant leakage through the hose wall. These new barrier hoses are also compatible for R-12 refrigerant systems **(Figure 7-12)**.

Suction Line

The term **suction line** refers to any kind of refrigerant line/hose between the evaporator outlet and the inlet or suction side of the compressor. This line carries the low-pressure, low-temperature refrigerant vapor to the compressor. The suction line is larger in diameter than the discharge line because it must allow the same volume of refrigerant to flow as the discharge, but at a greatly reduced pressure. When the air-conditioning system is operating properly, the suction feels cool or cold to the touch.

Discharge Line

The **discharge line** refers to the refrigerant line or hose that starts at the compressor outlet or discharge and allows the superheated refrigerant vapor to reach

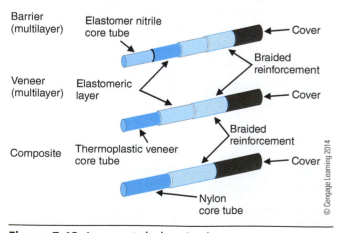

Barrier (multilayer)   Elastomer nitrile core tube   Cover

Braided reinforcement

Veneer (multilayer)   Elastomeric layer   Cover

Braided reinforcement

Composite   Thermoplastic veneer core tube   Cover

Nylon core tube

Figure 7-12 A new-style barrier hose.

the condenser. The discharge line is smaller in diameter than the suction line. In a properly operating system, the discharge line is very warm or even very hot to the touch.

Liquid Line

The **liquid line** is the refrigerant line that connects the condenser outlet to the receiver-drier, if so equipped, and then through to the thermostatic expansion valve or orifice tube. The refrigerant is high pressure and in a liquid state. For this reason, this line has the smallest diameter of any of the refrigerant lines/hoses on the air-conditioning system. This is because high-pressure liquid takes up less volume than the high-pressure vapor in the discharge line. This line is also very warm or hot to the touch on a properly operating air-conditioning system.

HOSE AND LINE REPAIR

Air-conditioning hoses and lines are commonly repaired by technicians. Refrigerant lines usually refer to nonflexible aluminum lines. These line assemblies may contain very complex bends and can be very expensive, not to mention difficult to obtain. Most line sets are dealership items that probably need to be ordered if they are not in stock.

Rubber hoses are used in areas where flexibility is required. This allows for the movement of the compressor as the engine twists in its mounts when it develops torque.

Hose repair techniques differ depending upon the type of refrigerant the system is using. Hoses are available in different sizes and styles **(Figure 7-13)**.

Simple Hose Repair

A leaking hose connection on an R-12 system may be repaired by first removing the refrigerant from the system. Next, cut the old connector off and discard. A new fitting with a barbed end is then pushed into the end of the hose until it bottoms. Lastly, a screw-type hose clamp is used to secure the fitting to the hose.

Hose splices are available for this type of repair. A hose splice may be used to repair a leaking line or to join two lines together. Nowadays, this hose clamp repair method is considered an adequate repair technique **(Figure 7-14)**.

Finger-Style Crimp

A better method of securing the fitting to the hose is with the use of a shell that is mechanically crimped

| Hose Size | Inside Diameter | | Outside Diameter (OD) | | | |
| | | | Rubber Hose | | Nylon Hose | |
	English	Metric	English	Metric	English	Metric
#6	5/16 in	7.94 mm	3/4 in	19.05 mm	15/32 in[1]	11.9 mm[3]
#8	13/32 in	10.32 mm	59/64 in[1]	23.42 mm[3]	35/64 in[1]	13.89 mm[3]
#10	1/2 in	12.7 mm	1–1/32 in[2]	25.8 mm[4]	11/16 in[1]	17.46 mm[3]
#12	5/8 in	15.87 mm	1–5/32 in[2]	29.37 mm[4]	NA	NA

[1] ±1/64 inch
[2] ±1/32 inch
[3] ±0.4 mm
[4] ±0.8 mm

© Cengage Learning 2014

Figure 7-13 Typical hose sizes for air-conditioning systems.

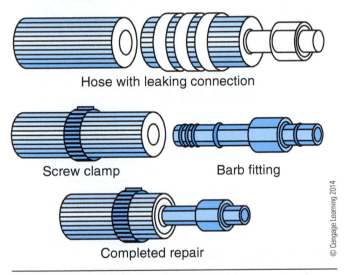

Hose with leaking connection

Screw clamp   Barb fitting

Completed repair

© Cengage Learning 2014

Figure 7-14 A simple repair of air-conditioning hose.

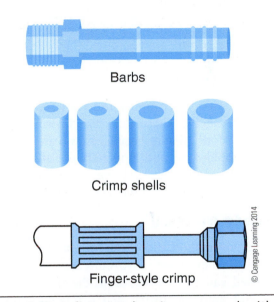

Barbs

Crimp shells

Finger-style crimp

© Cengage Learning 2014

Figure 7-15 A finger-style crimp, as used with R-12 air-conditioning systems.

onto the hose. The fittings have three raised barbs to grip the inside of the hose. These fittings are noncompatible with the new-style barrier hoses because the barbs can cut the inner liner, causing possible leaks. The crimp shells come in four popular sizes. The crimp shell is installed over the hose and then the three-barb fitting is pushed in until it bottoms. Then the shell is mechanically or hydraulically squeezed to lock the fitting securely **(Figure 7-15).**

Beadlock Fittings

Beadlock fittings are used with the modern-style barrier hoses. They have a smoother area to grip the inside of the hose so they will not cut the inner plastic liner. A captive shell is used to secure the fitting to the hose. A bubble-style crimp is used to reduce possible leaks between the hose and fitting **(Figures 7-16 and 7-17).**

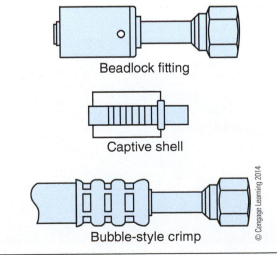

Beadlock fitting

Captive shell

Bubble-style crimp

© Cengage Learning 2014

Figure 7-16 A bubble-style crimp used with beadlock fittings.

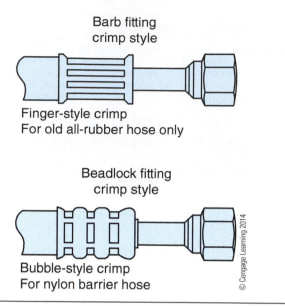

Barb fitting
crimp style

Finger-style crimp
For old all-rubber hose only

Beadlock fitting
crimp style

Bubble-style crimp
For nylon barrier hose

© Cengage Learning 2014

Figure 7-17 A comparison of a finger-style crimp and a bubble-style crimp.

Crimping Tools

Several styles of crimping tools are available to compress the shells. These tools use interchangeable dies to fit the four popular hose sizes and are available for both bubble-style and finger-style crimps. They may be mechanically squeezed like a vice, or hydraulic assist may be used to crush the shell **(Figure 7-18)**.

Alternate Method

This repair method allows the original fitting to be reused with a new hose. Two cuts are made so the factory shell can be removed from the fitting. A new shell is slid onto the fitting and a retaining ring is installed to hold the shell in place. A bubble-style crimp is used to lock the new shell and hose onto the old fitting **(Figure 7-19)**.

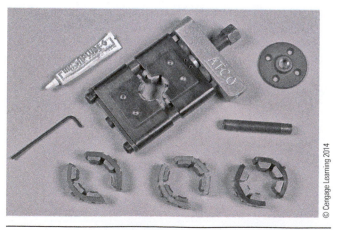

Figure 7-18 A mechanical crimping tool used to squeeze both finger-style or beadlock fittings.

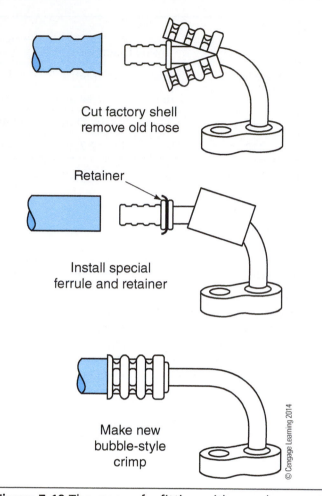

Cut factory shell
remove old hose

Retainer

Install special
ferrule and retainer

Make new
bubble-style
crimp

© Cengage Learning 2014

Figure 7-19 The reuse of a fitting with a replacement hose.

Evaporator Inlet Repair

Often, the inlet of the evaporator may be damaged by a technician attempting to replace an orifice tube assembly. Replacement tube kits are available to replace the damaged orifice tube line. The damaged tube is cut off, and the new orifice tube line is installed using a compression ferrule, nut, brass fitting, and O-ring. When the nut is tightened, the ferrule secures the new orifice tube line to the existing evaporator inlet.

Aluminum Line Repair

Small cracks and rub-throughs in aluminum tubing can be braze-repaired by skilled technicians capable of brazing aluminum tubing.

SERVICE VALVES

Service valves are used to allow the technician to install a manifold gauge into the system for service and/or diagnostic procedures. Generally, there will be a service valve on the high or discharge side of the system and on the low or suction side of the system. When manifold gauges are connected to the service valves, the gauge hose should be purged of air for accurate gauge reading and to prevent air from being accidentally introduced into the system. This is accomplished by cracking the manifold hose at the manifold fitting and bleeding a small amount of refrigerant from the line before retightening the hose connection. The refrigerant will expel any air that was in the line. There are two types of service valves used, stem-type and Schrader-type.

Stem-Type Service Valve

A **stem-type service valve** is sometimes used on two-cylinder reciprocating piston compressors. The valves are located at the compressor suction and discharge service ports and may also be located at various points throughout the refrigeration system for diagnostic or servicing procedures. The service valve allows the technician to install a manifold gauge set into the system for the purpose of testing, evacuating, recharging, or purging of the refrigeration system.

This valve not only allows access to the system for testing refrigerant pressures but can also be used to isolate the compressor from the rest of the refrigeration system. The service valve is normally left in the back-seated position, which blocks off the service port from the system. To place the valve in this position, the stem is rotated counterclockwise to seat the rear valve face and seal off the service gauge port **(Figure 7-20)**.

After the manifold gauge service line has been connected, the valve can be placed in the mid-seated position. This will allow the refrigerant to enter the service port, allowing the gauges to read. To mid-seat the valve for gauge readings, the stem need only be turned clockwise a half turn (cracked) or the discharge pulses of the refrigerant leaving the compressor can cause the gauge needle to vibrate, usually knocking the gauge out of calibration **(Figure 7-21)**.

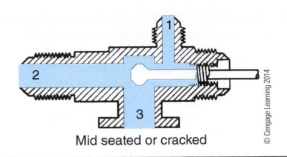

Mid seated or cracked

Figure 7-21 A stem-type service valve in the mid-seated or cracked position.

Front seating of the valve is accomplished by turning the stem clockwise all the way in. This can be done to allow the compressor to be serviced without removing the refrigerant from the rest of the system. The front-seated position is for service only and is never used while the air-conditioning system is operating **(Figure 7-22)**.

CAUTION *Care must be taken to **NEVER** front seat the discharge service valve while the compressor is operating. Even though the high pressure cut-off switch might be positioned below the valve, it would not operate fast enough to prevent major damage to the compressor and could cause personal injury.*

Schrader-Type Service Valve

The **Schrader-type service valve** is much like the valve used on pneumatic tires. One valve is usually located somewhere in the discharge line between the compressor and condenser inlet. The other valve is located in the suction line between the evaporator outlet and the compressor suction side. The Schrader valve allows the technician to install the manifold gauge set for service and diagnostic procedures. To activate the Schrader valve, the hoses from the

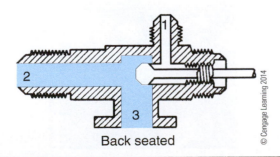

Back seated

Figure 7-20 A stem-type service valve in a back-seated position.

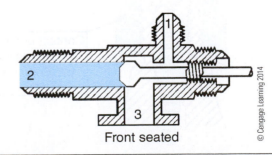

Front seated

Figure 7-22 A stem-type service valve in the front-seated position.

manifold must have a valve core depressor in them. As the hose is threaded onto the service port, the pin in the center of the valve is depressed by the valve core depressor, allowing refrigerant to flow to the manifold gauge set. When the hose is removed, the valve closes and seals automatically.

In earlier R-12 air-conditioning systems, the size of the Schrader valve fittings for high pressure and low pressure differed and required a special adaptor to install the high-side manifold gauge hose. The high-pressure Schrader was smaller to prevent mixing up of the high- and low-pressure sides of the system when the gauge set was attached. The different size fittings were originally introduced as a safety precaution, preventing do-it yourself mechanics from connecting the 1-pound (0.45 kg) disposable refrigerant cans to the high-pressure side of the system, which can cause the cans to explode **(Figure 7-23).**

> **Note:** When disconnecting the manifold service hoses from the system, confirm that the service valves are correctly seated and the Schrader valves are not leaking.

R-134 Service Valve

The R-134 service valves differ from the standard Schrader valve in that the gauge hoses are not threaded onto the fitting, but use a quick connect/disconnect system like that of a shop air line **(Figure 7-24).**

The hose end on the manifold gauge is attached as follows. The outer collar is spring-loaded and, once pulled backward, it can be installed. When the outer collar is released, it locks on to the fitting. Once the hose has been connected, the depressor is turned clockwise until the sealing ball is pushed off its seat, allowing refrigerant to flow through the hose to the gauge. Disconnection of this service is exactly the opposite of the installation. The depressor is turned counterclockwise all the way back and the collar of the spring-loaded fitting is pulled backward from the fitting, causing it to disconnect **(Figure 7-25).**

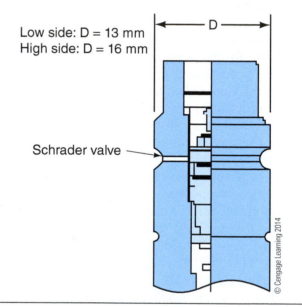

Low side: D = 13 mm
High side: D = 16 mm

Schrader valve

Figure 7-24 A service valve for R-134a systems.

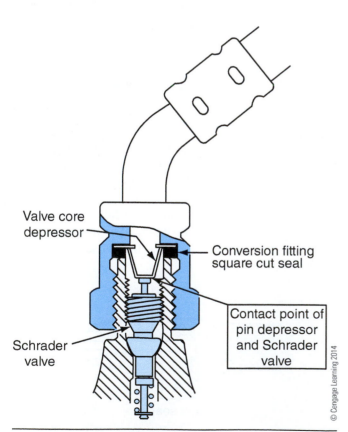

Valve core depressor

Conversion fitting square cut seal

Schrader valve

Contact point of pin depressor and Schrader valve

Figure 7-23 A Schrader valve with hose pin depressor.

Figure 7-25 A manifold gauge hose end for R-134a systems.

Installation of Refrigerant Management Center

P1-1 Remove the protective cap from the service valve.

P1-2 Pull back the spring-loaded outer collar and insert the hose end over the fitting. Once the collar is released, it will lock onto the fitting.

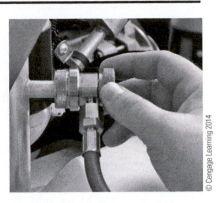

P1-3 Turn depressor clockwise until the sealing ball is pushed off its seat.

P1-4 Gauges should now read system pressure.

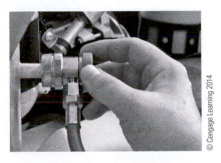

P1-5 Removal of the service lines is exactly the opposite of installation. Turn depressor counterclockwise all the way back.

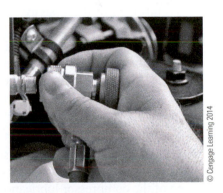

P1-6 Pull back the spring-loaded outer collar and remove from fitting. Don't forget to reinstall the protective caps of the service valves.

Note: In R-134a systems, the sizes of the gauge ports differ, with the high-pressure fitting being larger than the low-pressure side.

LEAK DETECTION

All air-conditioning systems leak refrigerant over time, and a system will eventually become depleted of refrigerant. Systems that are in good condition may lose up to half a pound per year and this is considered normal. If refrigerant losses are more than this, the technician must locate and repair the refrigerant leak.

Refrigerant leaks can occur anywhere throughout the system. This being said, here are a few locations that seem to cause more problems than others:

- Leaks around refrigerant lines at the compressor and at the compressor shaft seal.
- Leaks around various hose fittings and joints throughout the system.
- Condensers are notoriously susceptible to leaks as road debris is often driven through them at high speed, nicking or puncturing the coil.
- Refrigerant can also be lost through hose permeation.
- Also, don't rule out the evaporator coil itself.

Before performing one of the many leak detection methods, first perform a visual inspection by looking for signs of compressor oil. Whenever refrigerant is lost, some of the compressor lubrication oil is usually pushed out at the site of the leak. Look for wet, oily spots along refrigerant line hoses and components. Look for damage and corrosion of the refrigerant lines, hoses, and components. If a leak is suspected, it may be confirmed by using one of the methods described below.

Leak Detection Methods

1. One of the simplest methods of finding leaks in an air-conditioning system is with a **soapsuds solution**. The solution is sprayed or applied to the suspected leak location or fitting and, if the leak is large enough, it will cause bubbles to form at the point of the leak. This type of leak detection method is used when no other form of detection equipment is available. It may also be used to pinpoint a large leak that makes a sensitive electronic leak detector activate anywhere in the vicinity of the leak.

2. Another method along the same lines as the soapsuds solution is a commercially available **fluid leak detector**. This solution can find smaller leaks than the soapsuds solution. The fluid is sprayed or brushed onto the suspected leak area to be tested. If a leak is present, the liquid forms clusters of bubbles or large bubbles (depending on the size of the leak) at the site of the leak. This fluid leak detector may also be used to pinpoint a large leak that makes a sensitive electronic leak detector activate anywhere in the vicinity of the leak **(Figure 7-26)**.

3. **Flame-type leak detectors**, although not commonly used today, were very common for finding CFC refrigerant leaks. Another name for this device is a *halide torch*. The device is threaded on top of a disposable propane tank. The technician runs the inlet hose over any suspected areas and watches for any change in color, indicating a leak. The propane is burned in the presence of a copper element. If the detector finds a CFC refrigerant leak, the flame changes colors from blue propane flame to green or turquoise, depending upon the severity of the leak. Green indicates a small leak, while turquoise indicates a severe leak **(Figure 7-27)**.

> **CAUTION** *Flame-type leak detectors are a potential fire hazard, and when the refrigerant is burned, it can produce poisonous phosgene gas.*

4. **Ultraviolet (UV) leak detectors**, also called black light leak detectors, are used to locate very small and troublesome leaks. Some air-conditioning systems have phosphorous dye installed during production and are labeled on the compressor or receiver-drier, if so equipped. The phosphorous dye may also be installed into the low side of an air-conditioning system in a small specific quantity. The dye mixes with the compressor oil and is pushed out of the system with the

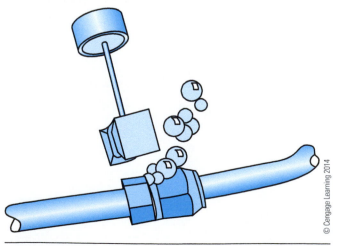

Figure 7-26 A fluid-type leak detector is applied to a suspected leakage point.

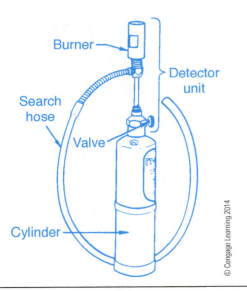

Figure 7-27 A flame-type leak detector.

escaping refrigerant. The technician can then broadcast a standard black light over the system, and any leak found will fluoresce bright yellow. This method of leak detection is used to pinpoint hard-to-find leaks, especially cold or hot leaks **(Figure 7-28)**. (Cold leaks occur only while the system is not operating, and hot leaks occur only when the air-conditioning system is operating.)

5. **Electronic leak detectors** should be used by any shop regularly servicing air-conditioning systems. They are the most sensitive of all the types of leak detectors. Some can find a leak of a half ounce per year **(Figure 7-29)**. The sensitivity of this device can make pinpointing of leaks in confined spaces very difficult. An electronic leak detector is a small hand-held instrument with a flexible probe that the technician runs over the refrigerant lines and fittings without actually making contact with the lines **(Figure 7-30)**. The end of the probe should be moved at a rate of 1 to

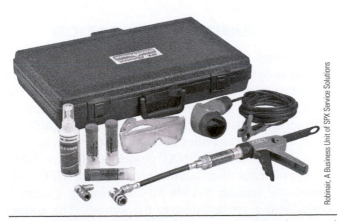

Figure 7-28 An ultraviolet (UV) leak detector kit, used to pinpoint hard-to-find leaks.

Figure 7-29 Electronic refrigerant leak detector for finding very small leaks.

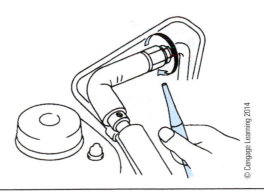

Figure 7-30 Tracing a suspected leak with an electronic leak detector.

2 inches per second (25–50 mm). When the probe senses a leak, the detector sets off an audible alarm or a visual light, depending which option the operator selects. The area being tested should be free of oil, grease, and residual refrigerant before the leak detection process is started. Suspected leakage areas should be cleaned using soap and water, not a solvent. A detected leak should be an active flow leak, not a residual condition caused by refrigerant trapped under a film of oil. The electronic leak detector can be specific as to the type of refrigerant it detects **(Figure 7-31)**.

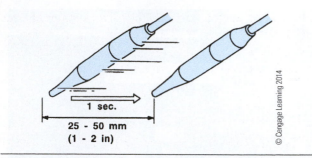

Figure 7-31 The rate of speed at which the probe of an electronic leak detector should be moved.

6. **Ultrasonic leak detectors** are another way of finding refrigerant leaks. This type of leak detector actually listens for leaks. This is accomplished with the use of an **ultrasonic tester**. This device can detect sounds in the ultrasonic frequency, which can't be detected by the human ear. The detector then converts and amplifies the sound so that the technician can hear it using a headset. Some detectors also display the sound/leak rate. Ultrasonic leak detectors are fast and accurate because they are not falsely triggered, as are electronic detectors. These detectors don't usually need to be recalibrated to compensate for background noise. Ultrasonic leak detectors can only be used on systems that are pressurized or in a vacuum because they sense the frequency of escaping gases or air entering the system. An ultraviolet detector can also be used when the system is depleted of refrigerant, as the dye will already have been pushed out of the leak site.

Ultrasonic leak detectors can also be used to find vacuum leaks on vehicles or can be used in a preventive maintenance program to detect bearings or solenoids that are starting to fail. To use the ultrasonic leak detector, the technician puts on the headset and turns the unit on. The technician can then adjust the sensitivity level and begin running the probe over any suspect leak areas. The sound of the leak will be amplified in the headset. The sound will get louder as the probe is positioned closer to the leak, allowing the technician to actually pinpoint it by the loudness of the sound. Most units automatically block out ambient noise, wind, stray gases, or other contaminants so that they will not detect a false leak **(Figure 7-32)**.

SERVICING AIR-CONDITIONING SYSTEMS

Before repair or replacement of any AC components (except compressor service if stem-type service valves are used), all refrigerant must be removed from the system. This is accomplished by using the manifold

Figure 7-32 An ultrasonic leak detector allows technicians to hear refrigerant or vacuum leaks in the ultrasonic range.

gauge set and a refrigerant reclaim/recycling machine to ensure that none of the refrigerant escapes to the atmosphere. These machines control the rate of refrigerant discharge, minimizing the amount of oil removed from the system. Years ago, when an AC system required service, the refrigerant was simply vented to the atmosphere. Today all refrigerant must be reclaimed and recycled. Heavy fines are imposed on technicians and shop owners noncompliant with Section 609 of the Clean Air Act. This act states that no one repairing or servicing a motor vehicle's air-conditioning system may do so without the proper refrigerant recycling equipment. It also states that only properly trained and certified persons can perform this service. For more information, refer to **Chapter 2**.

REFRIGERANT IDENTIFICATION

Although air-conditioning systems using R-12 look similar to systems using R-134a, the refrigerant within these two systems should never be mixed together. The EPA has made it illegal to do so and states that all refrigerants must be approved under the **Significant New Alternatives Policy (SNAP)**. At the present time, only HFC-134a meets SNAP guidelines **(Figure 7-33)**.

This means that anything other than HFC-134a introduced into an R-134a air-conditioning system is considered a contaminant to the system. The correct type of refrigerant should always be used when servicing air-conditioning systems. The introduction of

Robinair, A Business Unit of SPX Service Solutions

Figure 7-33 A refrigerant identifier, used to test the type and purity of refrigerant within the system.

nonapproved refrigerant into a system can cause both safety and performance problems. Contaminated refrigerant can also damage service equipment as well as contaminate the refrigerant in the storage vessel of a refrigerant management center. Contaminated refrigerant can make diagnostics difficult for the technician. For example, with only a 10% contamination of R-134a introduced to an R-12 system, the vapor pressure of the mixture at 100°F would change from approximately 115 psi to 135 psi. The operating pressure at any temperature would be much higher than specified, yet the evaporator cooling would be inefficient compared to the system operating with pure refrigerant.

Air-conditioning service equipment used to diagnose and recharge air-conditioning systems is specific to the refrigerant, so R-12 equipment should not be used to recover or recharge an R-134a system. Technicians would actually have to go out of their way to install R-12 equipment on R-134a systems and vice versa, because the hose connections and couplers are different. Vacuum pumps are generally interchangeable between systems as long as they have the capacity to pull deep vacuums.

Refrigerants should be tested before connecting the service equipment to the system. This can be accomplished with the use of a **refrigerant identifier**. These testers identify flammable refrigerants, operate with both R-12 and R-134a refrigerants, indicate the percentage of the base refrigerant, and indicate whether the refrigerant is at an acceptable level (98% pure).

This type of device helps protect shop equipment and refrigerant suppliers. It should be noted that in Mexico, propane is widely used as refrigerant, and if propane is drawn into a recovery station, it will disable it.

VACUUM PUMP

The vacuum pump is a tool used to remove air and moisture from the air-conditioning system. The principle behind moisture removal has to do with pressure and boiling point **(Figure 7-34)**. Recall from **Chapter 3** that if the pressure is lowered, the boiling point will also be lowered. The vacuum pump reduces pressure and will cause the moisture within the system to boil at room temperature, allowing vapor to be exhausted by the pump, as long as three conditions are met:

- The correct size of vacuum pump for the application is used.
- The correct vacuum pump oil is used.
- The vacuum pump is properly maintained.

Correct Size

Vacuum pumps are rated on the volume of air they can remove in one minute, in cubic feet per minute (CFM). Most truck applications require a vacuum pump in the 4 to 6 CFM range.

Correct Oil

Only oil that is specifically designed for use in vacuum pumps should be used to service the pump. This oil must be thermally stable so that it doesn't break down in the high heat caused by extended evacuation times. It must also be manufactured to have low moisture content. Moisture reduces the purity of

John Dixon © 2014 Cengage Learning

Figure 7-34 A vacuum pump, used to remove air, moisture, and contaminants for the air-conditioning system.

the oil, thinning it and reducing the vacuum pump's ability to pull a low vacuum level.

Vacuum Pump Maintenance

Vacuum pump performance relies on the quality and purity of the oil within the pump. Regular maintenance of the vacuum pump requires the technician to replace the oil on a regular basis. Moisture that is removed from an air-conditioning system comes in contact with the vacuum pump oil, causing it to become impure. The moisture thins the oil, which reduces the pump's efficiency. To maximize the pump's efficiency, change the vacuum pump oil when any of these conditions has occurred:

- An evacuation has been performed on a system known to have a high moisture content.
- A system that is known to have been contaminated is evacuated.
- The vacuum pump oil appears cloudy or milky.
- The pump will not pull to the desired level of vacuum.
- After every 10 hours of vacuum pump operation, regardless of the oil's appearance.

EVACUATING PROCEDURE

Evacuation of a system is required any time the system has been opened to the atmosphere for service or repair. The **evacuation process** removes air, moisture,

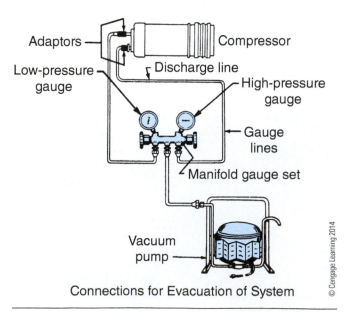

Connections for Evacuation of System

Figure 7-35 The vacuum pump is hooked up with the manifold gauge set to evacuate an air-conditioning system.

and contaminants that are able to enter the system when it is open to the atmosphere **(Figure 7-35)**.

The technician must first correctly install the manifold gauge set to the truck's air-conditioning system. The easiest way to tell the access ports apart is to follow the refrigerant lines from the compressor. The line with the smaller diameter is the high-pressure or discharge line. Follow the discharge line until you come to an access port and install the red hose of the manifold gauge to it. Take note that the red hose on the manifold gauge is also connected to the high-pressure gauge. Now from the compressor, follow the larger-diameter refrigerant line. This is the low-pressure or suction side of the system. Install the blue hose of the manifold gauge to the suction side access port. Again take note that the blue hose of the manifold gauge is connected to the compound gauge of the manifold. It is usually painted blue and has two scales capable of reading both vacuum and pressure. There is also a yellow line connected to the manifold. This is the service line and is used to add or remove refrigerant from the system. In the evacuation process, the yellow line from the manifold is connected to the port on the vacuum pump. The vacuum pump is used to draw air, moisture, and contaminants out of the system.

Before starting the vacuum pump, check to see that both the high- and low-side hand valves on the manifold gauges are closed and check that both gauges read zero; calibrate if necessary. Check the oil level of the pump. Follow the pump manufacturer's procedures for oil change frequency and specific maintenance required for that particular vacuum pump. Also look for particular vacuum pump starting instructions.

After starting the pump, open both the hand valves on the manifold. Watch the gauges and make sure that the low-side gauge does go into a vacuum. If it does not, there is a major leak that must be repaired before proceeding. If the gauge does go into a vacuum, run the vacuum pump for about 5 minutes. At this point, close both hand valves and note the position of the low-side gauge. Turn off the vacuum pump and wait another 5 minutes. Again, note the reading on the low-side gauge; any decrease in vacuum indicates a leak that must be repaired before continuing. First check that the gauge connections are tight and any seals are in good condition. Once you are satisfied that there are no leaks, turn the pump back on and open both the hand valves. Let the vacuum pump run for at least 30 minutes. Or refer to the vehicle manufacturer's specifications for evacuating time.

Once the evacuation time is up, you can close both hand valves on the manifold gauge set and turn off the vacuum pump. Follow any shutdown procedures for your particular vacuum pump. You can then disconnect the yellow line from the vacuum pump, knowing that all air and moisture have been successfully removed from the system.

Thermistor Vacuum Gauge

The **thermistor vacuum gauge** is used with a vacuum pump and manifold gauge set to measure the last and most critical inch of vacuum during the evacuation process. This unit accurately measures vacuum level in 10 segments from 25,000 microns to 50 microns (millionths of a meter). This allows the technician to know for sure that the system is free of air and moisture before recharging the system with refrigerant. The vacuum gauge also allows the technician to verify that there are no leaks in the air-conditioning system or that repairs have been performed correctly. The vacuum pump will not evacuate into the lower micron scale if there is a leak in the system **(Figure 7-36).**

Note: Vacuum pump service is essential to pulling a low vacuum level, and, in addition, the vacuum pump must be in good condition.

REFRIGERANT CHARGING

Once a proper evacuation has been achieved, it is time to install the refrigerant back into the vehicle. It is important that the correct amount of refrigerant be installed into the air-conditioning system. Too much refrigerant will create abnormally high system pressures, whereas too little refrigerant will not allow the system to cool properly and can cause compressor damage. Check for a manufacturer's label indicating the refrigerant type and weight required. If a label can't be found, check the manual for refrigerant type and quantity **(Figure 7-37).**

Always follow the instructions of the manufacturer. Here we provide a general overview of the refrigerant charging procedure.

Charging Procedure

Once the system has been successfully evacuated, both manifold hand valves are closed. The yellow (center hose) on the manifold gauge can then be disconnected from the vacuum pump and installed on the refrigerant cylinder.

The refrigerant cylinder may have one or two outlet valves. If the cylinder has two valves, you may choose between liquid refrigerant or vaporous refrigerant. If the refrigerant cylinder has only one outlet valve, vaporous refrigerant will be dispensed if the valve is on

Figure 7-36 A vacuum thermistor is used to measure the last, most critical inch of vacuum during the evacuation process.

Figure 7-37 A manufacturer's warning label with refrigerant capacity.

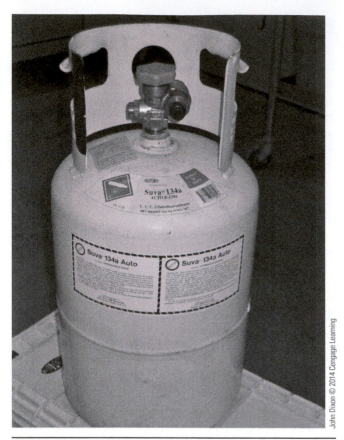

John Dixon © 2014 Cengage Learning

Figure 7-38 Refrigerant cylinder with one or two outlet valves.

top. If the cylinder is inverted, liquid refrigerant may be dispensed from the outlet valve **(Figure 7-38).**

The refrigerant cylinder should be placed on an accurate weigh scale to properly dispense the accurate amount of refrigerant by weight **(Figure 7-39).** Open the outlet valve on the refrigerant cylinder and **purge** the yellow hose on the manifold. This is done by quickly unscrewing the yellow hose at the base of the manifold. This action allows the refrigerant from the cylinder to push the air out of the yellow hose so that no air is able to enter the air-conditioning system with the refrigerant. Select liquid refrigerant and open the high-pressure hand valve (red) of the manifold gauge. This allows liquid refrigerant to flow into the high-pressure side of the system.

> **CAUTION** *NEVER charge liquid refrigerant into the low side of the system because the liquid refrigerant that may enter the suction side of the compressor can cause untold damage when the compressor starts up.*

TIF, A Business Unit of SPX Service Solutions

Figure 7-39 A refrigerant weigh scale, used to weigh in the correct amount of refrigerant.

The refrigerant will usually stop flowing before the weigh scale indicates that the proper weight of refrigerant has been dispensed. In that case, it is necessary to draw in the remaining refrigerant by operating the compressor. First, close the high-pressure hand valve on the manifold. Switch the refrigerant cylinder around so that vaporous refrigerant can then be dispensed from the refrigerant cylinder. Start the truck up and turn the air conditioner on. Slowly open the low-pressure (blue) valve on the manifold and allow the vaporous refrigerant to flow into the low side until the exact amount of refrigerant by weight has been installed into the system.

Partial Charge

Partial charging, sometimes referred to as a "**top up**," is not good practice as it cannot be known how much refrigerant is required to bring the system to its correct level by weight. Also, the condition of the refrigerant and the amount of moisture present in the system are not known. The correct service procedure that all technicians should practice is to evacuate, purge, and recharge the system with the correct weight of refrigerant.

Charging Cylinder

As previously discussed, refrigerant should be weighed into the system when an air-conditioning system is recharged. Another way to accurately measure the correct amount of refrigerant into the system is by using a refrigerant **charging cylinder**. The charging cylinder is transparent so that the technician can see the amount of refrigerant inside it. The outside of the cylinder has gradations marked in ounces. The charge cylinder allows the technician to adjust the scale to compensate for temperature, which has a direct impact on refrigerant volume within the cylinder. The multiple scales around the outside of the cylinder are marked with an ambient temperature scale. The technician can adjust the scale to the correct ambient temperature and then accurately dispense the correct refrigerant charge into the system through a manifold gauge set. Often during the charging procedure, the pressure of the refrigerant within the system will equal the pressure of the refrigerant within the charging cylinder. At this point, the technician operates the air conditioner and draws the remaining refrigerant into the low side of the system. Some charging cylinders incorporate an electric heater in the bottom of the cylinder to heat the refrigerant, thus increasing the pressure. This feature allows the technician to install the correct refrigerant charge without operating the system **(Figure 7-40)**.

REFRIGERANT RECOVERY/ RECYCLE

With the implementation of the Clean Air Act and EPA regulations, the **recovery** and **recycling** of refrigerant became mandatory by law. Refrigerant must be recovered, allowing nothing to be lost to the atmosphere. It would be very costly to just store this recovered refrigerant or have it disposed of properly, so most shops recycle the refrigerant. In recent times, refrigerant was routinely disposed of into the atmosphere; now, technicians can reuse the refrigerant. Over the past few years, refrigerant management centers have evolved into state-of-the-art equipment, making service work much easier.

Refrigerant Management Center

Refrigerant management centers incorporate the features of many different tools into one convenient package. Shops performing regular air-conditioning service will usually have a refrigerant center. Some of the tools incorporated into these machines are:

- Manifold gauges
- Vacuum pump

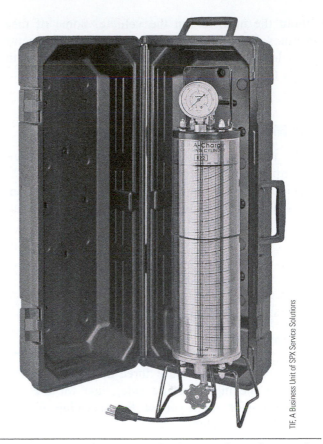

Figure 7-40 A charging cylinder used to charge the correct amount of refrigerant.

- Refrigerant weigh scale
- Internal refrigerant storage vessel

Today's refrigerant management centers make it much easier for technicians to accurately service air-conditioning systems. These units recover and recycle the refrigerant, evacuate the system, and weigh in the correct refrigerant charge. Once the hoses are connected to the vehicle, the technician may remove the entire amount of refrigerant charge from the system. This is known in the industry as *recovering the refrigerant*. The refrigerant is also recycled by passing the refrigerant through filters and separating the oil from the refrigerant. The refrigerant that is removed and filtered is known as recycled refrigerant and can be reused in the same vehicle. Once service work has been done, the technician can then select an evacuation time and the vacuum pump will pull the air-conditioning system into a deep vacuum. The pump will shut off automatically when the timed evacuation has been completed. The technician can then enter the amount of refrigerant by weight to be dispensed into the system. The compressor within the management center will pump the entire charge into the system without the need to run the vehicle. Some of these machines incorporate a feature called oil injection which lets the technician add compressor oil to the system **(Figure 7-41).**

ONLINE RESEARCH TASKS

For more information, check the following Web site: http://www.mastercool.com/

Go to technical support, and then view instructional videos. Check the videos for UV Leak Detector and Hydra Krimp.

Figure 7-41 Refrigerant management centers incorporate the features of many different tools into one convenient package.

Summary

- Manifold gauge sets use three color-coded hoses: red for the high-pressure side, blue for the low-pressure side, and yellow for servicing the air-conditioning system.

- All the components of the air-conditioning system are connected together by either refrigerant lines or hoses.

- The suction line refers to any kind of refrigerant line/hose between the evaporator outlet and the inlet or suction side of the compressor.

- The discharge line refers to the refrigerant line/hose that starts at the compressor outlet or discharge and allows the superheated refrigerant vapor to reach the condenser. The discharge line is smaller in diameter than the suction line.

- The liquid line is the refrigerant line that connects the condenser outlet to the receiver-drier, if so equipped, and then through to the thermostatic expansion valve or orifice tube.

- Soapsuds are one of the simplest ways to test for refrigerant leaks.

- Commercial liquid leak detectors are available and work better than soap bubbles.

- Flame-type leak detectors burn a different color in the presence of refrigerant.

- Electronic leak detectors are the most sensitive of all the leak detection methods.

- Ultraviolet leak detectors are used for hard-to-find leaks, often called hot or cold leaks.

- Ultrasonic leak detectors are able to listen and amplify the sound of leaking refrigerant.

- Before repairing or replacing any AC components (except compressor service if stem-type service valves are used), all refrigerant must be removed from the system.

- The correct type of refrigerant should always be used when servicing air-conditioning systems.

- Refrigerant identifiers are used to test refrigerant type and purity, preventing equipment contamination.

- The evacuation process removes air, moisture, and contaminants that are able to enter the system when it is open to the atmosphere.

- The thermistor vacuum gauge is used to measure the last and most critical inch of vacuum during the evacuation process.

- The correct refrigerant is critical to system performance.

- Topping up the air-conditioning system is never good practice because it is not known how much refrigerant is in the system already, or how much to add.

- Charging cylinders can be used to install the correct amount of refrigerant into the vehicle.

- Electronic scales may be used to measure in the correct refrigerant charge.

- Refrigerant management centers recover and recycle refrigerant as well as evacuate the system and install the correct refrigerant charge as selected by the technician. Some also allow compressor oil to be injected into the system.

Review Questions

1. If the low-side manifold hand valve is closed, the refrigerant path is between:
   - A. The service hose and the low-side hose
   - B. The service hose, the low-side hose, and the low-side gauge
   - C. The low-side gauge and the center hose only
   - D. The low-side gauge and the low-side hose

2. The low-side gauge may also be referred to as a:
   - A. High-pressure gauge
   - B. Compound gauge
   - C. Low-pressure gauge
   - D. Dual-pressure gauge

3. The method most effective in finding small leaks in air-conditioning systems is with the use of _____ leak detector.
   - A. a liquid
   - B. a fluorescent
   - C. an electronic
   - D. a halide

4. The burst pressure rating of a service should be in the range of:
   - A. 500 psig
   - B. 750 psig
   - C. 1000 psig
   - D. 2100 psig

5. Ultrasonic leak detectors detect refrigerant leaks by:
   - A. Sound
   - B. Sight
   - C. Vacuum loss
   - D. Pressure loss

6. The purpose of the vacuum pump is to remove _____ from the air-conditioning system.
   - A. Moisture
   - B. Air
   - C. Contaminants
   - D. All the above

7. Technician A states that the vacuum pump oil should be changed whenever the oil looks cloudy or milky. Technician B states that the vacuum pump oil should be changed after every 30 hours of operation. Which technician is correct?
   - A. Technician A is correct.
   - B. Technician B is correct.
   - C. Both technicians A and B are correct.
   - D. Neither technician A nor B is correct.

8. What air-conditioning tool is used to check the purity of refrigerants?
   - A. A manifold gauge set
   - B. An ultrasonic detector
   - C. A refrigerant identifier
   - D. A thermistor gauge

9. Technician A says that a refrigerant identifier is used to identify the amount of impurities in a sample of refrigerant. Technician B says that a refrigerant identifier will determine whether a sample of refrigerant is R-134a or R-12 and is better than 98% pure. Which technician is correct?
   - A. Technician A is correct.
   - B. Technician B is correct.
   - C. Both technicians A and B are correct.
   - D. Neither technician A nor B is correct.

10. When a technician removes the manifold gauge set from a system using a stem-type service valve, the stem should:

A. Be front-seated (turned clockwise all the way in)

B. Be mid-seated (turned in half way clockwise)

C. Be back-seated (turned counterclockwise all the way out)

D. Block the path of refrigerant from the gauge port

E. Both statements C and D are correct.

11. Which statement is correct?

A. The high-pressure gauge port on an R-12 air-conditioning system is larger than the low-pressure port.

B. The high-pressure port on an R-134a system is smaller than the low-pressure port.

C. The high-pressure port on an R134a system is larger than the low-pressure port.

12. In regard to refrigerant, which statement is correct?

A. Recovered refrigerant may be reused in the vehicle from which it was recovered.

B. Recycled refrigerant may be reused in the vehicle from which it was recovered.

C. Recovered refrigerant must go through a recycling process before it can be reinstalled in a vehicle.

D. Both statements A and B are correct.

E. Both statements B and C are correct.

13. What features may be found built into a refrigerant management center?

A. A weigh scale

B. A manifold gauge set

C. A vacuum pump

D. A refrigerant storage vessel

E. All of the above

14. Technician A states that the manifold gauge set is an indispensable tool used for the servicing of an air-conditioning system. Technician B states that the manifold gauge set can be used to add or remove refrigerant (in conjunction with other equipment) from an air-conditioning system. Which technician is correct?

A. Technician A is correct.

B. Technician B is correct.

C. Both Technicians A and B are correct.

D. Neither Technician A nor B is correct.

15. Technician A states that an electronic weigh scale can be used to measure either R-12 or R-134a. Technician B states that the same vacuum pump can be used to evacuate an R-12 system or an R-134a system. Which technician is correct?

A. Technician A is correct.

B. Technician B is correct.

C. Both Technicians A and B are correct.

D. Neither Technician A nor B is correct.

16. Partial charging of an air-conditioning system should not be performed because:

A. Too much refrigerant may be added to the system.

B. Too little refrigerant may be added to the system.

C. The condition of the refrigerant within the system is unknown.

D. Moisture and contaminants can't be removed from the system.

E. All of the above are correct.

8 Truck Engine Cooling Systems

Learning Objectives

Upon completion and review of this chapter, the student should be able to:

- Describe the purpose of a truck's cooling system.
- Define the terms *conduction, convection,* and *radiation.*
- List the three types of antifreeze used in today's diesel engines, and the advantages and disadvantages of each.
- Describe the need for a supplemental cooling additive package.
- Determine the freezing and boiling point of a coolant mixture, based on antifreeze and water ratios.
- Properly mix coolant using the correct proportions of water, antifreeze, and SCAs according to the OEM's recommendations and ambient temperature conditions.
- Measure the coolant strength, using the appropriate instrument.
- Describe the unfavorable effects of scaling in a cooling system and describe how to cure the situation.
- Describe how to test the SCA level and maintain it at the desired level.
- Identify the economic advantages of using extended life coolants as outlined by the manufacturers.
- Explain the role of a coolant filter and its service requirements.
- Explain the purpose of the radiator.
- Describe the coolant flow through three different styles of radiator construction.
- Explain the functions of a radiator cap.
- Test a radiator cap and determine its serviceability.
- Explain the operation of a cooling system thermostat.
- Describe the operation of a water pump and the principles of operation of a centrifugal pump.
- Explain the function of a heater core.
- Describe the role played by the heater control valve.
- Outline the roles played by the shutters and engine fan in controlling air flow through the radiator and engine compartment.
- Explain the operation of a thermatic-type viscous fan hub.
- Perform an engine coolant leak test.
- Diagnose basic cooling system malfunctions.
- Outline the coolant management requirements of some engine OEMs.

Key Terms

antifreeze	downflow	refraction
aqueous	fanstat	refractometers
by-pass circuit	heat exchanger	shutterstat
conduction	impeller	single pass
convection	kinetic energy	supplemental coolant additive
counterflow	litmus test	thermatic viscous drive
crossflow	non-positive	total dissolved solids
double pass	radiation	

INTRODUCTION

The cooling system is used to remove the excess heat energy that is a by-product of the engine as it burns fuel. If the engine were able to be 100% efficient at converting the energy of the fuel into **kinetic energy** (the energy of motion), there would be no heat left to dissipate. Since this has not yet been achieved, there must be a way of removing the heat to prevent the engine from overheating. This unwanted heat may be removed in one of two ways; some will be swept away with the exhaust gases, and the remainder will be removed by the engine's cooling system. If an engine is operating at a 40% thermal efficiency rating, the remaining 60% of the fuel's potential energy must be rejected. Approximately half of the rejected heat is discharged in the exhaust gas; that leaves the cooling system to transfer the remaining heat to the atmosphere. This is no easy task, given the extreme range of temperatures in North America. It is necessary to maintain a consistent engine operating temperature at all engine speeds and loads to ensure optimal performance and minimum emissions. Also, every engine has an ideal temperature at which optimum power is obtained. Cooling systems have the distinct task of maintaining the balance between temperature and power. Liquid cooling systems are universal in the North American truck industry **(Figure 8-1).**

SYSTEM OVERVIEW

The function of the engine's liquid cooling system is to:

- Absorb heat from the engine components
- Move the heat that has been transferred to the liquid

- Transfer the heat absorbed by the liquid to the ambient air by means of a heat exchanger (radiator)
- Maintain a constant engine operating temperature

The excess engine heat is removed by the cooling system by one of three methods, just as heat is transferred in an air-conditioning system:

1. **Conduction**: the transfer of heat from molecule to molecule through a solid object. Heat is transferred through the engine block in this method.
2. **Convection**: the transfer of heat by a flowing substance. Because solids don't flow, convection only occurs in liquids and gases. Heat is transferred from the engine compartment to the ambient air rushing through the radiator as the truck moves down the road.
3. **Radiation**: the transfer of heat from a source to an absorbent surface by passing through a medium (air) that is not heated. The turbine housing of a turbocharger radiates a significant amount of heat.

The truck's cooling system is a sealed system under pressure. Pressure produced within the system increases the boiling point of the coolant. Cooling systems are usually designed to maintain the temperature of the coolant at just below its boiling point. Most engine coolants are a mixture of water and **antifreeze**. The actual boiling point of the liquid depends upon the chemistry of the antifreeze and its concentration within the cooling system. The antifreeze performs two very important functions within the cooling system. First, it lowers the freezing point of the coolant, preventing damage in extremely cold weather. The antifreeze also acts as an anti-boiling agent, increasing the boiling point above that of water alone.

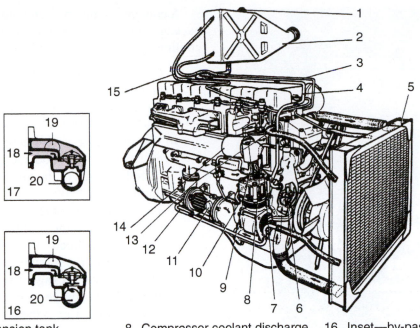

1. Coolant expansion tank pressure cap
2. Coolant expansion tank
3. Radiator gas relief hose
4. Feed line from coolant expansion tank water pump (suction side)
5. Radiator
6. Water pump impeller
7. Water conditioner discharge line
8. Compressor coolant discharge line
9. Coolant conditioner supply
10. Coolant feed to air compressor cylinder head
11. Oil cooler
12. Heater core supply line
13. Heater core discharge line
14. Coolant conditioner
15. Cylinder head gas relief line
16. Inset—by-pass circuit (thermostats closed)
17. Inset—radiator circuit (thermostats open)
18. Thermostat housing inlet
19. Thermostat housing outlet to radiator inlet
20. Thermostat housing outlet to water pump inlet

© Navistar International Corp.

Figure 8-1 The components of a truck's cooling system.

COOLANT

The coolant used to absorb excess heat from the combustion chamber of the engine is water-based. The heat absorbed by the coolant is transferred to the ambient air by the use of a **heat exchanger**. Coolant is circulated by the water pump, which causes it to flow through the engine's water jackets, absorbing heat as it flows **(Figure 8-2)**.

Engine coolant is a chemical mixture of water, antifreeze, and a **supplemental coolant additive** (SCA) package. If the only requirement of the coolant were to act as a heat transferring medium, water would work more efficiently than any currently used antifreeze mixture. The problems with using water only in a cooling system are that it has a low boiling point, a high freezing point, poor lubrication properties, and promotes oxidation and scaling **(Figure 8-3)**.

Most currently used truck engines use a mixture of ethylene glycol (EG), propylene glycol (PG), or carboxylate-type extended life antifreeze plus water to form what is known as coolant. Alcohol-based solutions are not used anymore because they evaporate at low temperatures. Engine coolant should always be kept at the correct proportions of water, antifreeze, and coolant additives. When EG or PG is used as the antifreeze component of the coolant, the additive package must be monitored and routinely replenished.

Extended life coolant is low maintenance because the life of the coolant is 6 years, with the supplemental coolant additive being replenished only once.

Antifreeze is an extremely important component of the coolant because water must not be allowed to freeze within the cooling system. Water expands about 9% in volume when it freezes and can distort or break the container holding it as it freezes. This can even occur if the container is a cast-iron engine block. Water takes up the least amount of space when it is in the liquid state close to its freezing point of 32°F (0°C). When water is heated from its near frozen state to close to its boiling point 212°F (100°C), it expands approximately 3%. If the water is contained in a 50/50 mixture of antifreeze, it expands even more than water alone to approximately 4% through the same temperature range.

John Dixon © 2014 Cengage Learning

Figure 8-2 An EG antifreeze container.

John Dixon © 2014 Cengage Learning

Figure 8-3 A supplemental coolant additive.

For this reason, the cooling system must be able to accommodate the expansion and contraction rate of the liquid coolant within the system. Antifreeze also increases the boiling point of the mixture.

A coolant mixture of water, antifreeze, and additive package should provide the following benefits to the cooling system:

- Corrosion inhibitors in both the antifreeze and additive package provide corrosion protection for the metal, plastic, and rubber compounds found within the cooling system.
- The antifreeze component of the coolant mixture protects the coolant from freezing and is directly related to the proportion of antifreeze in the mixture.
- The antifreeze component of the coolant mixture also raises the boiling point of the mixture and is again directly related to the proportion of antifreeze in the mixture.
- The additive package should also contain an anti-scaling additive to prevent hard water mineral deposits from adhering to the walls of the cooling system. This scale can contribute to poor heat transfer as well as reduced coolant flow.
- It should inhibit the formation of acid in the coolant by the use of a pH buffer. Acid, if present within the cooling system, will cause internal corrosion.
- Foaming suppression of the coolant prevents aeration of the coolant caused by the action of the water pump and the flow through the cooling system.
- An anti-dispersant prevents insoluble particles from coagulating and plugging the small passages of the cooling system.

The freeze protection and anti-boiling characteristics of glycol-based antifreeze do not break down over time, unlike the protective additives, which become depleted with engine operation. This additive package must be tested and restored at regular intervals to retain the protection level required for an efficient cooling system. Ethylene glycol and propylene glycol are both petrochemical products. Ethylene glycol has been the standard antifreeze for many years, but the federal Clean Air Act and OSHA regard it as a toxic hazard. When propylene glycol is in its virgin state, it is said to be less toxic than ethylene glycol. Consequently, propylene glycol is becoming more popular as a base antifreeze ingredient. This being said, leaks and spills of both ethylene and propylene glycol are to be regarded as dangerous to mammals (including humans) and plant life. The longer the coolant remains in the engine, the more toxic the mixture becomes. It is claimed that extended life coolants have lower toxicity than ethylene or propylene-based coolants.

Concentration of antifreeze by % volume	Freeze point of coolant			
	EG		PG	
0% (water only)	32°F	0°C	32°F	0°C
20%	16°F	−0°C	19°F	−7°C
30%	4°F	−16°C	10°F	−12°C
40%	−12°F	−24°C	−6°F	−21°C
50%	−34°F	−37°C	−27°F	−33°C
60%	−62°F	−52°C	−56°F	−49°C
80%	−57°F	−49°C	−71°F	−57°C
100%	−5°F	−22°C	−76°F	−60°C

© Cengage Learning 2014

Figure 8-4 The chart shows the freeze protection of EG and PG antifreeze when mixed with a percentage of water.

When propylene or ethylene glycols are mixed with water, they may be described as **aqueous**, meaning they mix easily and don't separate. The chart **(Figure 8-4)** compares the freezing point of ethylene glycol and propylene glycol as a percentage of the coolant mixture.

Ethylene and propylene glycol-based coolants should not be mixed. Mixing of the two will not harm the engine or cooling system, but measuring the strength of the coolant mixture will be impossible using either a refractometer or a hydrometer. If it is known that ethylene and propylene glycols have been mixed in a cooling system and it is impossible to flush the system immediately, measure using a refractometer with an ethylene glycol scale and a propylene scale. Then take the average of the two readings. To avoid problems down the road, the system should be drained completely and the coolant should be replaced with a mixture of ethylene glycol and water or propylene and water.

Extended life coolant is sold only in a premixed container and is dyed a red color. Only an extended life premix should be added to the cooling system,

though an extended life concentrate may be added in extremely cold conditions. Extended life coolant is noncompatible with ethylene or propylene glycol antifreeze.

Testing Coolant Strength

It is very important that technicians regularly test the strength of the engine coolant. Frequently, the cooling system is topped up with straight water while the vehicle is on the road and this has a major effect on the freeze protection of the coolant mixture. Standard antifreeze hydrometers are calibrated to measure the strength of ethylene glycol mixtures and tend to be fairly inaccurate, requiring calculations to correct for temperature. A **refractometer** is a more accurate measuring instrument to test coolant strength. These devices are recommended by most OEMs and the TMC (Truck Maintenance Council). Refractometers should be accurate to within 7° of the actual freeze point of the coolant throughout the temperature range in which the vehicle will be expected to operate. Refractometers designed for use with truck diesel engine coolant have a calibration scale for both ethylene and propylene glycol-based coolants. Some refractometers can also be used for checking the strength of electrolyte solution in batteries.

The technician is responsible for ensuring that the correct scale for the antifreeze base is being used, or measurement will be inaccurate. The refractometer measures how far light is bent as it passes through a liquid. The term used to indicate this is **refraction**. The refraction index of coolant increases as the concentration of antifreeze increases **(Figures 8-5 and 8-6)**.

Operation. The use of the refractometer involves taking a small sample of coolant and placing a drop of it on the prism surface. The lid is then lowered and the technician looks through the eyepiece while holding the lens up to the light. Ambient light is usually

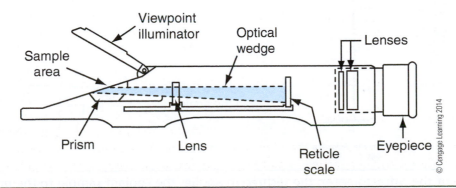

Figure 8-5 The parts of a typical handheld refractometer.

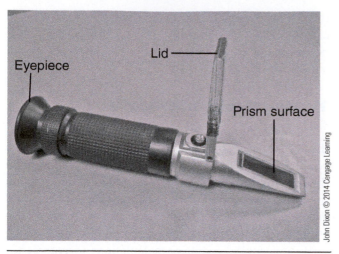

Figure 8-6 A refractometer, used to accurately measure the freeze protection of coolant mixtures.

sufficient to get a good view. Focusing the eyepiece provides a sharp, easy-to-read scale, which is designed with a semitransparent background. The view inside the eyepiece appears like a shade on a window being pulled down. The intersecting point of the light and the shade correspond with a temperature scale indicating the freeze protection of the coolant mixture. Although most refractometers can be used in this manner, always follow the operating procedures outlined by the manufacturer of the instrument **(Figure 8-7).**

Scaling

Scaling is caused by the minerals found in hard water that adhere to the surface of the cooling systems

where temperatures are highest. Scaling buildup insulates engine components designed to transfer heat. If scaling conditions are left unimpeded, engine overheating and consequent engine failure will occur.

De-scaling agents are commercially available to remove minor scale buildup, followed by flushing the cooling system. If the engine has already developed an overheating problem caused by scaling, the de-scaling agent will have little effect on the cooling system. In this case, the engine would have to be disassembled and the cylinder block and heads boiled in a soak tank.

Testing Supplemental Coolant Additives

The supplemental coolant additives (SCA) should be tested during regular PM services or when there is a substantial loss of coolant and the system has to be replenished. SCA testing is recommended by OEMs to ensure the coolant is protecting the internal components of the cooling system. The SCA is determined by the engine manufacturer and is specific to the materials used within the cooling system. Therefore, test kits of one manufacturer's engine cannot generally be used for other OEMs' products. The test kits usually contain test strips that must be stored in airtight containers. These test strips are also manufactured with an expiration date that if not adhered to may lead to inaccurate test results. The coolant test kits allow the technician to test the concentration level of the SCA, the pH level, and total dissolved solids (TDS). The pH level

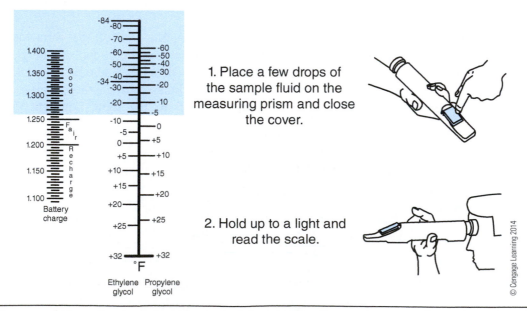

Figure 8-7 The use of the refractometer; the technician applies the coolant sample to the prism, then holds it up to the light and makes a reading.

PHOTO SEQUENCE 2 Testing Coolant Strength with a Refractometer

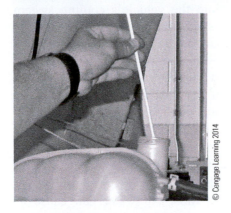

P2-1 Take a sample of coolant from either the recovery bottle or from the top of the radiator.

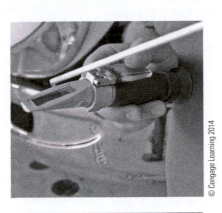

P2-2 Apply coolant to the prism surface.

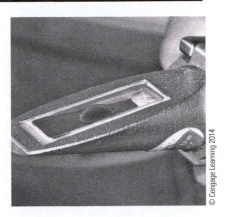

P2-3 One drop is all that is required.

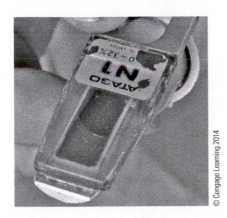

P2-4 Close the lid on the refractometer.

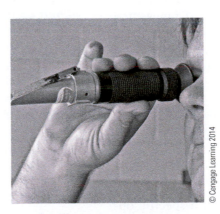

P2-5 Hold the refractometer up to the light and make a reading. It will appear as if a shutter has been pulled down.

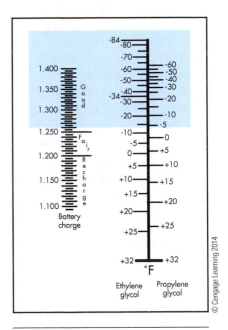

P2-6 The view through the refractometer.

determined indicates the relative acidity or alkalinity of the coolant. When coolant comes in contact with degrading metals (ferrous and copper based) or combustion gases, acids may form in the coolant. The test is known as a **litmus test** and is performed by immersing a test strip in a coolant sample. Once the test strip is removed, the color changes. The strip is then compared to a color chart provided with the kit. The best possible pH level is determined by each OEM, but generally falls between 7.5 and 11.0 on the pH scale. Higher acidity (readings below 7.5 on the pH scale) readings of the test coolant indicate the corrosion of ferrous and copper metals and exposure to combustion gases. In some cases, high acidity can indicate the breakdown of the coolant itself. Higher-than-normal alkalinity levels indicate aluminum corrosion and the possibility that a low silicate antifreeze has been used where a high silicate antifreeze is required. From the tables, the amount of additive to add to the system can be determined. The additive can be installed in the cooling system in a number of ways. It may be installed in the engine coolant filter, although this method generally results in higher-than-normal additive levels. The SCA may also be mixed with the coolant outside of the cooling system and then reinstalled. Most OEMs recommend testing additive levels in the coolant, then adding to adjust to the required additive levels. Unmeasured quantities of coolant additive should not be dumped into a cooling system at each PM service because engine damage could occur due to higher-than-normal acidity.

Note: Some litmus test kits can also read the freeze protection of the coolant.

The SCA may also be adjusted to suit a specific operating environment or set of conditions. Unusually hard water, for instance, requires a greater amount of anti-scale protection.

Testing for **total dissolved solids (TDS)** requires the use of a TDS probe. This probe measures the conductivity of the coolant by conducting a current between two electrodes. Distilled water does not conduct electricity, but as the total dissolved solids build up in the coolant, so does its ability to conduct electricity. This test is performed by inserting the probe into the coolant through the radiator cap. The TDS is measured in parts per million, and if the reading is higher than that specified by the OEM, the condition of the coolant may be conductive to scale buildup.

Mixing Heavy-Duty Coolant

It is preferable not to mix coolant within the vehicle's cooling system. The best method is to premix the coolant in an external container and then pour or pump the coolant into the vehicle's cooling system. High-quality water should be added to the container first. This water should not be excessively hard or have any iron content. Next, the correct amount of antifreeze is added, and finally the SCA package. Combine the contents of the container thoroughly before adding to the vehicle's cooling system.

Note: Tap water should not be used in a cooling system because the magnesium and calcium found in most tap water can cause scaling on cooling system components. Corrosion of engine and cooling system parts can also result from the presence of sulfates in tap water. For these reasons, distilled water should always be used when mixing coolant solutions.

CAUTION *Whenever antifreeze or a coolant solution contacts the skin, the affected area should be washed thoroughly without delay.*

Some OEMs recommend the use of commercially premixed coolant when a cooling system is topped up. This shifts the responsibility of mixing coolant away from the technician because the water, antifreeze, and SCA are already premixed in the correct proportions. It also eliminates the problems involved with poor quality water.

Note: Technicians should always use a refractometer to test the freeze protection of the coolant mixture and should make certain that the correct scale is used for the type of antifreeze being tested.

High Silicate Antifreeze

When aluminum is one of the internal components that must be protected, a coolant with high silicate concentrations should be used. However, many OEMs require the use of low silicate coolant in their engines. High silicate and low silicate antifreeze should not be mixed. High silicate antifreeze should not be used unless specified by the OEM. Extended life coolants (ELCs) do not use silicates and other chemicals to reduce scaling, but instead use a carboxylate base that, according to the manufacturer, drastically outperforms the complicated cocktail required of an ethylene or propylene glycol coolant.

Extended Life Coolants

Manufacturers of extended life coolants (ELC) suggest a service life of 600,000 miles (960,000 km) or 6 years, with one additive recharge at 300,000 miles (480,000 km) or 3 years. This is quite an advance compared to conventional EG or PG coolants, which had a service life of 2 years and could require up to 20 SCA adjustments. Extended life coolant is available in a premixed or full-strength solution. The price of ELC is comparable to that of EG or PG, and with its reduced maintenance and extended service life, it will probably become the coolant of choice for engine manufacturers. SCA test kits are not required for ELCs **(Figure 8-8)**.

Some of the advantages of extended life coolants are:

- Significantly extended service life of 6 years or 600,000 miles (960,000 km)

- No SCA testing required
- Longer water pump life because of reduced TDS content (TDS can be abrasive)
- Less hard water scaling
- Better corrosion protection
- Better cavitation protection
- Increased ability to transfer heat
- No gelling problems because no silicates are used (EG and PG get sludge buildup due to silicate dropout)
- Better protection for aluminum components
- Improved high-temperature performance over EGs and PGs

COOLANT FILTERS

Engine coolant filters are usually of the spin-on cartridge type and are connected in parallel to coolant flow. Some filters are charged with the SCA package, so it is important that the correct filter is used along with the correct coolant base, as specified by the OEM. This is also a good reason to avoid overservicing of the filter. When it is time to change a coolant filter, the technicians should familiarize themselves with the type of shutoff valve used on the vehicle, because some are automatic while others are of the manual type. These valves permit changing of the filter with minimal coolant loss. Once the filter is changed and valves are reopened, the filter will prime itself **(Figure 8-9)**.

Note: Some coolant filters may be available with different chemical strengths, allowing the technician to adjust the amount of SCA added.

Figure 8-8 Extended life coolant.

Figure 8-9 A typical engine coolant filter with hand shutoff valves.

Fluid flow through the filter is consistent with that of most other engine filters in that the coolant enters the canister through the outer ports and exits the center single port. Some coolant filters are equipped with a zinc oxide electrode to cancel out the electrolytic effect of the coolant. The electrode is sacrificed to the coolant, protecting internal metal components within the cooling system. A zinc electrode is sometimes referred to as a zinc plug and can be screwed into the engine block. These electrodes are more commonly found in marine applications **(Figure 8-10).**

Note: In severe winter conditions, a running truck circulates coolant through the engine and radiator, which are exposed to wind chill factors well below ambient temperatures. The coolant can ice up if the freeze protection is not sufficient. Make sure the coolant freeze protection is adequate for running wind chill factors.

If an engine is using an EG or PG coolant in its cooling system, it is permissible to change it to an ELC. All that needs to be done is to drain the old EG or PG and dispose of it correctly. Then the cooling system is flushed with fresh water. The ELC can then be added to the system. If the cooling system uses a coolant filter, it must be replaced with an SCA-free filter. The freeze protection of ELC may be strengthened for users in the northern United States and Canada. Check with the manufacturer of the ELC for a concentrate additive.

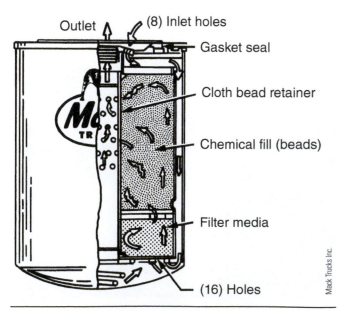

Figure 8-10 The flow of coolant through a typical engine coolant filter.

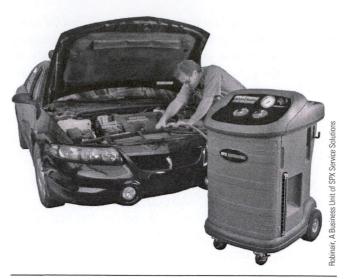

Figure 8-11 An engine coolant recycler.

Coolant Recycler

This is a compact unit that can be used for all coolant service work. It recycles on or off the vehicle, drains and refills a system, has a reverse flow option that provides back flushing, and includes a gauge to check system pressure for diagnostics **(Figure 8-11).**

COOLING SYSTEM COMPONENTS

The cooling system is made up of components that store, pump, condition, and manage engine temperature and coolant flow rate. These components may be very similar from one diesel engine OEM to the next. Nevertheless, when servicing and repairing cooling system components, always consult the appropriate service literature.

Radiators

A radiator is a form of heat exchanger. The engine horsepower rating usually determines the area the radiator takes up in the front of the truck. On average, this is in the range of 3 to 4 square inches per BHP unit. The cooling medium for the radiator is ram air, that is, ambient air forced through the radiator core as the truck is driven down the road. The radiator holds a large volume of coolant in close contact with a large volume of air so heat will transfer from the coolant to the air. The efficiency of the radiator is determined by the ambient temperature and the speed of the vehicle.

Note: As covered in **Chapter 3,** the rate of heat transfer depends on the temperature difference between coolant temperature and ambient temperature.

A fan shroud will also increase the heat transferring efficiency of the radiator as well as the efficiency of the fan.

Radiator Construction. Currently, most truck diesel engine radiators are constructed from copper and brass components, although the use of aluminum and plastics is gaining popularity. Normally, radiators consist of bundled rows of round or elliptical tubes to contain the flowing coolant. Fins are attached firmly to the tubes to increase the sectional area of the radiator, increasing its efficiency as the ram air picks up heat from both tubes and fins **(Figure 8-12)**.

The tubes are usually constructed from brass and the fins are copper in truck applications. OEMs are experimenting with aluminum, widely used in automotive radiator construction, due to its low cost and lighter weight. Copper, brass (an alloy of copper and zinc), and aluminum all have a high heat transfer coefficient and are ideal for use as base metals in the construction of radiators.

Aluminum is more vulnerable to corrosion than either copper or brass. This corrosion may form from the inside (coolant breakdown, poor water quality) or from the outside (salt from the air or road salt).

Plastics are widely used in the construction of radiator tanks, replacing bolted steel tanks. The plastic tanks are crimped onto the main radiator core enclosing the headers.

Radiators are equipped with a petcock (drain valve) that should be located at the lowest point of the cooling system so that the system can be drained by gravity. The inlet of the cooling system is usually sealed with a radiator cap. Most truck radiators are of the single pass variety, using a downflow principle.

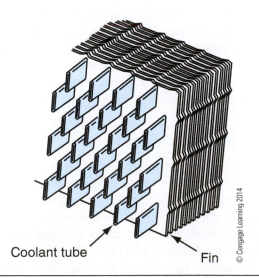

Coolant tube   Fin

© Cengage Learning 2014

Figure 8-12 Radiator tube and fin construction.

Truck radiators are classified according to the direction that coolant flows through them. The three styles of radiators are the downflow, the crossflow, and the counterflow.

- **Downflow** radiators are constructed with tanks on top and bottom, with the tubes running vertically between the tanks. Coolant enters the upper tank, where it is collected and distributed across the top of the core with the use of an internal baffle. The core is made up of numerous rows of small tubes that run vertically between the upper and lower tank assemblies. Sandwiched between the rows of tubes are thin sheet metal fins. The coolant passing through the tubes transfers heat to the fins, where air movement strips away the heat, expelling it to the atmosphere. The lower tank then collects the coolant from the core and discharges it to the engine through the outlet pipe. An overflow tube directs the flow of coolant to an overflow tank in the event that the pressure of the system is exceeded. This protects the cooling system components from damage. Downflow radiators have a **single pass** design because the coolant flows from the top tank to the bottom tank only once before it is returned to the engine to cycle through again.

- **Crossflow** radiators are constructed with the tanks on the sides of the core. The tubes run horizontally from tank to tank. Coolant enters the tank on one side of the radiator and then flows horizontally to the tank on the opposite side. The tank with the radiator cap is normally the outlet tank (the tank from which coolant flows back into the engine). A crossflow radiator can be constructed with a shorter height than a downflow radiator, allowing trucks to be built with a lower vehicle hood **(Figure 8-13)**.

- **Counterflow** radiators are similar in design to the downflow style. Coolant usually enters into the bottom tank, which is divided into inlet and outlet sides. The coolant flows vertically upward through the tubes from the inlet section of the bottom tank to the top tank. It then takes a second pass through another set of tubes back down to the outlet section of the bottom tank before it is returned to the inlet side of the engine. The efficiency of this style of radiator is increased because the coolant remains in the core section of the radiator for a longer period of time. The coolant flows through the core of the radiator twice; this is called a **double pass** radiator.

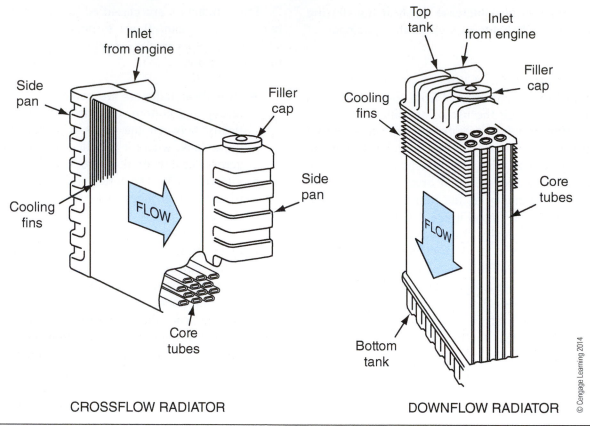

Figure 8-13 Radiator core construction.

This design was popular when liquid-cooled, charge-air heat exchangers were common.

Radiator Components

Core: The core is the center section of the radiator. It is made up of tubes and cooling fins.

Tanks: Tanks are the metal or plastic ends that fit over the core tube ends to provide storage for coolant and fittings for the hoses (**Figure 8-14**).

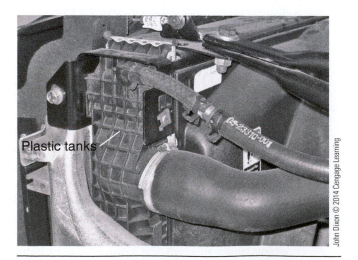

Figure 8-14 Plastic tanks on a crossflow radiator.

Filler Neck: The filler neck is the opening for adding coolant. It is also designed to accept a pressure cap and overflow tube.

Petcock: The petcock is the fitting located at the bottom of the tank, used for draining coolant from the system.

Radiator Hoses: Radiator hoses are used to direct coolant between the engine water jackets and the radiator. These hoses must be flexible to withstand the engine movement resulting from engine torque. The upper radiator hose normally connects the radiator inlet to the thermostat housing on the intake manifold or cylinder head. The lower radiator hose connects the radiator to the water pump inlet.

Radiator Servicing

Radiators are generally neglected by most technicians until the cooling system experiences some sort of failure because of leakage or overheating. The condition of the radiator should be checked on every PM service to ensure that it is clean and that there is no buildup of road dirt, insects, or debris that can restrict air flow and in turn hinder the radiator's heat transferability. Radiators should be washed frequently with either a low-pressure steamer, or a garden hose and detergent and a soft nylon bristle brush.

Note: High-pressure washers should never be used to clean a radiator because the high-pressure water can damage the delicate fins and can impede air flow through the radiator.

Radiator Testing

Frequently, coolant leaks are the result of damage and not corrosion. Some leaks may be located by white or reddish stains at the leakage point. To locate problem leaks, the radiator may also be pressure tested to around 10% above its normal operating pressure, but OEM recommendations must be followed, especially when plastic tanks are used in the construction. It is important that repairs to radiators are performed promptly to protect the engine from possible damage. If a leak is found in the radiator caused by external damage and the radiator is in good condition, it may be repaired by shorting out the affected tubes. This procedure usually involves removing the tanks and plugging the damaged tube at the header. If the leaking tube is accessible, it may be repaired by soldering with a 50/50 lead-tin solder, but the preferred method would be to braze the repair with silver solder, which requires considerably more heat, which may affect other soldering joints within the radiator core. Frequently, radiators are removed and sent out to shops with equipment and technicians that specialize in the repair of heating and cooling system components.

Radiator Cap

A very important component of a modern cooling system is the radiator pressure cap. The radiator cap locks onto the filler neck with a quarter turn. Usually the radiator cap must be pushed down before it can be turned on or off **(Figure 8-15)**.

The radiator cap performs four functions:

- It seals the top of the radiator's filler neck, preventing coolant leaks.
- It pressurizes the cooling system, thereby raising the boiling point.
- It relieves excess system pressure to protect the system from damage.
- It allows coolant to flow to and from the overflow tank to maintain a constant volume of coolant.

Radiator caps allow for the thermal expansion of the coolant to pressurize the cooling system. For each 1 psi

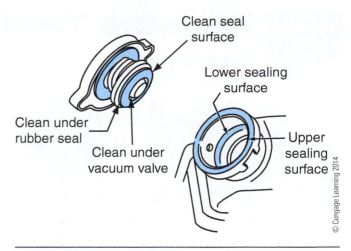

Figure 8-15 A typical radiator pressure cap and radiator filler neck.

(7 kPa) above atmospheric pressure, the boiling point rises by 3°F (1.67°C) at sea level. System pressures will hardly ever be designed to operate above 25 psi (172 kPa) and are more commonly designed to operate between 7 psi and 15 psi (50 kPa and 100 kPa).

A radiator cap is identified by the pressure the system is designed to maintain before the spring pressure of the cap is overcome, unseating the seal; when this occurs, the coolant is routed to an overflow tank **(Figure 8-16)**.

The overflow tank is at its highest level when the engine is operating at its hottest level. When the engine is shut off, it begins to cool. This will cause the coolant to contract (thermal contraction), which decreases the pressure to the point at which the system is in a partial vacuum. When this occurs, the vacuum valve within the cap unseats, allowing coolant from the

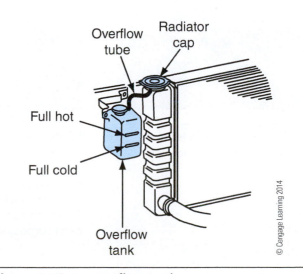

Figure 8-16 An overflow tank.

Pressure Testing a Cooling System

P3-1 Check radiator cap and make note of the pressure setting of the cap. Ensure it is the correct cap for the system you are working on.

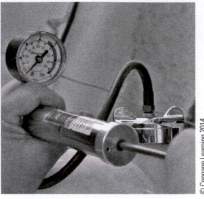

P3-2 Using the hand pump of the cooling system pressure tester, raise the pressure within the cooling system to system pressure, as indicated on the pressure cap.
Note: It may be permissible to raise the pressure 10% above the rating of the cap but always follow manufacturer's recommended procedures, especially when working with plastic tanks. Visually inspect cooling system for signs of external leaks.

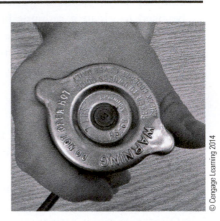

P3-3 When testing a radiator cap, make note of the pressure setting on the cap.

P3-4 Install adaptor to test the pressure setting of the radiator cap.

P3-5 Use the pressure tester pump to raise the pressure exerted on the radiator cap to system pressure indicated on the cap. The cap should not leak air until the pressure setting of the cap has been surpassed.

P3-6 Relieve pressure from cooling system or cap before removing cooling system pressure tester. This is accomplished on the tester pictured by rotating the upper collar.

overflow tank to be pushed by atmospheric pressure back into the radiator **(Figure 8-17).**

Some cooling systems do not use a coolant overflow tank, but instead use a coolant expansion tank **(Figure 8-18).** This tank has a sight glass that the

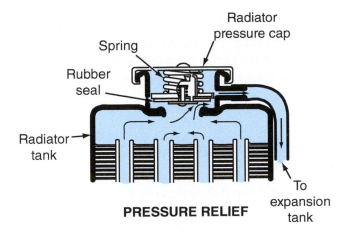

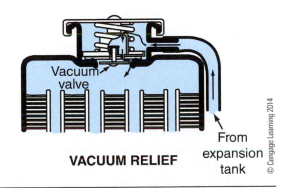

Figure 8-17 The flow of coolant through the radiator cap under pressure relief and vacuum relief.

Figure 8-18 A coolant expansion tank; coolant is filled until the sight glass is full, and the rest of the tank above the glass allows for coolant expansion.

technician can watch to make sure that coolant is at the correct level. This allows for space at the top of the tank for the coolant to expand. If pressure exceeds safe limits, the radiator cap bleeds pressure from the tank to maintain the intended pressure setting. As the system cools and the coolant contracts, the vacuum portion of the pressure cap allows atmospheric pressure to enter the cooling system, preventing the collapse of radiator and heater hoses.

Note: For the vacuum valve in the pressure cap to open, the atmospheric pressure applied to the surface of the coolant in the reservoir tank must be higher than that of the cooling system.

CAUTION *The radiator cap should NEVER be removed while the system is under pressure. Hot, pressurized coolant may be forced out the filler neck, causing serious burns to anyone close by. Most filler necks are fitted with double cap lock stops, preventing the removal of the radiator cap in a single counterclockwise motion. NEVER attempt to remove a radiator cap until the system has cooled and the pressure has equalized.*

Radiator Cap Testing

The radiator cap may be tested to ensure that it will contain the required pressure and also relieve any pressure above its rated capacity. This can be done with a standard cooling system pressure testing kit **(Figure 8-19).** The radiater cap is first installed on the correct size adaptor and then connected to the hand pump. The hand pump is used to pressurize the cap to 1 psi (7 kPa) above its rated capacity. The cap should open, allowing pressure to bleed off. Next, relieve the

Figure 8-19 Pressure tester and adapters for testing radiator caps.

pressure and again pump until the exact pressure rating of the cap is achieved. Stop pumping and observe the pressure gauge. A properly sealing cap should not drop more than 2 psi (15 kPa) in 1 minute.

WATER PUMP

The water pump, as the name implies, circulates coolant throughout the cooling system. Most water pumps are **non-positive** centrifugal pumps directly driven by gear or by belt. (Basically, pumps that discharge liquid in a continuous flow are referred to as non-positive displacement pumps and do not provide a positive seal against slippage, like a positive displacement pump.)

When the water pump is turned by the engine, an impeller spins inside the pump housing. This creates low pressure at its inlet, usually located at or close to the center of the impeller. The vanes of the impeller throw the coolant outward, and centrifugal force accelerates it into the spiraled pump housing and out toward the pump outlet. The fluid leaving the pump flows first through the engine block and cylinder head, then into the radiator, and finally back to the pump **(Figure 8-20).**

Due to the fact that the pressure of the coolant is at its lowest as it enters the water pump, boiling will always take place at this point first. This condition can speed up an overheating condition, as the impeller will not efficiently pump vapor. Water pumps are the main reason that coolants should contain some lubricating properties. The **impeller** is susceptible to the abrasive

wear of the coolant when TDS levels are high. Some of the reasons that water pumps fail are:

- Overloading of the pump's bearings and seals, caused by too much belt tension or belt misalignment.
- Erosion of the impeller caused by high TDS levels in the coolant.
- Mineral scale buildup on the pump housing.
- Overheating of the water pump caused by coolant boiling at the pump's inlet. If this happens during a hot shutdown, a vapor lock may occur.

Water Pump Replacement, Inspection

If a water pump is thought to be defective, it should be removed from the vehicle and inspected for the cause of the failure in order to avoid a repetition. Water pumps are rebuildable by the technician, but are most commonly replaced with a rebuilt unit. Rebuild centers are equipped with specialized tooling to perform the task of rebuilding the pumps quickly and efficiently. When a pump is rebuilt, inspect the components thoroughly; in many cases, especially where plastic impellers are used, only the housing and shaft are reused. Examine the shaft seal contact surface for wear, and pay special attention to the drive gear teeth. The OEM instructions should be observed, and where ceramic seals are used, great care is required during installation to avoid cracking them. One critical measurement is the impeller-to-housing clearance; failure to meet this will negatively affect the pump's efficiency.

THERMOSTATS

The thermostat is a valve in the cooling system that automatically adjusts the flow of engine coolant to maintain optimum engine operating temperature. A thermostat must perform the following functions:

- The thermostat must start to open at a specified temperature.
- The thermostat must be open fully at a set number of degrees above the start-to-open temperature.
- The thermostat must define a flow area through the thermostat in the fully open position.
- The thermostat must block coolant flow completely, or a defined small by-passed quantity, when the valve is fully closed.

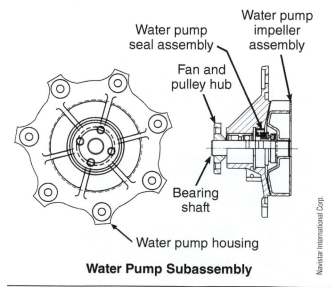

Water pump seal assembly

Water pump impeller assembly

Fan and pulley hub

Bearing shaft

Water pump housing

Water Pump Subassembly

Navistar International Corp.

Figure 8-20 A typical water pump assembly (cutaway view).

Thermostat Operation

The thermostat is usually located in the coolant manifold or housing attached to the manifold. When the engine is cold, the thermostat is closed, directing the coolant through the water pump and back into the engine (by-pass circuit). This allows the engine to warm up quickly. Once the engine has reached its normal operating temperature, the thermostat begins to open, allowing coolant to flow past it and circulate throughout the cooling system. Because the thermostat defines the flow area through which the coolant can pass, a system may use more than one thermostat (**Figure 8-21**).

The thermostat incorporates a heat-sensing element, which actuates a piston that is attached to the seal cylinder. When the engine coolant is cold, it is by-passed from the radiator and routed directly to the water pump, where it is circulated through the engine. When the engine has sufficiently warmed to operating temperature, the seal cylinder blocks off the passage to the water pump and routes the coolant to the radiator. The heat-sensing element consists of a hydrocarbon or wax pellet into which the actuating shaft of the thermostat is immersed. As the coolant in contact with the heat-sensing element heats up the medium inside, the element heats up and expands. This forces the actuating shaft outward in the pellet, which in turn opens the thermostat.

By-Pass Circuit

The **by-pass circuit** describes the flow of coolant during startup before the thermostat opens. This allows the water pump to circulate coolant through the engine cylinder block and head. This flow of by-pass coolant allows the engine to heat up faster to its proper operating temperature (**Figure 8-22**).

Operating Without a Thermostat

The practice of running a truck engine without a thermostat in not recommended by engine OEMs and may void warranty. Running a truck in this condition also violates the EPA requirements in regard to tampering with emission control components. Removal of the engine thermostat will result in the engine running too cool. This can cause water vapor in the crankcase to condense and cause corrosive acids and sludge in the crankcase. In addition, when an engine is run at low temperature, hydrocarbon emissions increase, causing the engine to fail EPA requirements.

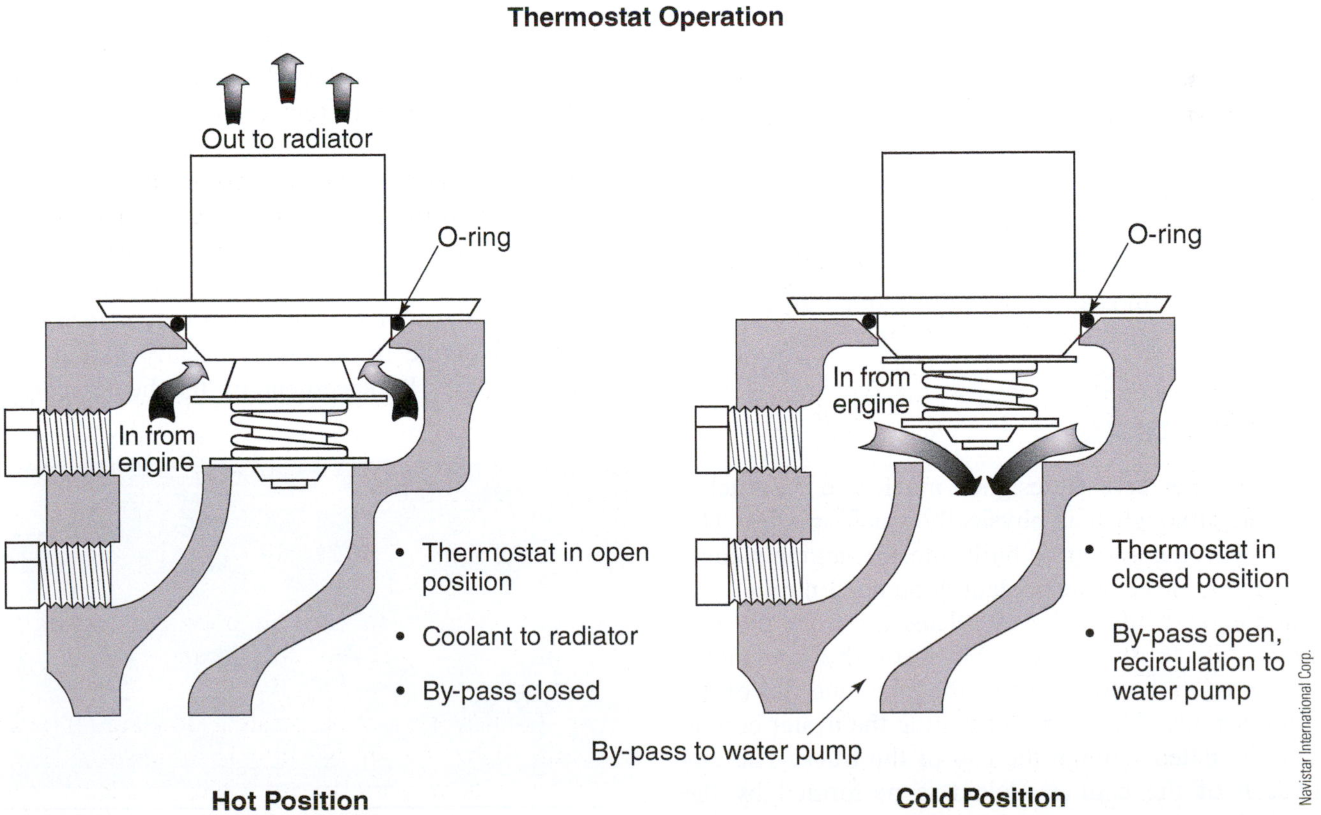

Thermostat Operation

Figure 8-21 A thermostat, closed in by-pass mode and open at operating temperature.

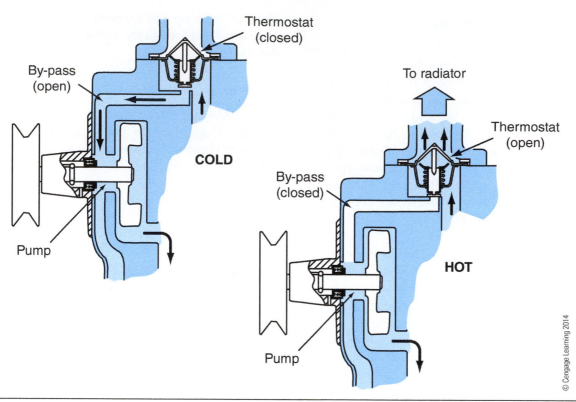

Figure 8-22 A thermostat closed and a thermostat fully open.

Thermostat Testing

Thermostats may be tested using a specialized tool that essentially consists of a tank, a heating element, and an accurate thermometer. This apparatus can also be duplicated with a pot of water on a hot plate. Always consult the OEM specifications for the thermostat you are testing. Also remember that there is a difference between start-to-open and fully open temperature values. The number stamped into the thermostat is the temperature at which it must start to open. The thermostat must be fully open at approximately 20°F above start-to-open temperature.

HEATER CORE

The heater core is designed much like the truck's radiator, although it is physically much smaller. The truck's heating system is built into the engine's cooling circuit. Hot engine coolant is pumped through the heater core, while a fan circulates air from the cab through it. In this way, heat from the coolant is transferred to the cooler air by the tubes and fins of the core assembly. The warm air exiting the heater core is then circulated through the cab of the truck. The circulation of the engine coolant is performed by the truck's water pump. Control doors, sometimes referred to as vent doors, route the air to specific areas of the

cab compartment to perform the heating and defrosting requirements of the truck. The vent doors may be actuated by vacuum, cable, air, or electric motor. The motors and cables of the system are controlled by either one or two control levers or by rotary dials that vary the function and temperature of the system **(Figure 8-23)**.

The heater core itself consists of inlet and inlet tanks connected by headers to a heat exchanger core. The heater core tank, tubes, and fins may become blocked in time by scale, rust, and mineral deposits circulated by the coolant. The heater core is located on the firewall of the truck and is usually buried deep within the dash. Replacement of the heater core is

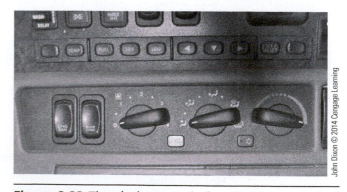

Figure 8-23 The dash controls for climate control of the truck cab.

Figure 8-24 The case containing heater core, evaporator, blower fan, and mixing door and vents.

usually a time-consuming task, so it is well worth the effort to leak-test the replacement core before installing it into position. Engine coolant kept in good condition is essential in prolonging the life of both the heater core and the radiator **(Figure 8-24)**.

HEATER CONTROL VALVE

The heater control valve may also be referred to as a coolant valve or water valve. This valve's function is to control the flow of engine coolant into the heater core. If the valve is in a closed position, no hot coolant is permitted to circulate through the heater core, allowing it to remain cool. If the valve is open fully, hot coolant will be allowed to circulate through the heater core, providing maximum heat. A variety of valve positions between fully open and fully closed allow the intensity of heat to be controlled.

As previously stated, the engine thermostat that helps to control the engine coolant temperature plays a significant role in providing heat for the truck's cab. A malfunctioning thermostat can cause the engine to overheat or not allow the truck engine to reach its desired operating temperature at all. In both of these instances, the performance of the truck's heating system will be compromised.

The climate control system for the cabin of the truck consists of the heater core and heater control valve for all the truck's heating and windshield defrosting needs, and the evaporator for all vehicle cooling needs. The heater core and evaporator are compact units that are usually mounted below and to the right of the instrument panel. A variable speed blower motor is mounted with the heater and evaporator core. It circulates air from the cab through both coils, and supplies heated or cooled air to the ductwork behind the dash panel.

BUNK HEATER AND AIR CONDITIONING

Bunk or sleeper climate control systems use remotely mounted components, usually located in the storage compartment below the bunk. These systems consist of a secondary heater core, evaporator core, and control valves. The bunk air-conditioning system is dependent on the vehicle's regular air-conditioning system, sharing the same compressor, condenser, and refrigerant. The bunk heater is plumbed directly to the engine and is usually dependent on the truck's cab heater **(Figure 8-25)**.

The operations of the bunk's heating and cooling systems are identical to those of the cab system. A blower motor circulates air through the coils and from there to the duct in the bunk's ventilation system **(Figures 8-26, 8-27, and 8-28)**.

Figure 8-25 Two sets of heater hoses from the engine, one to the main heater core, and the other to the bunk heater core.

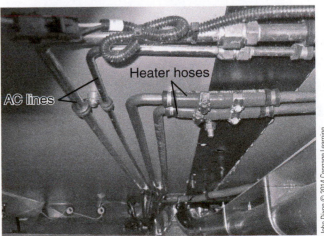

Figure 8-26 Heater hose lines and air-conditioning lines for the bunk climate control system.

Figure 8-27 A case containing the heater core and the secondary evaporator for the bunk climate control.

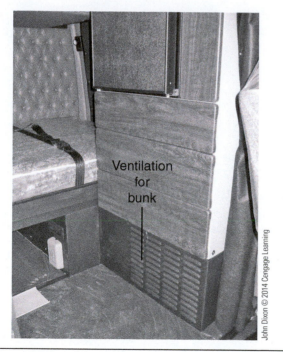

Ventilation for bunk

Figure 8-28 Inside the bunk ventilation for heating and cooling.

SHUTTERS

Shutters are located in front of the truck's radiator. Their purpose is to control the air flow through the radiator and engine compartment. Shutters are sets of louver-like slates that pivot on interconnected shafts and may be rotated in unison from fully closed to fully open positions, much like a window blind. The movement of the shutters is controlled by a **shutterstat**. The shutterstat is a temperature actuated control mechanism located in the coolant manifold. It receives

a feed of system air pressure, which it allows to pass through to the shutter cylinder until a predetermined temperature is reached.

When air from the shutterstat is delivered to the shutter cylinder, the plunger extends to actuate a lever and close the shutters. The shutters are spring-loaded to open, so if there is a failure with the shutter system they will be held open by spring force. During engine warm-up, the shutters are normally held closed once there is enough air pressure delivered to the cylinder to overcome the opposing spring force. With advances in engine technology, shutters are not nearly as common today. Commonly, engines are turbo charged with air-to-air coolers requiring constant air circulation **(Figure 8-29)**.

WINTER FRONTS

Winter fronts installed on the hood grille are used in extremely cold climates to limit the amount of ram air that can be driven through the radiator and engine compartment. This device causes an increase in coolant temperature, allowing the system to reach normal operating temperature. These devices should never cover the entire grille opening because 25% of the grille area must remain open; otherwise, coolant charge air and engine oil temperature become extreme. The recommended opening size varies according to these additional factors: ambient air temperature fluctuation, increase or decrease in load, vehicle groundspeed, altitude, ice and snow buildup, and wind chill factor. The problem with some of these devices is that the opening can be offset to the fan center, creating an unbalancing effect whenever the fan is operating. This uneven loading of the blades can cause them to flex, which over time can cause blades to break. Many engines produced after 1990 do not require the use of a winter front. If one is to be installed, it should be one approved by the OEM **(Figure 8-30)**.

COOLING FANS

OEMs generally use two basic types of engine cooling fans: suction-type and pusher-type fans. Suction fans are designed to pull outside air through the cooling system and engine compartment. Pusher-type fans, like those used on buses, push heated air out of the engine compartment while pulling in fresh air from the outside. Highway trucks that use the ram air effect typically use a suction-type fan. Depending upon engine size, radiator design, and vehicle application, ram air is capable of performing the required cooling 95%

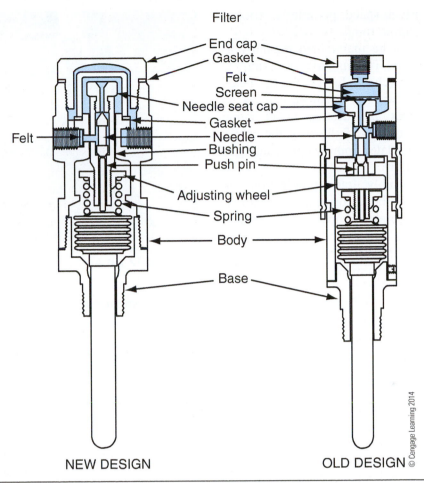

Filter

End cap
Gasket
Felt
Screen
Needle seat cap
Gasket
Felt
Needle
Bushing
Push pin
Adjusting wheel
Spring
Body
Base

NEW DESIGN

OLD DESIGN

© Cengage Learning 2014

Figure 8-29 The shutterstat, which controls the opening and closing of the shutters, if equipped (cross-section view).

John Dixon © 2014 Cengage Learning

Figure 8-30 The winter front limits the amount of ram air that can be driven through the radiator and engine compartment.

John Dixon © 2014 Cengage Learning

Figure 8-31 A fiberglass engine cooling fan.

of the time. The design of the fan is an important consideration because a typical modern fiberglass fan blade may draw around 6 BHP (brake horse power) from the engine, whereas a large steel fan may require

twice that amount. As a result, many OEMs have chosen the fiberglass blade over the steel or aluminum blade **(Figure 8-31).** Some of the current blade designs use flexible pitch blades that alter the fan's efficiency proportionally to its driven speed, permitting efficient fan operation at low engine rpms.

The cooling fan must be maintained regularly as part of the cooling system. This includes checking the condition of the blades (check for cracks or nicks) because fan blades that are not balanced can set up harmonic vibrations that can destroy water pump bearings, create torsional stress failures of the crankshaft, and damage other related components, leading to total cooling system failure.

Because the fan assembly draws power from the engine while the truck is being driven, most current truck engines use lightweight, temperature-controlled fans. Fanstats controlling air, electric, or oil pressure engaged clutch fans of the on/off-type are used, as well as thermo-modulated, viscous drive fans.

On/Off Fan Hubs

On/off fan hubs use clutches, which are actuated electrically or by oil pressure or air pressure. These clutches are controlled by a fanstat or the engine management system. As discussed in an earlier chapter, a vehicle's air-conditioning system may also control the fan hub cycles independently from the ECM, using electric over pneumatic switching. The fanstat is usually located in the coolant manifold. It is a temperature-activated switch that activates the clutch to lock up or to freewheel, using either an electrical signal or an air pressure signal as a means of engaging the clutch. Fan hubs are usually spring-loaded to keep the clutch in the engagement mode so the electrical or air pressure signal disengages the fan clutch, allowing it to freewheel. In this way, if there is a malfunction in the system the fan will be engaged, preventing an overheating situation. On/off-type fan hubs rob the least amount of power from the engine when the coolant temperatures are below the trigger value, which can be as much as 95% of the time. This percentage may be altered as ambient temperatures increase and the vehicle's air-conditioning system locks up the fan hub **(Figure 8-32)**.

Engine oil pressure may also be used as the medium to couple the fan hub (common in bus applications). The **fanstat** controls the flow of oil to the hub assembly. The fanstat sends oil pressure to disengage the fan hub, allowing it to freewheel. Once the fanstat has reached its set temperature limit, it stops the flow of oil, allowing string pressure to lock up the fan hub and engaging the fan. Oil pressure that is directed to the fan hub acts as a fluid coupling, so there is always some slippage. Some electronically managed engines directly control the

Figure 8-32 An air-actuated fan hub.

fan cycle and use it as a retarding mechanism as well as a cooling aid.

Thermatic Viscous Drive Fan Hubs/ Thermo-Modulated Fans

The **thermatic viscous drive** fan hub is an independent component and is not externally controlled. It uses silicone fluid as the drive medium between the drive hub and the fan drive plate. The hub assembly may be broken down into three sections: the drive hub (input section), the driven fan drive plate (output section), and the control mechanism. This fan drive system does not use a mechanical connection between the drive and the driven members. During minimum slip operation, torque is transmitted through the internal friction of the silicone drive fluid in the working chamber that couples the input and output sections. A wiper attached to the driven member continually wipes the fluid, and centrifugal force returns it to a supply chamber, where an open valve cycles it back to the working chamber. A bimetal strip located in the control mechanism senses temperature, and when the temperature within the engine compartment drops, the bimetal strip contracts, closing the valve that supplies the fluid medium to the working chamber, and trapping it in the supply chamber. One of the advantages of a viscous drive fan hub is its ability to increase or decrease its efficiency through the variable amount of slippage allowed by the bimetal strip.

Fan Shrouds

A fan shroud is usually a molded fiber device bolted to the radiator assembly. The fan shroud may partially enclose the fan, providing some degree of safety for technicians working around running engines

Figure 8-33 The fan shroud increases air flow through the condenser and radiator by channeling their entire surface area to the inlet side of the fan.

while the fan is engaged. The shroud increases air flow through the condenser and radiator by channeling the entire surface area of these components to the inlet side of the fan. The shroud also shapes the air flow through the engine compartment. When weather conditions become hot, temperature management problems can occur if the shroud is missing or damaged because the shroud has an extreme effect on the efficiency of the cooling fan **(Figure 8-33).**

Fan Belts and Pulleys

Fan belts and pulleys should be checked on every PM service. Fan pulleys use bearings that may be roller-, tapered roller-, or bushing-type. The pulleys use external V or poly-V groove belts. Belt tension should be adjusted with the use of a belt tensioner to ensure that the belt is neither too loose nor too tight. If the belt is adjusted too tightly, it applies excessive loads on the pulley bearings, shortening both bearing and belt life. If the belt is adjusted too loosely, the slippage will cause the belt to be destroyed even more quickly than if it were too tight.

Fan belts should be replaced when they show signs of being glazed, cracked, or nicked. It is much more cost effective to replace belts if they show any of these early signs of failure than it is for the truck to experience a roadside breakdown due to belt failure.

Cooling System Leaks

It is not uncommon for cooling systems to leak, and they should be inspected by the operator daily. Some leaks may be caused by the contraction of mated components at joints, especially around hose clamps.

These leaks are referred to as cold leaks and often disappear as the engine comes up to operating temperature. The actual cause of a cold leak is that as temperatures rise in the cooling system, the water necks to which the coolant hoses connect expand from the rising heat. The problem with this is that the hoses and clamps expand at different rates. The clamp does not allow the hose to expand as much as the water neck. The elastomer in the hose wall conforms to the size of the expanded neck. As the coolant temperature drops after engine shutdown, the water neck contracts and the seal between the neck and the hose diminishes. As the repetitive action occurs, the hose-to-neck seal is completely lost, resulting in water leakage at the hose connection as the system cools down.

Coolant hoses may be made of a rubber compound or of silicone. Silicone hoses are more expensive than rubber compound hoses, but have longer service life. Silicone hoses require the use of special hose clamps that will not bite into the hose. These hoses are sensitive to being over tightened, so they should be torqued to specification **(Figure 8-34).**

Most external leaks can be found with the use of a pressure tester. A typical cooling system pressure-testing kit consists of a hand actuated pump and a gauge assembly calibrated from 0 to 25 psi (170 kPa), plus various adaptors to allow installation to the different sizes of radiator fill necks and radiator caps. Some testers are capable of testing for vacuum as well.

Testing for Leaks

The coolant pressure tester is installed into the filler neck of the radiator in place of the radiator cap. The hand pump is operated until the pressure on the gauge indicates the maximum operating pressure, as noted in

Figure 8-34 Hose clamp used with silicone hoses.

the OEM manual, and ensures that the radiator cap is also correct for its application. The pressure on the gauge should not fall off. If the gauge does lose pressure, you may be able to locate the leak by following the trail of coolant on the floor. If there is a loss of pressure but no external coolant, this may indicate the presence of an internal leak. Some external leaks, especially cold leaks, may be hard to locate and may be found by increasing the pressure by 10%, but never by more (**Figures 8-35** and **8-36**).

Note: Never exceed the cooling system pressure by more than 10% because internal sealing points may be damaged, causing internal leaks. Also, before pressure testing, make sure the radiator cap is the correct unit for the application.

Figure 8-35 The pressure gauge of a coolant pressure tester.

CAUTION *The use of products that claim to stop leaks should be avoided even in a situation that may be deemed an emergency. These products may work temporarily, but in doing so may plug thermostats, radiator/heater cores, and oil cooler bundles. They may also insulate the components of the cooling system, reducing their ability to transmit heat. In short, they may cause more problems than they cure.*

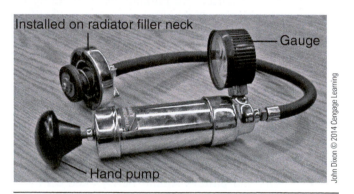

Figure 8-36 A pressure tester hand pump.

COOLING SYSTEM MANAGEMENT

Most truck cooling systems are designed to manage temperature by first opening the thermostat. Next, the shutterstat opens the shutters. And finally, the engine cooling fan engages. The fan is the last in the chain because it uses power from the engine (around 6–12 hp at rated speed); therefore, most OEMs subscribe to the theory that the engine fan should be run as little as possible.

Summary

- Engine cooling systems are required to absorb heat from the engine components.

- Transfer heat is absorbed to the ambient air by means of a heat exchanger.

- Engine cooling systems maintain a constant engine operating temperature.

- Heat may be removed by convection, conduction, or radiation.

- Most engine coolants are a mixture of water and antifreeze.

- Coolant should not be mixed in the cooling system but in an external container.

- The boiling point of coolant depends on the chemistry of the antifreeze and its concentration within the cooling system.

- Antifreeze lowers the freezing point and increases the boiling point of the coolant mixture, while protecting the cooling system from corrosion, scaling, and foaming, and inhibiting acid buildup.

- EG- or PG-based coolants must have the additive package routinely replenished.

- ELC can have a service life of up to 6 years with only one SCA recharge required.

- ELC is only available in premixed solution, ensuring that the water is at the mandatory level.

- The TDS level of the coolant is an indication of electrical conductivity. The higher the TDS level, the more conductive the coolant becomes.

- The most accurate instrument for measuring the freeze protection of a coolant mixture is a refractometer.

- The SCA used with EP and PG coolants must be tested and adjusted on a regular basis because the additives diminish over time.

- Radiators are usually positioned to take full advantage of the ram air effect for improved heat transfer.

- Radiators are defined by the way coolant flows through them, either downflow, crossflow, or counterflow.

- Radiator caps are designed to safely seal system pressure from the atmosphere, and a vacuum valve allows for the expansion and contraction of the coolant.

- Water pumps are non-positive, centrifugal-type pumps that may be belt- or gear-driven.

- Water pumps are lubricated by the coolant flowing through them and are defenseless to high TDS levels that become abrasive.

- Coolant filters should be changed according to OEM recommendations because some contain SCAs. Too much SCA in a cooling system can cause problems, just as low levels can.

- The job of the thermostat is to maintain the engine temperature exactly to ensure the best performance, fuel economy, and minimum noxious emissions.

- When the engine is cold, the thermostat routes the coolant through the by-pass circuit to quickly warm up the engine.

- When the engine warms sufficiently, the thermostat opens, directing coolant to flow through the radiator.

- The purpose of shutters is to control the air flow through the radiator and engine compartment.

- Shutters are controlled by the shutterstat, which directs system air pressure to open or close the shutters, depending upon coolant temperature.

- Shutters are usually designed to fail in the open position.

- Winter fronts installed on the hood grille are used in extremely cold climates to limit the amount of ram air that can be driven through the radiator and engine compartment.

- There are two basic types of engine cooling fans: suction-type and pusher-type.

- Most engine fans are temperature-controlled by a fanstat located in the water manifold. Another style of fanstat senses the temperature of the engine compartment in order to control the fan.

- Some engine fans use flexible pitch blades that alter the fan's efficiency proportionally to its driven speed, permitting efficient fan operation at low engine rpms.

- Viscous-type thermatic fans sense the temperature of the engine compartment and are driven by a fluid coupling designed to produce minimum slip at normal engine operating temperature.

- Engine fans are driven by V or poly-V belts whose tension should be adjusted with a belt tensioning gauge to avoid premature bearing failure or belt slippage.

- External leaks of cooling systems may be found using a handheld pressure tester. With the use of adaptors found in the kit, the radiator cap may be tested as well.

Review Questions

1. Of the following coolant types, which is thought of as the most toxic?

   A. Pure water

   B. Ethylene glycol

   C. Propylene glycol

   D. Extended life coolant

2. Of the following products, which has the best ability to transfer heat?

   A. Pure water

   B. Ethylene glycol coolant mixture

   C. Propylene glycol coolant mixture

   D. Extended life coolant

3. Which of the following components could cause cooling system hoses to collapse when the truck is left parked overnight?

   A. Defective thermostat

   B. Improper coolant

   C. Defective radiator cap

   D. This is normal.

4. What would be the most likely outcome if a radiator cap pressure valve fails to seal?

   A. Cooler operating temperatures

   B. Coolant boiling in the cooling system

   C. Increased hydrocarbon emissions

   D. Coolant from the overflow tank being drawn into the radiator

5. Which section of a downflow radiator would be the warmest when the truck is up to operating temperature?

   A. The bottom tank

   B. The overflow tank

   C. The top tank

   D. The center portion of the radiator core

6. Which of the following best describes the type of water pump typically found on truck diesel engines?

   A. Positive displacement

   B. Non-positive displacement centrifugal

   C. Constant volume

   D. Gear-type pump

7. When the thermostat is closed and the coolant is flowing through the by-pass circuit, what is actually taking place?

   A. The coolant is being circulated through the radiator.

   B. The coolant is being circulated through the engine.

   C. The coolant is being circulated through the coolant filter.

8. Which of the following components controls the drive efficiency of a thermatic viscous drive fan hub?

   A. A bimetal strip

   B. The viscosity of the silicone drive medium

   C. The fanstat

   D. The solenoid

9. At what speed does a fiberglass flexible pitch fan blade produce its greatest efficiency?

   A. At high speed

   B. At low speed

   C. At all speeds

10. If a diesel engine starts to overheat, at what location will the coolant begin to boil first?

    A. Within the thermostat housing

    B. Within the top radiator tank

    C. At the engine cooling jackets

    D. At the inlet to the water pump

11. Of the following instruments, which is the most accurate to test the degree of freeze protection of a diesel engine coolant?

    A. A spectrographic analyzer

    B. A color-coded test coupon

    C. A hydrometer

    D. A refractometer

12. Two technicians are having a discussion about ELC, and Technician A says that some ELCs have a service life of up to 6 years while Technician B says that water should never be added to ELC even if the water is distilled. Which technician is correct?

    A. Technician A is correct.

    B. Technician B is correct.

    C. Both Technicians A and B are correct.

    D. Neither Technician A nor B is correct.

13. Two technicians are discussing the operation of the shutters on a truck and Technician A says that if the shutterstat fails, the shutters will be left in the open position. Technician B says that if the air pressure is not able to enter the shutter cylinder, the shutters will be left in the closed position. Which technician is correct?

    A. Technician A is correct.

    B. Technician B is correct.

    C. Both Technicians A and B are correct.

    D. Neither Technician A nor B is correct.

14. Two technicians are discussing the freeze protection of antifreeze. Technician A says the protection should be 10 degrees below the expected ambient temperature. Technician B says that due to the wind chill factor that trucks experience while traveling down the road, the freeze protection must always be factored to the lowest expected wind chill factor. Which technician is correct?

    A. Technician A is correct.

    B. Technician B is correct.

    C. Both Technicians A and B are correct.

    D. Neither Technician A nor B is correct.

15. Two technicians are discussing the coolant mixtures. Technician A says that ELC should always be mixed at a ratio of 40% coolant with 60% distilled water. Technician B says that a 50/50 mixture of EG and water has better freeze protection than straight EG. Which technician is correct?

    A. Technician A is correct.

    B. Technician B is correct

    C. Both Technicians A and B are correct.

    D. Neither Technician A nor B is correct.

16. Two technicians are discussing radiator construction. Technician A says that a double pass radiator has better cooling capacity than a single pass radiator of the same size. Technician B says that a crossflow-style radiator is chosen by manufacturers designing trucks with low hood and grille designs. Which technician is correct?

    A. Technician A is correct.

    B. Technician B is correct.

    C. Both Technicians A and B are correct.

    D. Neither Technician A nor B is correct.

17. Which of the following components is not a heat exchanger?

    A. The radiator

    B. The heater core

    C. The evaporator

    D. The blower motor

Cab Climate Control/ Supplemental Truck Heating and Cooling

Learning Objectives

Upon completion and review of this chapter, the student should be able to:

- Explain how temperature is controlled in a blend air system.
- Explain how temperature is controlled in a water-valve controlled system.
- Describe the function of the fan switch.
- List the different features controlled by the air selection switch.
- Explain the function of the temperature control switch.
- Describe the function of the air-conditioning switch.
- Explain what happens when the recirculation button is activated or deactivated.
- Describe the purpose of the bunk override switch.
- Explain the purpose of the air louvers used in truck HVAC systems.
- Explain why it is important not to shut off manual coolant hand valves.
- Explain the importance of ventilation in the cab of the vehicle.
- Explain what stepper motors are used for in an HVAC system.
- Describe the function of a temperature sensor.
- List three types of supplemental cab climate controls, and explain the differences between them.
- Explain the advantages of truck stop electrification.

Key Terms

auxiliary power units (APUs)

blend air system

Constant Discharge Temperature Control (CDTC)

greenhouse gases (GHGs)

hydronic heater systems

pulse modulating water valve

pulse width modulation

recirculation

set point

shore power systems

stand-alone systems

stepper motor

thermistor

truck stop electrification

water-valve controlled system

INTRODUCTION

This chapter is written to help technicians understand how the climate control system operates. Technicians must understand what is supposed to happen when the HVAC controls are placed in their many different positions and what is happening within the dash in order for the system to maintain the desired temperature requested by the vehicle operator. These systems can be very challenging even for the most skilled technician to master. If the technician does not know exactly how the system operates, the possibility of misdiagnosis is drastically increased.

Most truck manufacturers use one of two methods of climate control in their vehicles. The two systems are the blend air system and the water-valve controlled system. To describe the systems, in this chapter we will use a Freightliner Century Class as our example vehicle.

This chapter will also examine supplemental truck heating and cooling systems and the advantages these units provide the vehicle owner and the environment.

SYSTEM OVERVIEW

The Blend Air HVAC System

In a **blend air system**, air circulated through the heater core and fresh air or air circulated through the evaporator coil are blended in correct proportions to maintain the desired temperature as selected by the vehicle operator.

Water-Valve Controlled System

The **water-valve controlled system** uses a water valve to vary the amount of coolant flowing through the heater core. When the valve fully opens, coolant flow through the heater core increases, so more heat is available to be circulated through the cab. When the valve closes down, restricting coolant flow, the amount of available heat decreases. The Freightliner Corporation refers to one of its systems as **Constant Discharge Temperature Control** (CDTC). These systems are able to maintain a constant air flow temperature within the cab of the vehicle, regardless of ambient conditions, selected fan speed, engine coolant temperature, or coolant flow. Once the operator selects the desired temperature with the control switch, no other adjustment by the driver is necessary. This is accomplished with a **pulse modulating water valve**. These valves control coolant flow very accurately, maintaining the exact temperature or **set point**, as selected by the vehicle operator **(Figures 9-1 and 9-2).**

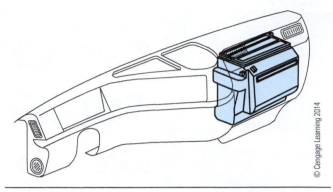

Figure 9-1 The HVAC system for both the blend air system and the water-valve controlled system are contained in a package beneath the dash on the passenger side of the cab.

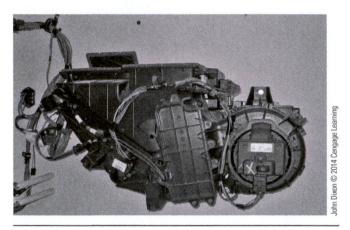

Figure 9-2 A typical HVAC unit containing the evaporator, heater core, blower fan, ducting, and circulating doors.

Note: Pulse modulating valves also known as pulse width modulation (PWM) are controlled by an electronic control unit. The volume of fluid that the valve will allow to pass can be controlled in very small increments.

Supplemental Heating and Cooling Systems

These systems allow the vehicle operator to maintain climate control in the cab and sleeper compartment when the engine is not running. These systems are proven to save huge amounts of diesel fuel compared to idling the truck's engine to maintain cab and sleeper temperature. Due to the reduced idling times, the environment is spared a great deal of **greenhouse gases (GHGs)**.

WATER-VALVE CONTROLLED SYSTEMS

For this section, we will be looking at the climate control feature of a water-valve controlled system in a Freightliner Century Class truck. The climate control panel, as shown in **Figure 9-3,** allows the vehicle operator to control all the functions of the HVAC system at one convenient center.

The Fan Switch

The fan switch is located at position 2 in **Figure 9-3.** This switch gives the operator the ability to control the fan speed. The air is forced by the fan through the selected air outlets. The switch shown here has eight fan speeds and an off position. If the switch is at the 0 position, the fan will not operate. If the operator wishes to increase the air flow, the switch is turned clockwise or to a higher setting. To decrease the air flow, the operator may turn the switch counterclockwise or to a lower setting. Setting the switch to the off position disables the air conditioner and places the air source in the fresh air mode.

Note: If the fan switch on the climate control panel for the cab is in the off position and the air conditioning for the bunk is on, the cab fan operates at low speed. It is necessary to keep this fan running to protect the cab evaporator from freezing because the refrigerant is still flowing through the main evaporator when the remote evaporator within the sleeper is being operated.

When the engine is started, there is a 2-second delay before the blower fan begins to operate and it can take an additional 4 seconds for the blower motor to reach its high-speed setting. The blower motor performs a self-diagnostic test once the engine has started, which accounts for the delay.

Air Selection Switch

The air selection switch, located at position 4 in **Figure 9-3,** allows the operator to control the flow of air from, the defrost (windshield) outlets, the face outlets, the floor outlets, or a combination of these outlets, making a total of nine different selection modes **(Figure 9-4).**

Refer to **Figure 9-5** for air selection switch positions. The different selection modes are as follows:

Position 1 is the **face mode.** In the face mode, all air from the blower is forced through the face or instrument panel vent outlets **(Figure 9-6).**

Position 2 is **between the face mode and the bi-level mode.** In this position, approximately 75% of the air flow is distributed through the face vents and the remaining 25% is directed through to the floor outlets.

Position 3 is the **bi-level mode.** In this position, the discharge air is divided equally between the face outlets and the floor outlets.

Position 4 is the position **between the bi-level mode and the floor mode.** In this position, 25% of discharge air is forced through the face vents and 75% is directed through the floor outlets.

Position 5 is the **floor mode.** In this position, all the discharge air is directed to the floor outlets **(Figure 9-7).**

Position 6 is the position **between the floor mode and the floor/defog mode.** In this position, 75% of the discharge air is forced out the floor outlets and the remaining 25% is directed through the defrost vents at the windshield.

Position 7 is the **defog mode.** In this position, the discharge air is divided equally between the

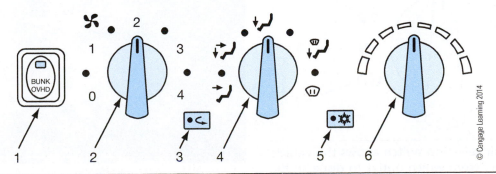

Figure 9-3 A climate control panel featuring CDTC.

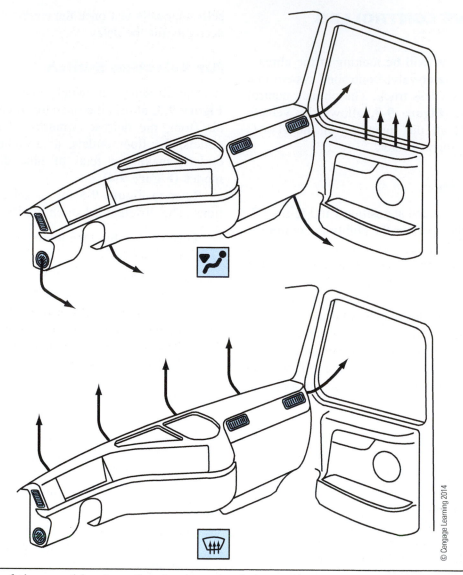

Figure 9-4 Some of the possible air outlets in the cab of the truck.

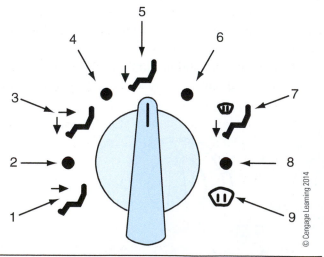

Figure 9-5 The air selection switch allows the vehicle operator to select from which outlet to disperse the conditioned air (whether heated or cooled).

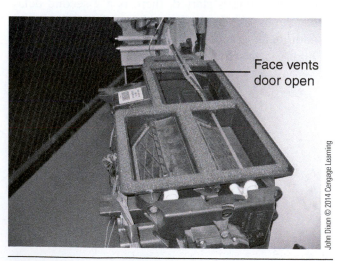

Face vents door open

Figure 9-6 The vent door for the face mode is open.

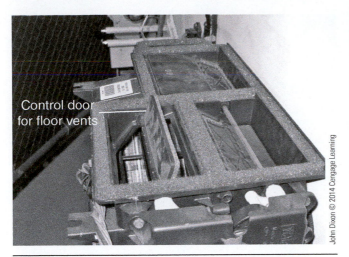

Figure 9-7 The vent door is open to direct air to the floor vents.

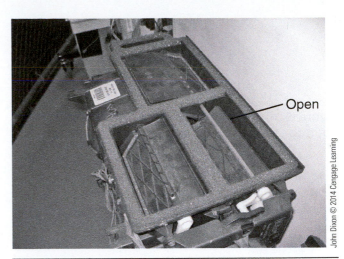

Figure 9-9 The defrost vent door is open, and all others are closed, supplying all air to the windshield.

floor outlets and the defrost vents at the windshield (**Figure 9-8**).

Position 8 is the position **between the defog mode and the defrost mode.** In this position, 75% of the discharge air is forced though the defrost outlets and the remaining 25% through the floor outlets. When the air selection switch is in this position, the air conditioner automatically turns on. The recirculation button does not operate when the switch is in this mode.

Position 9 is the **defrost mode.** In this position, all of the discharge air is forced through the defrost outlet and onto the front windshield. When the air selection switch is in this position, the air conditioner automatically turns on. The recirculation button does not operate when the switch is in this mode (**Figure 9-9**).

Temperature Control Switch

The temperature control switch is found in position 6 on the control panel in **Figure 9-3**. This switch allows operators to select the desired temperature to suit their own personal comfort level. On both the manual water-valve system and the CDTC climate control

panel, if the switch is turned clockwise into the red zone, the temperature increases (becomes warmer). If the switch is rotated counterclockwise into the blue zone, the discharge temperature decreases (becomes cooler).

Air-Conditioning Switch

The purpose of the air-conditioning unit is to cool and dehumidify the interior of the cab. The control button for the air conditioner is in position number 5 on the control panel in **Figure 9-3**. This switch, when pushed in, operates the air conditioner, and if the button is again pushed in, the air conditioner is turned off.

The air-conditioning button contains an amber indicator lamp that is on whenever the demand for air conditioning has been selected from either the cab climate control panel or the sleeper climate control panel. If the instrument panel lights are on, then the snowflake indicator on the air-conditioning button will be on.

The air-conditioning function will be disabled when:

1. The ambient air temperature is so low that the air-conditioning function would be ineffective.
2. The engine is not running at a high enough rpm.
3. Certain conditions cause the heater and air-conditioning system to go into a protection mode (**Figure 9-10**).

Recirculation

The **recirculation** button is located at position 3 on the control panel in **Figure 9-3**. The recirculation mode restricts the proportion of outside air entering the cab. Often, the recirculation button is used to prevent

Figure 9-8 The vent doors are open, distributing air to defrost vents and floor outlet vents.

Figure 9-10 The air-conditioning switch allows the operator to turn air conditioning on or off.

dusty or smoky air from entering the cab. The recirculation feature can also be used to decrease the time required to heat or cool the interior of the cab when ambient temperature conditions are extreme. Whenever the climate control system is in recirculation mode, the amber light is illuminated. When the air selection switch is in certain positions, the recirculation feature will not operate:

1. When the air selector switch is in the defog mode.
2. When the air selector switch is in between the defog mode and the defrost mode.
3. When the air selector switch is in the defrost mode.

Note: The recirculation feature is controlled by a timer and will automatically switch the system to partial recirculation mode after 20 minutes of operation. This prevents the buildup of fumes or odors and also prevents oxygen depletion within the cab. This feature alone will help the operator remain alert. If conditions continue to be extremely dusty or smoky, the partial recirculation mode can be overridden back to full recirculation mode by simply pressing the recirculation button twice. This allows the timer to reset for 20 minutes **(Figures 9-11** and **9-12).**

Optional Bunk Override Switch

The bunk override switch (BUNK OVRD) gives the vehicle operator the ability to control the temperature and fan speed of the sleeper compartment from the cab climate control panel. This switch is located at position 1 on the climate control panel in **Figure 9-3.** If the operator depresses the top half of the switch, it allows the cab settings to override the sleeper control settings. Whenever the bunk override mode is in operation, the amber light is illuminated. If the operator makes any adjustment to the controls within the

Figure 9-11 The recirculation door is open, allowing fresh air to be introduced into the HVAC.

Figure 9-12 The recirculation door is closed, preventing dust and contaminants from entering the cab as well as reducing cooling time when air conditioning has been selected.

sleeper compartment, the override mode feature will be canceled, enabling the bunk climate control settings to take over.

When the climate control system is in the bunk override mode, the climate control panel of the cab can be adjusted without affecting the settings of the sleeper compartment. If it is necessary to make adjustments to the sleeper climate settings, set the fan speed and temperature settings on the cab climate control panel to the desired sleeper settings and depress the upper half of the bunk override switch again. Once again, the cab climate control panel can be adjusted for the cab settings without affecting the new sleeper settings.

When the operator wishes to cancel the bunk override mode, the lower half of the bunk override

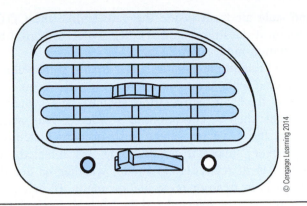

Figure 9-13 Louvers allow the operator to direct the discharge air where it is wanted.

switch is depressed. Once the override mode has been terminated, the sleeper climate control system is controlled from the sleeper climate control panel, and the amber indicator on the override button is off.

Air Outlet Vents

On the instrument panel, outlets for the face have louvers that may be directed so the discharge air from them can be made to flow up, down, left, or right. The operator can point these vents wherever desired or close the vents completely **(Figure 9-13)**.

Defrost (windshield) vents and the vents that are directed at the doors are fixed and therefore nonadjustable. The sleeper compartment may have one or more vents that may be positioned up, down, or closed.

SLEEPER CLIMATE CONTROL PANEL

As with the climate control system of the cab, the sleeper climate control system also features a watervalve controlled system with constant discharge temperature control (CDTC) **(Figure 9-14)**.

The CDTC maintains constant air flow temperature from the vents, regardless of ambient temperature conditions, selected fan speed, engine coolant temperature, and coolant flow. Just like the cab climate control panel, once the temperature control switch is rotated to the preferred temperature, no other adjustments are necessary.

Fan Switch

The fan switch is located at position 1 in **Figure 9-14**. The fan switch in the sleeper controls the fan speed of the blower for the heater and air conditioning. The fan switch on the sleeper control panel has eight fan speeds as well as an off position. Rotating the switch in

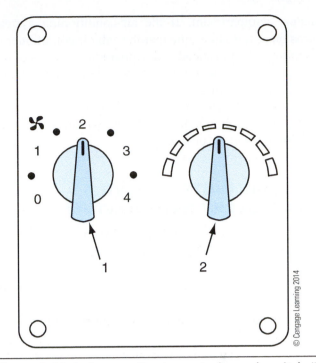

Figure 9-14 A sleeper climate control panel: switch #1 controls fan speed and switch #2 controls temperature.

a clockwise direction or to a higher number increases the fan speed, and turning the switch counterclockwise or to a lower number decreases the fan speed.

> **Note:** If the fan switch on the climate control panel for the sleeper is in the off position and the air conditioning for the cab is on, the sleeper fan operates at low speed. It is necessary to keep this fan running to protect the bunk evaporator from freezing, because refrigerant is still flowing through the remote evaporator when the main evaporator within the cab dash is being operated.

Temperature Control Switch

The temperature control switch for the sleeper is found in position 2 in **Figure 9-14**. This switch allows operators to select the desired temperature to suit their own personal comfort level within the sleeper compartment. Like the temperature control switch for the cab climate control, if the switch is turned clockwise into the red zone, the temperature increases (becomes warmer). If the switch is rotated counterclockwise into the blue zone, the discharge temperature decreases (becomes cooler).

The air-conditioning system automatically turns on in order to maintain the chosen temperature in the sleeper compartment. If the air-conditioning system turns on at the same time that the cab air conditioner is operating, the fan speed and temperature settings of the sleeper compartment will be overridden by the cab air-conditioning settings.

The CDTC will not function whenever the temperature control switch is in the maximum cooling or maximum heating position.

MANUAL WATER-VALVE CONTROLLED HVAC SYSTEM

The water-valve controlled HVAC system operates much like the CDTC previously mentioned, with a few small differences.

This system is also from a Freightliner Century Class truck.

In position 1 in **Figure 9-15** is the fan switch. This switch allows the operator to circulate fresh or re-circulated air at four different speeds or to shut the blower fan off entirely. Turning the switch clockwise to higher numbers increases the speed of the blower fan. When the switch is in the 0 position, the fan is off.

In position 2 in **Figure 9-15** is the air selection switch, which is used to direct the flow of discharge air on the face and feet, face only, feet only, feet and windshield, or windshield only.

In position 3 in **Figure 9-15** is the temperature control switch. This switch allows operators to select the desired temperature to suit their personal comfort level. When the switch is turned clockwise into the red zone, the temperature increases (becomes warmer). If the switch is rotated counterclockwise into the blue zone, the discharge temperature decreases (becomes cooler).

In position 4 in **Figure 9-15** is the maximum air-conditioning button. When this button is depressed, the ventilation system is placed in the recirculation mode. This mode of operation should not be used for more than 20 minutes to prevent the buildup of fumes, odors,

and stale air from inside the cab. Unlike the CDTC system, it does not automatically switch to partial re-circulation mode after 20 minutes.

On Freightliner Century Class trucks, if the vehicle is ordered without air conditioning, it will be equipped with a fresh air/recirculation switch. This switch is optional when air conditioning is ordered. This switch allows the driver to select fresh or recirculated air. Whenever "max air conditioning" is selected, the system is in recirculation mode, whether or not the switch is in fresh or recirculation mode. If air conditioning has been selected and the fresh air recirculation switch is in recirculate mode, the system will function the same as it does when the air selection switch is set at maximum air conditioning.

The control panel in **Figure 9-15** may contain a fan bunk switch. This switch on the cab control panel overrides the sleeper climate control panel. This is a rocker switch with three positions. To increase the ventilation, the switch is put in the HI position. For normal fan speed, the switch can be moved to the mid-position, and if operators wish, they may turn off the fan in the bunk entirely by moving the switch to the off position.

Sleeper Climate Control Panel

The sleeper heater and air conditioning used in a water-valve controlled HVAC system operates much like the CDTC system, with a few minor changes. The bunk HVAC system uses a separate evaporator coil, expansion valve, and blower fan for the air conditioning side, and a separate heater core and water valve that operates independently from the cab HVAC system **(Figures 9-16** and **9-17).**

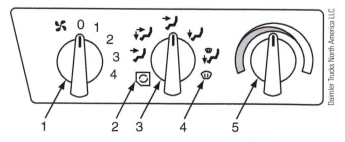

Figure 9-15 An HVAC control panel for a water-valve system.

Figure 9-16 A bunk HVAC system containing the evaporator, heater core, blower fan, and ducting.

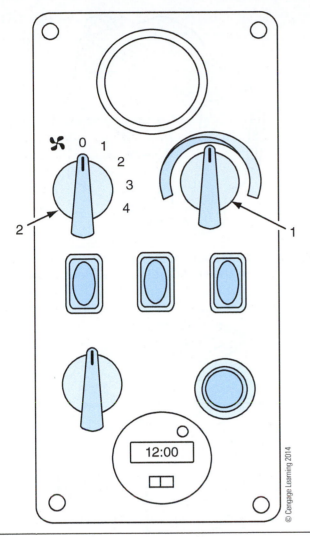

Figure 9-17 A sleeper control panel for a water-valve controlled system; the temperature control switch is in position 1 and the fan control switch is in position 2.

Temperature Control

The water-valve controlled system controls the intensity of the heating system by controlling the flow of coolant through the heater core. More coolant flow through the heater core means more heat at the vent outlets. When coolant flow is restricted, the temperature at the vent outlets decreases. When the air conditioning is on, the cold air from the evaporator is mixed with warm air from the heater core. Again, the water-control valve controls the temperature of the air conditioning by controlling the temperature of the heater core.

There are two types of water valves used on trucks. One is a manual valve, which is cable activated, and the other is electrically operated **(Figure 9-18)**.

The manual valve is opened and closed by a cable, which is linked to the control panel or it may be linked to the heater control stepper motor.

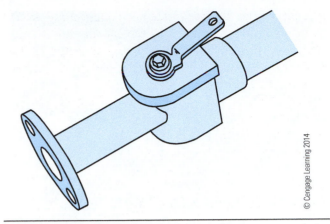

Figure 9-18 A manual water valve operated by movement of a cable.

The Volvo Corporation calls their water valve a solenoid water valve, whereas Freightliner calls theirs pulse modulated. No matter what these companies call these valves, they are both **pulse width modulation** controlled **(Figure 9-19)**.

These valves are used in systems that are electronically controlled by electrical impulses from the control unit. The solenoid water valve is constantly supplied with 12 V and is switched on the ground side.

Note: If the water valve does not have the ability to close all the way, the air conditioning will not work efficiently when set to the lowest temperature because it is working against heat generated at the heater core.

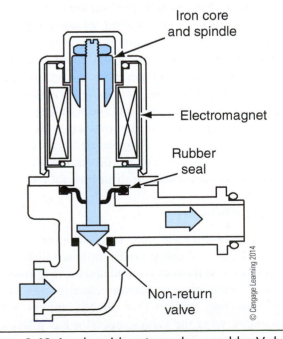

Figure 9-19 A solenoid water valve used by Volvo.

Figure 9-20 Manual water shutoff valves on the engine.

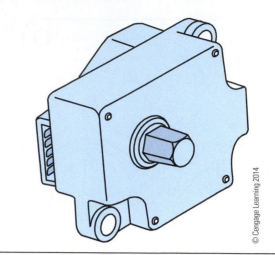

Figure 9-21 Stepper motors are used to move vent doors within the HVAC unit into various positions.

It is important not to shut off the valve at the engine for coolant to the heater core **(Figure 9-20)**. This will severely hinder the system's ability to control the air-conditioning temperature in both the cab and sleeper compartments. When the valve is shut off, the warm coolant will not be able to circulate through the heater core, making the outlet air temperature controls unable to change the air temperature.

BLEND AIR SYSTEM

The control panels for blend air HVAC systems are virtually the same as the control panels for water-valve controlled systems. The real change is in the way that the outlet air temperature is controlled. With the blend air system, both the heater core and the evaporator coil can be operating at their highest intensity (if AC has been selected). The air conditioning is on and there is no restriction in the amount of coolant circulating through the heater core. The temperature is controlled by a blending door. This blend door may be automatically or manually positioned to allow more or less air to be directed through each of the two coils in order to maintain the desired vent temperature. If air conditioning has not been selected by the driver, the air from the heater core is blended with fresh air brought into the cab.

Stepper Motor

Stepper motors are used to position doors within the HVAC unit, thereby controlling air distribution, temperature, and fresh or recirculated air **(Figure 9-21)**.

These small devices convert electric signals to a stepped rotating motion that directly moves a door according to what the vehicle operator has selected on the control panel. The stepper motor consists of a rotor

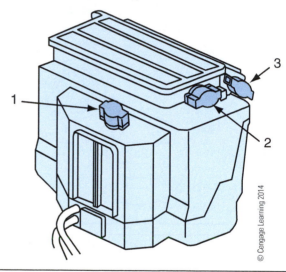

Figure 9-22 Placement of three stepper motors on a Volvo climate control system.

with permanent magnets, four coil windings, and four in series diodes **(Figure 9-22)**. Stepper motors are controlled by an ECU and can be moved in very precise increments.

The stepper motor in position 1 controls the shutter, which mixes fresh air with recirculated air. The stepper motor in position 2 controls the air outlet distribution. And the stepper motor in position 3 controls air bypass on electronically controlled systems and controls the water valve and air mixing on the manual air-conditioning systems or systems without A/C.

Ventilation

An exhaust system is included with sleeper cabs. When the climate control panel is in the fresh air mode of operation, approximately 300 cubic feet of air per minute is brought into the cab and exhausted through a vent in the rear area of the sleeper. This will exhaust the stale air, allowing fresh air to take its place and slightly pressurize the cab. To prevent the buildup of fumes, odors, and stale air, the HVAC system should not be run in the recirculation mode for more than 20 minutes at a time.

HVAC GENERAL INFORMATION

The controls for other truck manufacturers tend to vary, but all operate on much of the same premises **(Figure 9-23)**.

The control panel for one of Western Star's trucks has a digital LED display for the cab temperature. There is also a rocker switch for temperature control. Arrows above the rocker switch indicate the direction the temperature will change when the switch is depressed. If the switch is pressed to the left side, the setting for the cab temperature decreases. If the switch is pressed to the right side, the temperature increases. The vehicle operator can change the display from degrees Fahrenheit to degrees Celsius. This may be accomplished by turning the ignition switch "ON" and the fan speed switch to the "OFF" position. The rocker switch is then depressed to the side with the temperature down arrow and held for 5 seconds. The units of temperature will be displayed on the LED screen. Press the down side of the rocker switch again and the display units of temperature will change **(Figure 9-24)**.

If the unit of temperature display is changed on the cab control panel, the unit of temperature display on the sleeper compartment also changes. Western Star trucks use a system called Automatic Temperature Control or ATC. When the auto setting is engaged, the system engages full automatic control mode. When in this mode of operation, the fan speed varies as the system maintains the temperature set on the display.

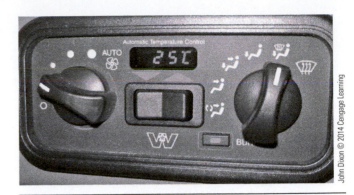

Figure 9-24 A control panel from a Western Star truck; the temperature display has been changed from degrees Fahrenheit to degrees Celsius.

Most truck manufacturers offer vehicles with some type of electronically controlled HVAC system. These systems require a control unit that may be on its own or integrated into the vehicle's engine management system. The HVAC control unit (computer) controls the temperature within a set of values. It receives input signals from temperature and pressure sensors, and delivers output signals to control compressor operation, water valve operation, and stepper motor positions. The vehicle operator sets the control panel to the desired temperature, fan speed, ventilation position, and fresh or recirculated air position.

Temperature Sensors

Temperature sensors are used at various points throughout the vehicle to send signals back to the electronic control unit for the HVAC system. These sensors contain a temperature-dependent resistor. This type of temperature sensor is known in the industry as a **thermistor**. These thermistors are referred to as negative temperature coefficient (NTC) thermistors. As the temperature increases, the resistance goes down, and vice versa when the temperature increases. The electronic control unit monitors these changes in resistance and responds accordingly **(Figure 9-25)**.

Operator Maintenance

It is important that the operator run the air conditioning compressor at least once every month for approximately 5 minutes. Doing this helps prevent the drying and cracking of tubing seals, thereby reducing refrigerant leaks in the system. The air conditioner should only be operated after the engine compartment is warm and the cab's interior is 70°F (21°C) or higher. If ambient conditions are cold, the heater may also be operated to avoid driver discomfort.

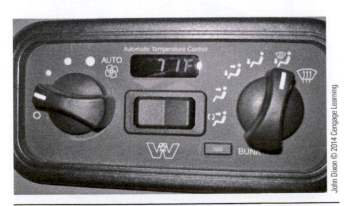

Figure 9-23 A control panel from a Western Star truck.

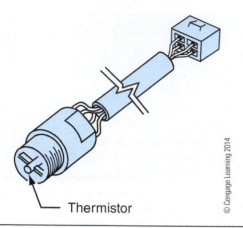

Thermistor

Figure 9-25 Temperature sensors, called thermistors, allow the system to monitor the temperature at various points to aid in controlling the set point.

© Cengage Learning 2014

General Maintenance

Freightliner recommends that the fresh air filter be replaced every six months, regardless of mileage, to permit proper operation of the HVAC system. If the filter is not replaced, damage to the heater and air-conditioner components could occur.

Western Star recommends that their foam-type air filter be cleaned every 3 months, or replaced annually. These foam-style filters can be cleaned a maximum of three times before they should be replaced. It is also recommended that the optional dust/odor filter be replaced rather than cleaned every 3 months. The air filter assembly for the sleeper compartment should also be replaced or serviced at the same interval as the cab air filter.

....

Note: The HVAC system should not be operated without the fresh air filter installed.

....

Follow the instructions in the service literature for filter replacement, making sure that the new filter is oriented properly. The filter label has an arrow that indicates the air flow direction through the filter.

SUPPLEMENTAL CAB CLIMATE CONTROL

In order for the heating and air conditioning to function, the truck engine must be running. Both the United States and Canadian governments are aware that with the number of trucks across North America, this has become a huge environmental problem.

According to a Canadian study, a typical heavy-duty, freight-hauling, Class 8 truck idles about 1840 hours

per year when parked overnight at truck stops. The main reasons for idling are to:

- heat or cool the truck cab and/or sleeper,
- keep the fuel warm in winter,
- keep the engine warm in winter so that the engine is easier to start.

In addition, bus operators operate their engines for similar reasons.

A Class 8 truck/bus engine consumes approximately 1 gallon (4 L) of diesel fuel per hour when idling at 900 rpm. Because each quarter gallon (liter) of diesel fuel consumed by an engine produces 6.1 lbs (2.8 kg) of GHGs (Greenhouse Gases), one hour of idling by a Class 8 engine produces 24.7 pounds (11.2 kg) of GHGs, and one year's worth of idling (1840 hours) produces 22.6 tons (20,496 kg) of GHGs.

Such engine idling can be avoided through the use of devices such as fuel-fired interior heaters, engine coolant heaters with truck cab or bus interior heating apparatus, and auxiliary power generators **(Figure 9-26).**

Fuel-Fired Interior Heaters

Fuel-fired heaters that use diesel fuel from the truck's fuel tank are incredibly more efficient at heating the truck's interior, than idling the engine, because heat is provided directly from the combustion flame to the heat exchanger. The vehicle's engine uses some of its heat energy to overcome internal friction

Espar Heater Systems

Figure 9-26 A supplemental truck heating system.

and to drive engine accessories such as alternators and power steering pumps. Fuel-fired interior heaters are available with outputs between 5100 and 50,000 British thermal units per hour (BTU/h). These devices can only be used for heating the cab's interior but not for cooling it.

According to the study previously referenced on the idling habits of North American truck drivers, Natural Resources Canada has decided that, for Canadian conditions, it would assume 1450 hours of engine idling for heating and 390 hours for cooling. Manufacturers of fuel-fired interior heaters have published figures for their fuel consumption of between 0.07 and 0.46 gallons per hour (0.27 to 1.80 liters per hour). At these rates, a fuel-fired interior heater could save from 0.58 to 0.98 gallons per hour (2.2 to 3.7 liters/hour) for approximately 1450 hours per year. This would equate to the approximate consumption of 843 to 1417 gallons (3190 to 5365 liters) of diesel fuel, representing a GHG reduction of 8.8 to 16.5 tons (8.0 to 15 tonnes) per year. These figures indicate an enormous reduction in fuel consumption and GHG production that would otherwise result from idling the vehicles' engines **(Figure 9-27)**.

Operation

For this explanation, we will look at an air heater manufactured by Espar Corporation. The heater may be turned on by means of a switch, a thermostat, or a timer. A control unit signals the blower motor to draw combustion air into the heat exchanger; simultaneously, fuel is drawn from the fuel tank by the heater's fuel pump and mixed with the air. As an ignition plug ignites the air/fuel mixture, a controlled flame is established inside the sealed heat exchanger. Combustion gases then pass safely through the exhaust and into the atmosphere. Clean air is drawn over the surrounding walls of the heat exchanger and circulated by the heater fan. A flame sensor, a temperature regulating sensor, and an overheat sensor ensure safe operation of the unit. These units cycle through four heat settings (boost, high, medium, and low) in order to maintain the temperature selected by the vehicle operator. In some cases, less heating is required than the heater produces in the low mode of operation. When this occurs, the heater switches to a standby mode. When the unit is shut off by any method, it starts a controlled cool-down sequence. The fuel pump stops delivering fuel. The ignition plug is then energized for 15 seconds. The blower will continue to run for 3 minutes and will then automatically switch off.

Engine Coolant Heaters with Truck Cab or Bus Interior Heating

This category of supplemental truck climate control includes products that both supply heated air to the truck or bus interior, as well as heat and circulate engine coolant. These units are fuel-fired, using diesel fuel from the truck's fuel tank. They can be designed as a single unit, or they can be a combination of air heater and coolant heater, as long as both operate independently of the vehicle's engine. Fuel-fired coolant heaters are available with outputs between 13,700 and 120,000 British thermal units per hour (BTU/h). These devices can only be used for heating the engine coolant and cab's interior and provide no means for cooling.

According to a recent study, Natural Resources Canada has decided that, for North American winter conditions, it would assume 1450 hours of engine idling for heating. Depending on the application, these devices could save up to 0.92 gallons per hour (3.5 liters/hour). At this rate of idling, approximately 416 gallons (5075 L) of diesel would be consumed, representing a GHG reduction of 15.7 tons (14.2 tonnes). Again, these figures indicate an enormous reduction in the fuel consumption and GHG production caused by idling the vehicle's engine. Preheating the engine also eliminates white smoke engine emissions that occur when starting a cold engine. Warm starts also reduce engine stress and wear and reduce battery demand.

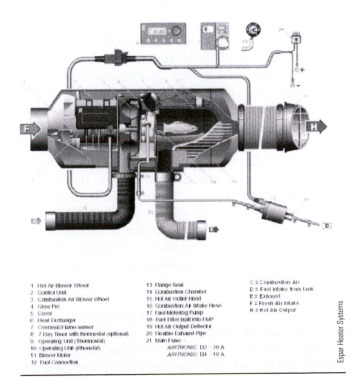

1 Hot Air Blower Wheel	13 Flange Seal	C = Combustion Air
2 Control Unit	14 Combustion Chamber	D = Fuel Intake from tank
3 Combustion Air Blower Wheel	15 Hot Air Outlet Hood	E = Exhaust
4 Glow Pin	16 Combustion Air Intake Hose	F = Fresh Air Intake
5 Cover	17 Fuel Metering Pump	H = Hot Air Output
6 Heat Exchanger	18 Fuel Filter built into FMP	
7 Overheat/Flame sensor	19 Hot Air Output Deflector	
8 7 Day Timer with thermostat (optional)	20 Flexible Exhaust Pipe	
9 Operating Unit (Thermostat)	21 Main Fuse	
10 Operating Unit (Rheostat)	AIRTRONIC D2 - 20 A	
11 Blower Motor	AIRTRONIC D4 - 10 A	
12 Fuel Connection		

Espar Heater Systems

Figure 9-27 A diesel-fired cab interior heater manufactured by Espar.

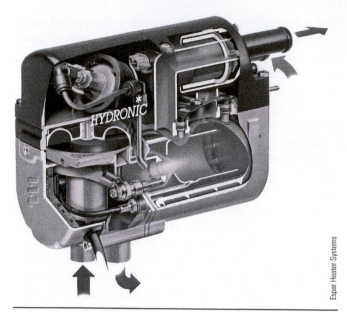

Espar Heater Systems

Figure 9-28 A coolant heater, called a hydronic heater, by Espar.

Because these coolant heaters are capable of maintaining coolant temperatures in the 185°F (85°C) range, heat is produced for the cab and sleeper compartment. This heat may be extracted by the OEM heater cores, and a thermostatically controlled fan or the coolant may be plumbed through another remote heat exchanger **(Figure 9-28).**

Operation

For this explanation, we will look at a **hydronic heater** manufactured by Espar Corporation. These heaters are connected to the engine's cooling system. The heater may be turned on by means of a switch or timer. A control unit signals the integrated coolant pump to start circulating coolant; simultaneously, a blower motor draws combustion air into the heat exchanger. At the same time, fuel is drawn from the fuel tank by the heater's fuel pump and mixed with the air. The mixture is then ignited by a glow pin. Once a flame is established, the glow pin shuts off. Combustion gases pass safely through the exhaust and into the atmosphere. The heater now runs in high heat mode and the temperature is monitored at the heat exchanger. An integrated water pump moves the coolant over the heat exchanger and heats the coolant, and then circulates the coolant thorough the engine's plumbing system. These units have the capacity to raise the temperature of the coolant 100°F in one hour. Once the coolant reaches 185°F (80°C), the heater automatically switches to low heat mode and continues to run. If the coolant drops down to 167°F

(75°C), the heater automatically switches back to the high heat mode of operation. If the temperature of the coolant continues to rise, the heater shuts down when coolant temperature reaches 187°F (86°C). The water pump continues to operate, allowing the unit to monitor the engine temperature. Once the coolant temperature falls to 167°F (75°C), the unit automatically restarts. The unit automatically continues to operate as described, unless it is switched off either manually, automatically by a timer, or by a heater malfunction. When the unit is shut off by any method, it starts a controlled cool-down sequence. The fuel metering pump stops delivering fuel, extinguishing the flame. The combustion air blower and water pump continue to operate for 130 seconds to cool the unit down. The unit then shuts down.

Note: If the unit shuts down due to a flame-out while running, it attempts to restart once. If the restart is successful, it continues to run; if not, the unit shuts down completely after a cool-down cycle. Also, if the battery voltage drops below 10 VDC or rise above 16 VDC, the unit shuts down after a cool-down mode.

AUXILIARY POWER UNITS

Auxiliary power units, or APUs, are portable, truck-mounted systems that can provide climate control (heating and cooling) as well as electricity for electrical appliances, battery charging, and/or engine preheating (block heater). These systems generally consist of a small internal combustion engine (usually diesel) equipped with a generator and heat recovery system to provide electricity and heat. For the air-conditioning function, an electrically powered air-conditioning unit is normally installed in the sleeper, although some systems use the truck's air-conditioning system. The more sophisticated APUs consist of a generator powered by a diesel engine, a compressor, an alternator, and an inverter/charger. These units are fully integrated into the truck's heating, ventilation, and air-conditioning system. In addition, the inverter/charger allows the system to provide electric power to the cab and sleeping compartments for appliances such as a television or microwave.

Output ratings for these units range from 6000 to 25,000 BTUs of cooling capacity. The system heating capacity ranges between 6000 and 18,000 BTUs. In addition, these units also provide varying

Figure 9-29 An auxiliary power unit mounted to the frame rail of a Class 8 truck.

amounts of AC and/or DC electricity to support engine-off use of truck cab/sleeper accessories **(Figure 9-29).**

According to one study, Natural Resources Canada has decided that, for North American climate conditions, it would assume 1450 hours of engine idling for heating and 390 hours for cooling. Manufacturers of APUs have published figures for their fuel consumption of between 0.01 and 0.2 gallons per hour (0.38 to 0.78 liters per hour). At these rates, a fuel-fired interior heater could save from 0.87 to 0.95 gallons per hour (3.2 to 3.6 liters/hour) for approximately 1840 hours per year. This would approximately equate to the consumption of 1555 to 1750 gallons (5888 to 6624 liters) of diesel fuel, representing a GHG reduction of 18.2 to 20.4 tons (16.5 to 18.5 tonnes) per year. Again, these figures indicate an enormous reduction in the fuel consumption and GHG production caused by idling vehicle engines.

Note: Supplemental heating and cooling systems also reduce noxious emissions such as hydrocarbons, NOX, and carbon monoxide.

TRUCK STOP ELECTRIFICATION

The U.S. Department of Transportation (DOT) estimates that approximately 5000 truck stops in the United States offer parking and other services, including fueling stations, restaurants, stores, and showers. Truck stops are vital to America's over-the-road transportation system **(Figure 9-30).**

Figure 9-30 Truck stop electrification.

Truck stop electrification allows the driver to plug the vehicle in to operate necessary systems without idling the truck's engine. In some cases, a **stand-alone system** can provide heating, ventilation, and air conditioning directly to the sleeper compartment.

Options for truck stop electrification include:

- stand-alone systems that are owned and operated by the truck stop
- combination systems that require both onboard and off-board equipment

Both of these systems eliminate idling of the truck's engine, thereby greatly reducing the amount of harmful GHGs produced.

Stand-Alone Systems

The stand-alone systems for HVAC operations are contained in a structure above the truck parking spaces. A hose from the host HVAC unit is connected to the truck window and a computer touch screen allows the operator to make payment for the service provided. Stand-alone systems are owned and maintained by private companies that charge an hourly fee. To accommodate the HVAC hose, a window template must be installed in the truck **(Figure 9-31)**.

Onboard or Shore Power Systems

These systems provide electrical outlets that the truck can plug into. (The shipping and boating industries have been doing this for many years.) To use **shore power systems**, the truck must be equipped with an

Figure 9-31 Picture of an HVAC stand-alone system contained in a structure above the truck parking space.

inverter to convert 120 volt AC power. This power can then be used to run onboard HVAC equipment or any other appliance the operator wishes to plug into the vehicle's outlets (TV, refrigerator, microwave, etc.).

The truck stop outlets are owned by private companies that regulate use and fees. Onboard equipment is owned by the owner of the vehicle or trucking company.

Summary

- A water-valve controlled system uses the valve to control the amount of coolant flowing through the heater core.

- Blend air systems combine air from the heater core with air from the evaporator in the correct proportions to maintain the desired outlet temperature.

- The fan switch gives the operator control over the fan speed.

- The air selection switch allows the operator to choose which vents discharge air from the system.

- The temperature switch allows the operator to select the desired temperature for personal comfort level.

- The air-conditioning switch allows the operator to select air conditioning or turn it off.

- Recirculation mode restricts the proportion of outside air entering the cab.

- The bunk override switch gives the vehicle operator the ability to control the temperature and fan speed of the sleeper compartment from the cab climate control panel.

- Pulse width modulated water valves are electronically controlled by electrical impulses from the control unit.

- Supplemental climate control systems provide driver comfort without idling the truck's engine.

- Supplemental climate control systems save large quantities of fuel and wear and tear on the truck's engine, and create big reductions in GHGs compared to idling the truck's engine.

- Truck stop electrification can provide a total HVAC solution for the vehicle while it is parked, or provide electricity to run onboard electrical appliances for driver comfort and HVAC needs.

Review Questions

1. What component is responsible for controlling the temperature at the outlets in a blend air system?

   A. A water valve

   B. A thermistor

   C. A pulse width modulated solenoid valve

   D. A blend door

2. If the fan is in the off position in the cab and air conditioning has been requested in the sleeper compartment, the cab fan will:

   A. Not run

   B. Run in low speed

   C. Run in medium speed

   D. Run in high speed

3. The air selection switch allows the vehicle operator to:

   A. Select desired fan speed

   B. Select desired temperature

   C. Select where discharge air is discharged

   D. Override the climate control settings in the bunk

4. When the HVAC unit is placed in the recirculation mode, air is:

   A. Discharged through the windshield vents

   B. Filtered and introduced into the vehicle through the recirculation door

   C. Blocked from entering the cab by the closed recirculation door

   D. Discharged through the face vents

5. The bunk override switch allows the vehicle operator to:

   A. Control the cab's climate control system remotely from the sleeper compartment

   B. Control the climate control settings remotely from the cab's control panel

   C. Control both the cab climate settings and the bunk settings from the bunk control panel

   D. Control the ventilation system in the bunk

6. What would happen if a water control valve did not close completely?

   A. There would be no adverse effect to the air-conditioning system.

   B. The air-conditioning unit would cool more than was requested by the temperature switch.

   C. The air-conditioning unit would not cool to its maximum setting because it would be working against the heater core.

   D. The heating capabilities of the system would be severely affected.

7. Thermistors contain a temperature-dependent resistor. The resistance value _____ as the temperature goes up.

   A. Goes up

   B. Goes down

   C. Stays the same

8. Which of the following expel the most GHGs into the environment?

   A. Fuel-fired air heaters

   B. Fuel-fired coolant heaters

   C. Truck stop electrification

   D. Idling the truck's engine

9. Which products supply power to run electrical appliances in the truck's cab?

A. Fuel-fired air heaters   C. APUs

B. Stand-alone systems   D. Fuel-fired coolant heaters

10. Which of the following products can provide the air-conditioning component to the cab/sleeper when the truck's engine is not running?

A. Fuel-fired air heaters   D. Both A and B are correct.

B. APUs   E. Both B and C are correct.

C. Stand-alone systems

Troubleshooting and Performance Testing

Key Terms

catastrophic	hygroscopic	residue
combustible	non-combustible	stabilize
equalized	polyalkylene glycol (PAG)	sweep
evacuated	purging	
flushing	recovered	

INTRODUCTION

Technicians working on air-conditioning equipment must know how to work safely and competently in order to efficiently diagnose problems with the A/C system. Any technician can play hit and miss, changing components until the A/C system operates effectively. On the other hand, a good technician listens to the complaint, verifies the complaint, troubleshoots the system, and repairs it without any guesswork involved.

SYSTEM OVERVIEW

Troubleshooting air-conditioning systems is really not as hard as some make it out to be. The real trick to it is following a sequential order when diagnosing. The human hand is an incredibly sensitive tool in determining temperatures for the diagnostic procedures. The hands, along with a good serviceable gauge manifold set, are often the only tools required to make an accurate diagnosis. Start with a performance test (discussed later in this chapter). Observe the operating pressures on the gauge set. Compare your measurements and reading with what they are supposed to be on a properly operating system. From the examples found later in this chapter, identify what problems exist and repair as required.

CAUTION *Air-conditioning systems are constantly under pressure. Before internal repairs are made to any air-conditioning system, refrigerant must be recovered (removed). Once the repair has been completed, the system must be evacuated and recharged with refrigerant before it can be put back in service.*

Always wear safety goggles and non-leather gloves when recovering, evacuating, charging, or leak testing an air-conditioning system. Leather or rubber gloves should not be worn because the material can stick to the skin if it comes in contact with refrigerant liquid or gas.

Care must always be taken when handling refrigerant to prevent contact with the skin or eyes. If liquid refrigerant is exposed to the atmosphere, it evaporates very quickly, instantly freezing skin or eye tissue. Serious injury, including blindness, can occur from contact.

If refrigerant does come in contact with the eyes, they should be flushed with lukewarm water. Do not rub the eyes. Apply a light bandage and seek medical attention immediately. If refrigerant comes in contact with the skin, again flush with lukewarm water, apply a light bandage, and seek medical attention immediately.

Refrigerant 134a is **non-combustible** at atmospheric temperature and pressure; however, it can be **combustible** at pressures as low as 5.5 psi (38 kPa) at 350°F (177°C) when in a mixture of 60% air. For this reason, the air-conditioning system should never be pressure tested or leak tested with compressed air because a fire or explosion could result in personal injury or property damage. The air-conditioning system can be safely pressurized with dry nitrogen.

It is imperative that all refrigeration work be performed in a well-ventilated area. R-134a vapors have a slightly sweet odor that is difficult to detect. To ensure a safe working environment, performing frequent leak checks and using air monitoring equipment are recommended.

Storage of R-134a containers should be at temperatures less than 125°F (52°C) to prevent explosion of the containers.

Polyalkylene glycol (PAG) oil is used to lubricate the compressor in air-conditioning systems that use R-134a. When handling PAG oil, take the following precautions:

- Keep the oil in a clean container to prevent contamination.
- Do not leave the air-conditioning system or PAG oil container open to the atmosphere. PAG oil is **hygroscopic** (has a high moisture absorption capacity).

- Take precautions when handling PAG oil because it can damage painted surfaces, plastic parts, and other parts (drive belt) if it is spilled on them.
- Don't mix PAG oil with any other refrigerant oil.

SERVICING

Whenever a technician performs any internal service work to the air-conditioning system, the following rules should be adhered to:

- When the system is open, always cap the disconnected lines, preventing foreign material from entering the system.
- Whenever air-conditioning lines are disconnected, new O-rings lubricated with refrigerant oil should be installed before reassembly.
- The receiver-drier/accumulator should be replaced any time the system is opened up for service or repair.
- Manifold gauge lines should be kept connected to the manifold when not in use to keep foreign material from entering the service lines.
- Replace any lost refrigerant oil.

Before any service work is performed, the technician or service manager should talk to the driver to obtain any problematic history. Questions should include:

1. Did the air conditioning stop working all at once or did it gradually lose its effectiveness?
2. Does the air conditioner work occasionally or not at all?
3. Has the air-conditioning system been worked on or serviced recently?

The technician should then performance test the air-conditioning system to verify the operator's complaint.

PERFORMANCE TEST

- Start by running the engine at a speed of 1500 rpm. Close all cab doors and windows.
- Turn the air-conditioning system on, setting the system for maximum cool.
- A thermometer should be inserted into the center duct and the system should be run for approximately 10 minutes to allow the system and thermometer to **stabilize** (to keep from fluctuating).
- Check that the air flow is coming from the appropriate vent for the position of the air selection switch.

- If the fan is operating but air flow is low, check for leaves or debris that may be obstructing the air inlet.
- During the performance check, visually inspect the compressor drive belt for signs of slippage.
- Check the compressor clutch to make sure it is engaged, paying attention to any abnormal compressor noises.
- Feel the discharge line at the compressor; it should be hot to the touch. If it is warm or cold, that may indicate a faulty compressor or low refrigerant charge.
- Next, feel the inlet and outlet of the condenser. You should be able to feel a large change of temperature between the inlet and outlet.
- Feel the liquid line from the condenser outlet to the receiver tank; it should be warm. Cool spots in the line indicate restrictions. If the line is hot, that may indicate that the condenser is not transferring enough heat to the ambient air (check condenser fan).
- Check the inlet and outlet line at the receiver-drier; they should be the same temperature. If they are not, this is an indication that the filter-drier may be blocked or restricted.
- Feel the liquid line from the outlet of the receiver-drier to the inlet of the TXV. It should feel warm to the touch. Any cold spots in the liquid line indicate a restriction.
- Check the suction line from the evaporator outlet to the compressor inlet. It should feel cold. If humidity conditions are right, this line may be sweating and may even have some frost covering it. If the line is warm, it may indicate that the expansion valve is not metering enough refrigerant into the evaporator or that the refrigerant charge is insufficient. If the line is covered in ice, it can indicate an overcharged system or a TXV that is metering too much refrigerant into the evaporator.
- If the engine is running hot, the system will not operate as efficiently. If the water valve does not close completely, the heat from the heater core will enter the cab of the truck.
- Check the temperature of the discharge air from the dash vent to make sure the air is cool. The discharge air temperature will vary according to ambient temperature and humidity conditions.
- Now shut the unit down.
- After the engine is shut down, feel across the surface of the condenser. Any cold spots indicate crimped or restricted coils.

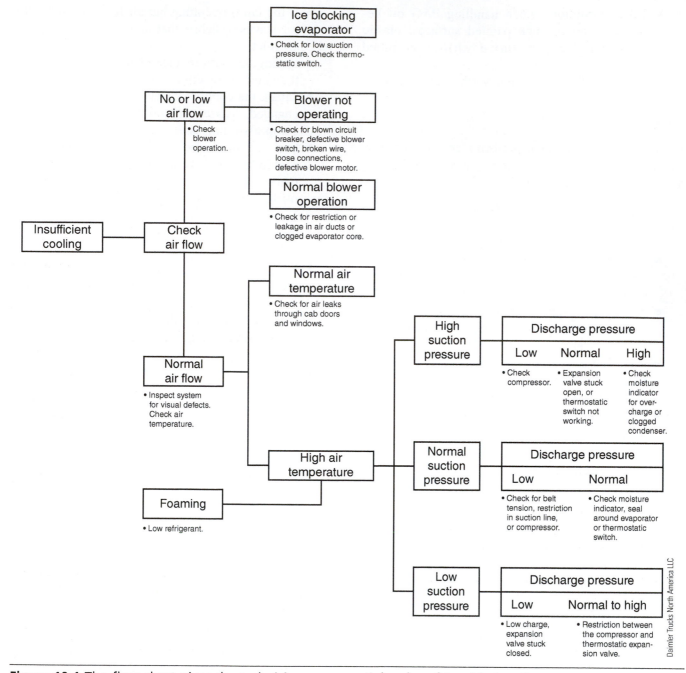

Figure 10-1 The flow chart gives the technician a sequential order of troubleshooting.

If a problem is found with the system at this point, follow through the fault analysis flow chart found in **Figure 10-1.**

GAUGE TESTING

To gauge test an air-conditioning system, the manifold set should first be installed. The gauges on a reclaiming station will also work for this procedure. Gauges should be installed like those in **Figure 10-2.**

The static readings on the gauges should be **equalized** (be at the same pressure) under static conditions. The pressure readings on the gauges should be equal to the ambient temperature found on the chart in **Figure 10-3.**

The actual refrigerant charge, whether it be over- or undercharged, cannot be determined at this point. A system would have to be extremely undercharged to have lower readings than those found in the tables. If the unit is overcharged, the pressure will not be higher than that found in the tables. If the refrigerant charge is

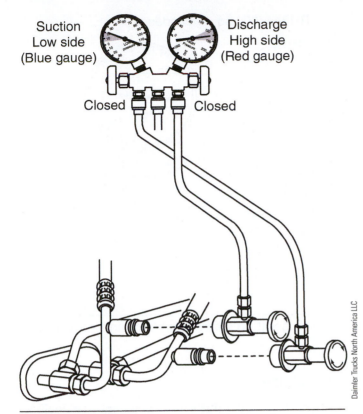

Suction
Low side
(Blue gauge)

Discharge
High side
(Red gauge)

Closed   Closed

Daimler Trucks North America LLC

Figure 10-2 Proper gauge installation for pressure testing of an air-conditioning system.

impure (there is air in the system or a mixed refrigerant is installed), the pressure may be higher than that found in the tables. If all appears to be normal and there is a complaint that the system is operating poorly, the technician should perform a gauge test. Begin by starting the engine, and set it to run at 1500 rpm. Close all windows and doors, and set the air-conditioning system to max cooling and high fan. Let the system run for 5 to 10 minutes and check the readings on the manifold gauges.

Freightliner recommends typical gauge readings for R-134a depending upon the size of the condenser. Using a 270-square-inch condenser, pressures should be in the range of:

- Suction side 25–35 psi
- Discharge side 175–225 psi

If the air-conditioning system uses a 600-square-inch condenser, the pressures should range from:

- Suction side 15–25 psi
- Discharge side 100–150 psi

Note: Higher-than-normal ambient temperature and humidity levels create higher pressure.

R-134a Temperature-Pressure Chart

Evaporator range		Condenser range	
Temperature °C (°F)	Pressure kPa (psi)	Temperature °C (°F)	Pressure kPa (psi)
−9 (16)	106 (15)	38 (100)	857 (124)
−7 (20)	124 (18)	40 (104)	917 (133)
−4 (24)	144 (21)	42 (108)	980 (142)
−2 (28)	166 (24)	44 (112)	1047 (147)
0 (32)	188 (27)	47 (116)	1114 (162)
2 (36)	212 (31)	49 (120)	1185 (172)
4 (40)	238 (35)	51 (124)	1260 (183)
10 (50)	310 (45)	53 (128)	1337 (200)
16 (60)	392 (57)	57 (135)	1481 (215)
21 (70)	487 (71)	63 (145)	1704 (247)
27 (80)	609 (88)	68 (155)	1948 (283)
32 (90)	718 (104)	71 (160)	2079 (301)

R-12 Temperature-Pressure Chart

Evaporator range		Condenser range	
Temperature °C (°F)	Pressure kPa (psi)	Temperature °C (°F)	Pressure kPa (psi)
−9 (16)	127 (18)	38 (100)	808 (117)
−7 (20)	145 (21)	40 (104)	859 (125)
−4 (24)	165 (24)	42 (108)	917 (133)
−2 (28)	190 (28)	44 (112)	969 (140)
0 (32)	207 (30)	47 (116)	1027 (149)
2 (36)	230 (33)	49 (120)	1087 (158)
4 (40)	255 (37)	51 (124)	1150 (167)
10 (50)	322 (47)	53 (128)	1215 (176)
16 (60)	398 (58)	57 (135)	1334 (194)
21 (70)	484 (70)	63 (145)	1519 (220)
27 (80)	580 (84)	68 (155)	1721 (250)
32 (90)	688 (100)	74 (165)	1940 (281)

© Cengage Learning 2014

Figure 10-3 A pressure-temperature chart showing the saturation pressure for the refrigerant at the current temperature.

Match the readings that have been taken with those found in the following section to identify the symptoms, cause, and cure.

SOME AIR AND MOISTURE

Refer to the gauges in **Figure 10-4**:

- Suction pressure normal
- Discharge pressure normal

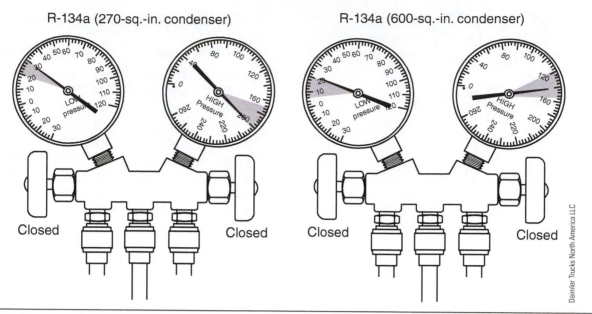

Figure 10-4 A manifold gauge set installed on an operating refrigeration system that contains some air and moisture.

Symptoms

Symptoms under this set of conditions may include:

1. Not cooling effectively.
2. Moisture indicator shows moisture level is high in the system.
3. Low-pressure gauge does not respond to the on/off cycle of the thermostatic switch.

Cause

The cause of this is probably air and moisture in the system, generally caused by an external leak on the low-pressure side of the air-conditioning system.

Cure

1. Leak test the air-conditioning system.
2. Remove refrigerant and repair any leaks found, as required.
3. Flush and purge the system, if required.
4. Top up refrigerant oil as required.
5. Remove and replace the receiver-drier.
6. Perform an evacuation on the system.
7. Charge the system with the recommended weight and type of refrigerant.

EXCESSIVE AIR AND MOISTURE

Refer to the gauges in **Figure 10-5**:

- Suction pressure high
- Discharge pressure high

Symptoms

Symptoms under this set of conditions may include:

1. Not cooling effectively.
2. Moisture indicator shows moisture level is high in the system.
3. Suction pressure is high in the beginning but as the system operates, may drop into the low range or even into a vacuum.
4. Discharge pressure is high.

Cause

Excessive amounts of air and moisture.

1. The receiver-drier is saturated beyond its capacity, allowing the remaining moisture to freeze at the restriction caused by the TXV.
2. The high pressure is caused by an excessive amount of air that was able to enter the system.

Cure

1. Leak test the air-conditioning system.
2. Remove the refrigerant and repair any leaks found, as required.
3. Flush and purge the system, if required.
4. Top up refrigerant oil as required.
5. Remove and replace the receiver-drier.
6. Perform an evacuation on the system.
7. Charge the system with the recommended weight and type of refrigerant.

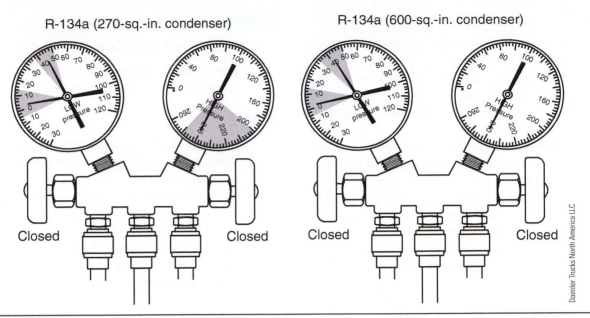

Figure 10-5 A manifold gauge set installed on an operating refrigeration system that contains excessive air and moisture.

CONDENSER AIR FLOW OBSTRUCTION OR OVERCHARGED

Refer to the gauges in **Figure 10-6:**

- Suction pressure high
- Discharge pressure high

Symptoms

Symptoms for this set of conditions may include:
1. Not cooling effectively.
2. Suction line is warm to the touch.
3. Discharge line is hot.

4. Truck engine may be hot or overheating.
5. High-pressure relief switch may occasionally open, breaking the electrical circuit of the clutch.

Cause

Air is not able to flow through the condenser or radiator. Foreign material is not allowing air to circulate, drastically affecting heat transfer.
1. The fan belt is slipping or the fan clutch mechanism is faulty.
2. Refrigerant is excessively overcharged.

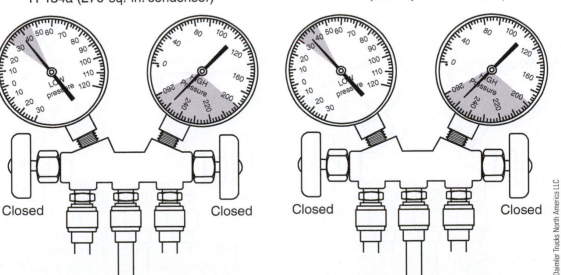

Figure 10-6 A manifold gauge set installed on an operating refrigeration system with some sort of an obstruction to the air flow through the condenser or that is overcharged with refrigerant.

Cure

1. Wash the condenser and radiator area. Avoid the use of pressure washers for this process because the fins may be damaged by the high pressure. Instead, use a soap and water solution with a soft brush and flush with clean water from a garden hose.
2. Check the condition and tension of the fan belt. Replace it if necessary. Check the fan speed and make repairs to the clutch fan as required.
3. Recover the refrigerant, top up refrigerant oil if required, evacuate, and charge the system with the recommended weight and type of refrigerant. (This is the only way of knowing exactly how much refrigerant is in the system.)
4. Again, test system pressures. If the discharge pressure is still too high, perform a test for a restriction in the high side.

LOW REFRIGERANT CHARGE

Refer to the gauges in **Figure 10-7**:

- Suction pressure low
- Discharge pressure low

Symptoms

Symptoms under this set of conditions may include:

1. Not cooling effectively.
2. Slightly lower than normal suction and discharge pressures.

Cause

Low refrigerant charge usually caused by:

1. A very small leak in the air-conditioning system. (It may have leaked out over a long period of time.)

Cure

1. Leak test the air-conditioning system.
2. Remove refrigerant and repair any leaks found, as required.
3. Flush and purge the system, if required.
4. Top up refrigerant oil as required.
5. Remove and replace the receiver-drier.
6. Perform an evacuation on the system.
7. Charge the system with the recommended weight and type of refrigerant.

VERY LOW REFRIGERANT CHARGE

Refer to gauges in **Figure 10-8**:

- Suction pressure low
- Discharge pressure low

Symptoms

Symptoms under this set of conditions may include:

1. Not cooling at all.
2. Discharge air from the evaporator may be warm.
3. Compressor may not be running or is frequently cycling on and off.
4. Low pressure switch is consistently opening (if equipped).

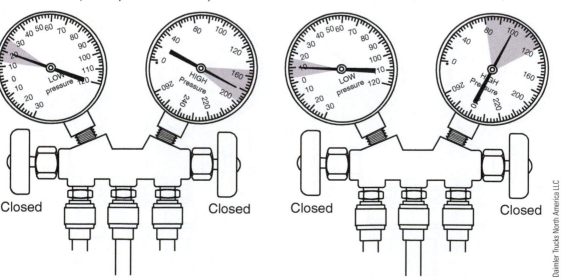

R-134a (270-sq.-in. condenser)   R-134a (600-sq.-in. condenser)

Closed   Closed   Closed   Closed

Daimler Trucks North America LLC

Figure 10-7 A manifold gauge set installed on an operating refrigeration system that has a low refrigerant charge.

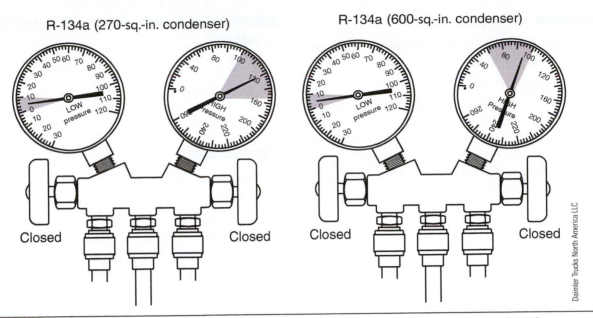

Figure 10-8 A manifold gauge set installed on an operating refrigeration system that has a very low refrigerant charge.

Cause

Extremely low refrigerant charge, usually caused by:

1. A leak in the air-conditioning system.

Cure

1. Leak test the air-conditioning system.
2. Remove refrigerant and repair any leaks found, as required.
3. Flush and purge the system, if required.
4. Remove refrigerant and flush, if required.
5. Top up refrigerant oil as required.
6. Remove and replace the receiver-drier.
7. Perform an evacuation on the system.
8. Charge the system with the recommended weight and type of refrigerant.

RESTRICTION IN THE HIGH SIDE OF THE SYSTEM

Refer to the gauges in **Figure 10-9**:

- Suction pressure low
- Discharge pressure normal to high

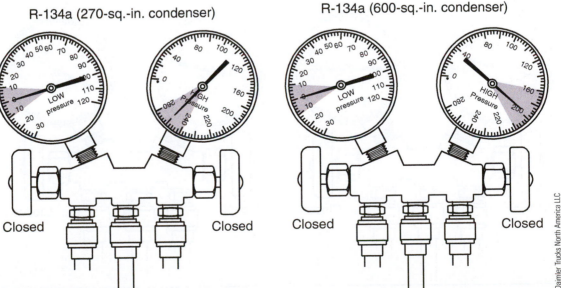

Figure 10-9 A manifold gauge set installed on an operating refrigeration system that has a restriction in the high side of the system.

Symptoms

Symptoms under this set of conditions may include:

1. Not cooling effectively.
2. Frosted discharge line just beyond the restricted section.
3. If the restriction is between the compressor discharge and the inlet of the receiver, the discharge pressure can be excessively high.
4. High pressure switch is open.

Cause

There is some kind of restriction between the compressor discharge and the inlet of the TXV.

Cure

1. Feel the discharge line for cold spots to identify the point of the restriction.
2. Remove refrigerant and flush, if required.
3. Remove and replace the defective component if it can't be flushed successfully.
4. Top up refrigerant oil as required.
5. Remove and replace the receiver-drier.
6. Perform an evacuation on the system.
7. Charge the system with the recommended weight and type of refrigerant.

EXPANSION VALVE NOT OPENING ENOUGH

Refer to the gauges in **Figure 10-10:**

- Suction pressure low
- Discharge pressure low

Symptoms

Symptoms under this set of conditions may include:

1. Not cooling effectively.
2. Sweat or frost build-up on the expansion valve.
3. Suction and discharge pressure return to normal if heat is applied to the TXV.

Cause

1. The TXV does not open enough.
2. The TXV is stuck in the closed position.
3. Problem with the capillary tube or sensing bulb (not firmly attached).
4. Blockage at the orifice tube on an orifice tube air-conditioning system.

Cure

1. Repair the TXV if possible; if not, go to Step 2.
2. Remove refrigerant and flush, if required.

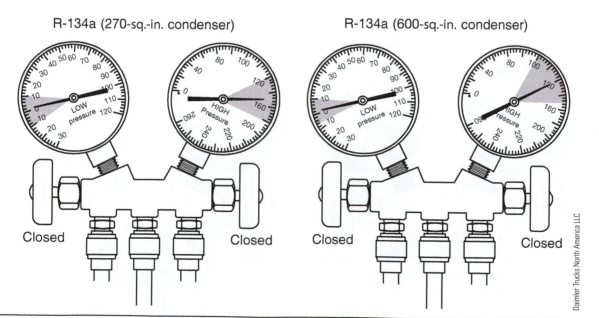

Figure 10-10 A manifold gauge set installed on an operating refrigeration system that has a thermostatic expansion valve that is not opening or a restriction at the orifice tube.

3. Replace TXV, being careful not to damage or crimp the capillary tube.
4. Top up refrigerant oil as required.
5. Remove and replace the receiver-drier/accumulator.
6. Perform an evacuation on the system.
7. Charge the system with the recommended weight and type of refrigerant.

EXPANSION VALVE HELD OPEN

Refer to gauges in **Figure 10-11**:

- Suction pressure high
- Discharge pressure low to normal

Symptoms

Symptoms under this set of conditions may include:

1. Not cooling effectively; cools well in the beginning, but in time the evaporator freezes up, blocking air flow.
2. Sweat or frost build-up on the expansion valve.
3. Suction line sweats heavily.
4. System pressures return to normal if the evaporator is allowed to defrost completely. (Much water should flow from the evaporator drip tube.)

Cause

1. The TXV is stuck open, causing the evaporator to operate in a flooded state.

2. There is a problem with the capillary tube or sensing bulb (not firmly attached).

Cure

1. Repair the TXV, if possible; if not, go to Step 2.
2. Remove refrigerant and flush, if required.
3. Replace the TXV, being careful not to damage or crimp the capillary tube.
4. Top up refrigerant oil as required.
5. Remove and replace the receiver-drier.
6. Perform an evacuation on the system.
7. Charge the system with the recommended weight and type of refrigerant.

DEFECTIVE THERMOSTATIC SWITCH

Refer to the gauges in **Figure 10-12**:

- Suction pressure low-vacuum
- Discharge pressure low

Symptoms

Symptoms under this set of conditions may include:

1. Not cooling effectively. Compressor runs constantly and compressor may become noisy.
2. Evaporator freezes up, blocking air flow.

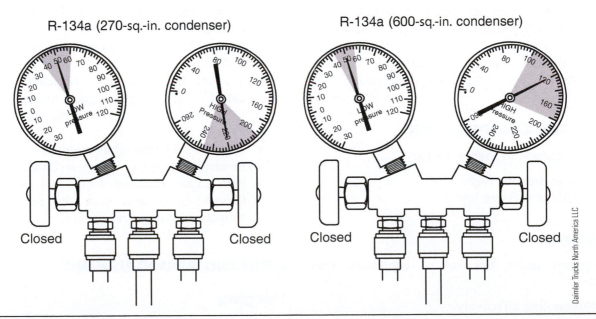

Figure 10-11 A manifold gauge set installed on an operating refrigeration system with the thermostatic expansion valve held open.

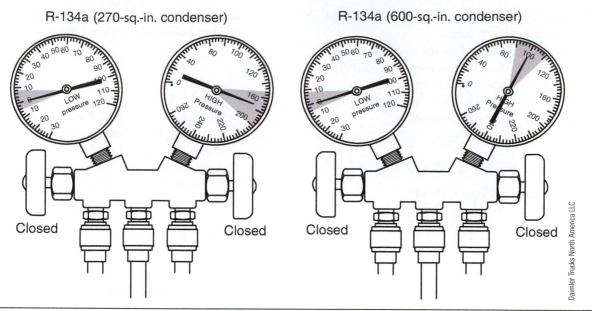

Figure 10-12 A manifold gauge set installed on an operating refrigeration system with a defective thermostatic switch.

Cause

1. A defective thermostatic switch is not disengaging the clutch at temperatures below 32°F (0°C).
2. A bent, kinked, or broken capillary tube is causing the thermostatic switch to malfunction.

Cure

1. Inspect the condition of the capillary tube for the thermostatic switch. Check that it is in the proper location.
2. Perform an electrical test on the switch.
3. Rectify any problems found with the capillary tube.
4. If problems with the capillary tube can't be corrected, replace the thermostatic switch.

DEFECTIVE COMPRESSOR

Refer to gauges in **Figure 10-13:**

- Suction pressure high
- Discharge pressure low

Symptoms

Symptoms under this set of conditions may include:

1. Not cooling effectively.
2. Noisy compressor operation with no compressor cycling.

Cause

1. The compressor drive belt is loose or worn.
2. The compressor clutch is not operating.
3. There is internal compressor failure.
4. There is a defective trinary or low pressure switch.

Cure

1. Inspect or replace the compressor drive belt, as necessary.
2. Inspect the engine drive pulley; replace it if necessary.
3. Inspect the compressor clutch assembly; replace if defective.
4. Test the trinary switch electrically. If found defective, remove and replace trinary switch.
5. If the compressor has to be replaced, the system must be flushed.
6. Top up refrigerant oil as required.
7. Remove and replace the receiver-drier.
8. Perform an evacuation on the system.
9. Charge the system with the recommended weight and type of refrigerant.

PURGING AND FLUSHING

Purging

The process of **purging** an air-conditioning system involves blowing out wet air, refrigerant, and loose dirt

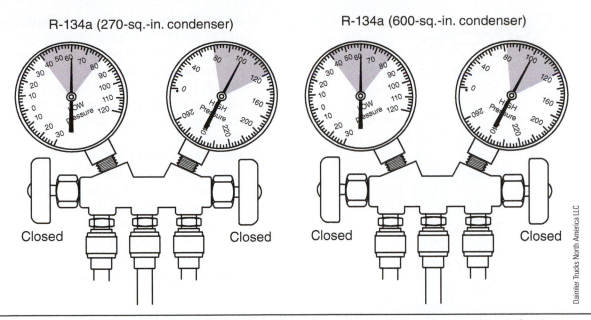

R-134a (270-sq.-in. condenser)   R-134a (600-sq.-in. condenser)

Closed   Closed   Closed   Closed

Figure 10-13 A manifold gauge set installed on an operating refrigeration system with a defective compressor.

from refrigerant hose lines and components with the use of a dry gas such as compressed nitrogen.

Note: Generally, dry nitrogen gas should be used for purging procedures, although any safe, inert, dry gas such as carbon dioxide or argon could also be used. Nitrogen can't be dispensed without the use of a regulator. A nitrogen cylinder contains a pressure in excess of 2000 psi (13,789 kPa). Purging procedures are performed at 200 psi (1378 kPa); therefore, a regulator capable of maintaining pressures between 0 and 200 psi (0 to 1378 kPa) is required.

Flushing

Flushing is used to clean extremely contaminated systems. Flushing solvents are commercially available that are compatible with both R-12 and R-134a systems and are also compatible with mineral, ester, paraffin, and PAG oils. One flushing product approved by several compressor manufacturers is HCFC-141b. This chemical is marketed under a number of brand names, including "Dura Flush 141" and "Acc-U-Flush." HCFC-141b has excellent cleaning properties, evaporates quickly, and leaves almost no **residue** (matter that remains after something has been removed) in the system (less than 4 ppm). It is non-flammable, safe to use, and is compatible with both R-12 and R-134a systems.

Note: Freightliner, Sterling, and Western Star do not endorse flushing with any chemical. Flushing procedures are not covered under any of their warranties. Their recommendation is that if flushing is necessary, it is caused by some type of **catastrophic** (extremely harmful) failure. Under warranty, lines and components are replaced instead of being flushed. Most times the blockage occurs at the TXV or orifice tube location. Once the vehicle's warranty has expired, some maintenance facilities will perform flushing procedures because the cost of replacing components can become too expensive for the vehicle owner.

GUIDELINES FOR PURGING AND FLUSHING

When contamination of an air-conditioning system is suspected, all sections of the system should first be purged or flushed. Once purging or flushing is complete, the receiver-drier or accumulator should be replaced before evacuation and recharging of the system.

The system should be purged or flushed in sections to prevent contaminants from being forced into other parts of the system. Flushing or purging the compressor, receiver-drier, or accumulator is never recommended. Always purge or flush in a reverse direction to the normal refrigerant flow.

If the system is heavily contaminated or the desiccant from the accumulator or receiver-drier has

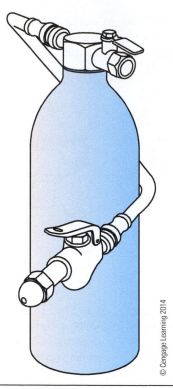

Figure 10-14 A pressurized container of flushing solvent with a rubber tip blow gun attached.

circulated through the system, the compressor, thermostatic expansion valve, or orifice tube should be replaced. The new HFCF-141b (hydrofluorocarbon flush) is recommended by most after-market air-conditioning part suppliers such as Everco, Murray, and Four Seasons. HFCF-141b is safe, nonflammable, and can also be used to flush electronic components. It is the flush intended to replace CFC-based flushing solvents R11 and R113, but has one-tenth the ozone depletion potential. HFCF-141b should not be flushed through compressors, driers/accumulators, refrigerant control valves, or other items that would trap solvent. Flushing with higher pressure may not provide enough liquid contact between the solvent and components. After-market suppliers recommend flushing at about 60 psi with dry air **(Figure 10-14)**.

PURGING AND FLUSHING PROCEDURES

Purging

1. When purging, disconnect both ends of the component or line, and tightly cap the rest of the system to prevent foreign material from entering.
2. Connect a hose with a rubber tip blow gun attachment to the outlet of the pressure regulator.

3. Adjust the regulator from the nitrogen cylinder to 200 psi (1379 kPa).
4. Place the rubber tip blow gun in one end of the component, and allow nitrogen to flow. (Make sure that you are blowing through in the reverse direction to the refrigerant flow.)
5. Allow the nitrogen to flow at this pressure for 30 seconds.
6. Back the regulator off to a constant flow of approximately 4 psi (28 kPa) for 1 to 2 minutes. If system is very wet, allow for a longer **sweep** time. (Sweep time allows the dry nitrogen to absorb moisture from the system.)
7. Close the valve on the nitrogen cylinder.
8. Tightly cap the component that has just been purged and go on to the next component.

Flushing with HFCF-141b

Flush solvents are available in bulk containers. Dispense the solvent into a container that will allow you to pour it into the hose or component to be flushed, then blow it through with a rubber tip blow gun. Solvent is also available in pressurized containers for flushing the system. Use dry air or nitrogen at about 60 psi to blow through the system. Remember to always flush in a reverse direction to refrigerant flow first to dislodge any material caught inside. Hold a clean rag over the end of the hose or component being flushed to identify the contaminants that are pushed out of the system. Add more solvent as necessary until the exiting residue appears clean. Once you are satisfied that the component is clean, pop the components dry in both directions.

How to Pop Components Dry

1. Hold a clean shop rag over the end of the component to be pop dried.
2. Place the tip of the blow gun over the opposite end of the component and apply pressure to seal the tip.
3. Hold a finger over the other end with the shop rag and pressurize with nitrogen until pressure builds up.
4. Release your finger, letting the pressure blow off into the shop rag. (Remember to wear your safety goggles for this procedure.)
5. Repeat Steps 2–4 until nothing comes out into the shop rag.
6. Whenever a system has been flushed, a long evacuation should be given to the system before refrigerant is reinstalled.

Summary

- Air-conditioning systems are constantly under pressure. Before internal repairs, refrigerant must be recovered.

- Always wear safety goggles and non-leather gloves when recovering, evacuating, charging, or leak testing an air-conditioning system.

- R-134a can be combustible under the right conditions.

- A/C systems should never be pressure or leak tested with compressed air because a fire or explosion could result in personal injury.

- PAG oil containers should always be kept tightly closed because PAG is hygroscopic.

- When the system is open, always cap the disconnected lines, preventing foreign material from entering the system.

- Whenever air-conditioning lines are disconnected, new O-rings lubricated with refrigerant oil should be installed before reassembly.

- The receiver-drier must be replaced any time the system is opened up for service or repair.

- Manifold gauge lines should be kept connected to the manifold when not in use to keep foreign material from entering the service lines.

- Add refrigerant oil to replace what may have been lost.

- Always question the vehicle operator to find out what the problem is and the history of the problem.

- A performance test allows the technician to identify problems that may exist in an air-conditioning system.

- To check the refrigerant level of an R-134a, performance and gauge tests must be performed. If there is an indication that the system is low on refrigerant, the leak must be repaired and the correct amount of refrigerant reinstalled. Never top up a system.

- Purging of an air-conditioning system involves blowing out wet air, refrigerant, and loose dirt from refrigerant hose lines and components with the use of a dry gas such as compressed nitrogen.

- Flushing is much like purging, but a liquid flushing solvent is used to flush the contaminants out of the refrigerant hose lines and components.

- "Popping components dry" is a process of allowing dry nitrogen gas to absorb moisture by allowing it to build up and then releasing the pressure quickly, allowing the moisture and residue to be forced out of the system.

Review Questions

1. If an air-conditioning system is operating correctly, the discharge line from the compressor to the condenser should feel:

   A. Hot to the touch, approximately 110°F–180°F (43°C–82°C)

   B. Warm to the touch, approximately 80°F (27°C)

   C. Cold to the touch, approximately 32°F (0°C)

   D. None of the above

2. A system that has had some type of catastrophic failure that has heavily contaminated the system should:

   A. Be flushed

   B. Be purged

   C. Have the compressor, receiver-drier/accumulator, and TXV or orifice tube replaced

   D. Have all the above done

3. A driver complains that the cab air conditioning is working intermittently. What is the most likely cause?

   A. There is moisture in the system.

   B. The receiver-drier is plugged.

   C. The expansion valve is stuck closed.

   D. The expansion valve is stuck open.

4. A driver complains that the cab air conditioning works well when he first starts it up, but after some time it starts to operate ineffectively. What is the most likely cause?

   A. There is moisture in the system.

   B. The receiver-drier is plugged.

   C. The expansion valve is stuck closed.

   D. The expansion valve is stuck open.

5. Technicians working on air-conditioning equipment should do so wearing:

   A. Goggles and rubber gloves

   B. Goggles and cotton gloves

   C. Goggles and leather gloves

   D. No goggles or gloves are required when servicing air-conditioning equipment.

6. If an air-conditioning system is operating correctly, the suction line from the expansion valve to the compressor should feel:

   A. Hot to the touch, approximately 110°F–180°F (43°C–82°C)

   B. Warm to the touch, approximately 80°F (27°C)

   C. Cold to the touch, approximately 32°F (0°C)

   D. None of the above

7. A technician is troubleshooting a truck's air-conditioning system and finds normal gauge readings and the temperatures of all lines feel to be correct as well, but the temperature from the vents is not nearly cool enough. What would you suspect to be the problem?

   A. The condenser has some sort of internal obstruction.

   B. The system is undercharged.

   C. The heater valve is stuck open.

   D. The system is overcharged.

8. If an air-conditioning system is operating correctly, the line from the receiver-drier to the expansion valve should feel:

   A. Hot to the touch, approximately 110°F–180°F (43°C–82°C)

   B. Warm to the touch, approximately 80°F (27°C)

   C. Cold to the touch, approximately 32°F (0°C)

   D. None of the above

9. A technician is inspecting a truck's air-conditioning system and finds a heavy buildup of frost at one point in the liquid line between the condenser and the receiver-drier. The technician should immediately identify:

   A. That there is a buildup of moisture within the liquid line

   B. That there is a buildup of liquid refrigerant at that point of the hose

   C. That the hose is restricted at that point

   D. None of the above

10. When the technician installs the manifold gauge set to a static air-conditioning system, the gauges should read:

   A. Higher pressure on the high side

   B. Higher pressure on the low side

   C. The pressures should read the same on both gauges.

   D. The pressure of the refrigerant just as it was turned off

11.   The first thing a technician must do while dealing with a driver's air-conditioning complaint is to:

   A. Hook up the manifold gauge set   C. Verify the complaint

   B. Calculate the total cost of the job   D. Discharge the refrigerant, and then recharge it and see if that works

12.   A technician is performing an evacuation of an air-conditioning system. When the technician shuts off the gauges and vacuum pump, the gauges rise rapidly and continue to rise until they read 0 psig (0 kPa). What could be the problem?

   A. The condenser is restricted.   C. The system has excessive moisture in it.

   B. The evaporator is restricted.   D. The system has an external leak.

APAD/ACPU A/C Control Systems

Learning Objectives

Upon completion and review of this chapter, the student should be able to:

- Explain how an APAD system improves the reliability of an air-conditioning system.
- Describe the functions of an APAD system.
- List the inputs to an APAD or ACPU module.
- List the outputs from an APAD or ACPU module.
- Explain what happens when the control module senses low system refrigerant pressure.
- Explain what happens when the control module senses a complete loss of refrigerant in the air-conditioning system.
- Describe what the module does when it senses a high refrigerant pressure condition within the system.
- Describe what the module does if it senses low supply voltage.
- Describe what the module does if it senses high supply voltage.
- Locate the diagnostic information in order to interpret the various blink codes the module can display.
- List, in order of priority, the faults the APAD control module can display.
- Describe how an ACPU communicates diagnostic faults, and where those faults are displayed.

Key Terms

ACPU	continuity	MOSFET
algorithms	data bus	nominal
APAD	interface	potted
bidirectional	light emitting diodes (LEDs)	

INTRODUCTION

This chapter will deal with APADs (air-conditioning protection and diagnostic systems) and ACPU (air-conditioning protection unit) systems. Although there are many versions of this product, the CM-813 controller and the CM-820 control module will be studied in detail. The CM-813 module is for use in TXV style A/C systems, whereas a CM-816 module can only be used in orifice tube systems. The CM-820 is used on current model trucks with more modern **data bus** (the linking of separate computers together) capabilities.

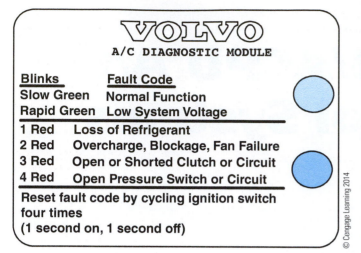

Figure 11-1 Volvo APADs unit mounted on the truck's firewall.

These products are designed to reduce maintenance costs associated with HVAC systems by increasing the life of system components. These systems incorporate an added feature to help the technician with air-conditioning diagnostics.

SYSTEM OVERVIEW

APAD or ACPU systems may be installed into the truck as stand-alone systems or may be incorporated into the truck's data bus. The stand-alone system uses a ruggedly manufactured module designed for survival under the hood of a heavy-duty vehicle. This unit is **potted** in an epoxy compound that provides a complete environmental seal, which also enhances heat sink for electrical components. (A heat sink is a device that cools an electrical component by dissipating the heat to the surrounding air.) These modules have two **light emitting diodes (LEDs)** built in to indicate the state of the air-conditioning system. One LED is green and indicates that the system is operating correctly; the other LED is red and flashes to indicate fault codes, otherwise known as blink codes **(Figure 11-1)**.

The **APAD** control system is composed of an electronic control module, two pressure switches, and a conventional evaporator thermostat (TStat). The module receives input from the two pressure switches and the TStat, and may also receive numerous parameters read from the vehicle SAE J1587 data bus. The input signals are interpreted by control laws, which derive outputs to the compressor clutch coil, fan actuator circuit, and diagnostic codes. In the APAD system, the controller becomes the only device through which power is switched to the compressor clutch coil.

The fan output is designed to **interface** (point of interaction or communication between a computer and any other entity) with electronically controlled diesel engines, but it can be used on mechanical engines with the addition of an appropriate engine fan temperature control.

Air-conditioning reliability is improved by actively monitoring system conditions and controlling the refrigerant compressor and the on-off fan drive. Air-conditioning system maintenance is reduced because the APAD control **algorithms** (step-by-step problem-solving procedures) do not allow operation in unstable and self-destructive modes. In addition, diagnostics aid technicians in servicing systems by communicating specific fault codes that warn of existing or impending problems.

ACPUs are the second generation of APAD systems. They connect via an L-1922 connector to the J1587 data bus. The ACPU functions in much the same way as the earlier APAD units, but also has the ability to display faults on the truck's dash panel.

COMMON AIR-CONDITIONING PROBLEMS

The resilience of the compressor and refrigerant hoses is adversely affected by operation of the air-conditioning system with reduced (or excessive) refrigerant charge. Operation with a partial refrigerant charge can cause compressor lubricant starvation, compressor overheating due to rapid cycling, seal failure, and can lead to damaged hoses and fittings due to exposure to excessive temperatures. Operation with an excessive refrigerant charge can severely stress the entire air-conditioning system, especially the compressor, and can cause hoses and fittings to leak. Low system voltages can cause the compressor clutch to slip and overheat, and loose electrical connections can cause erratic compressor cycling, which leads to clutch failures. Traditional electrical control systems based upon a series of simple pressure switches are unable to prevent or detect these harmful modes of operation.

THE APAD SYSTEM

The control module operates as the main air-conditioning system control. The module provides an on/off output to the compressor clutch coil and the fan actuator circuit trigger. Both of these outputs are a function of the APAD control laws. As stated previously, the module is the only device in which power is switched to the compressor clutch coil. The CM-820 module incorporates its own high current (high side, sourcing) drive relay, thereby eliminating external relays previously used with the CM-813 module to connect the compressor clutch circuit. The APAD module receives inputs from two smart or semi-smart switches and the evaporator thermostat. (A smart switch can communicate directly with the data bus, while the semi-smart switch must do so indirectly.) Fault diagnostic communication is accomplished through the J1587 data bus on the CM-820 controller and via blink code diagnostics on the CM-813 controller. The module is designed to interface with electronically controlled diesel engines. The APAD system is capable of reducing air-conditioning component failures that can damage one of the most expensive systems to be maintained on heavy commercial vehicles. This system protects the air-conditioning components from self-destruction by placing limits on the dynamic response of the system under certain environmental and undesirable conditions. APAD systems reduce air-conditioning operating costs by:

- Preventing rapid cycling of the compressor clutch due to high- or low-pressure conditions.
- Preventing rapid cycling of the engine fan at idle.
- Preventing slippage of the compressor clutch due to low voltage.
- Relieving stress on the starting system by holding off the A/C until 15 seconds after startup.
- Lubricating the A/C compressor and components year round by cycling the compressor on for 15 seconds after startup.
- Showing fault code indicators of potential A/C problems to aid in troubleshooting.

ELECTRICAL I/O DEFINITION

(INPUT AND OUTPUT)		
Name	**Function**	**Description**
A/C Drive	Output	A/C compressor clutch coil drive
Fan	Output	Fan actuator circuit
GND (3)	Power	Module ground, pressure switch returns (2)
HPx	Input	High-pressure switch
LPx	Input	Low-pressure switch
TStat	Input	Evaporator thermostat and A/C on/off input
Vign	Power	Module supply voltage, ignition switch

INPUTS FOR ACPU CM-813 CONTROLLER

The CM-813 module receives inputs from the following sensors:

- HPx, the high-pressure switch, which is normally closed
- LPx, the low-pressure switch, which is normally closed
- TStat, the evaporator thermostat switch (**Figure 11-2**).

The two pressure switches are semi-smart switches that use an internal resistor installed parallel to the electrical contacts, allowing diagnostics of sensor, wiring, and connector faults. Both the high- and low-pressure switches are designed to provide a current path to ground when they are closed (**Figures 11-3 and 11-4**).

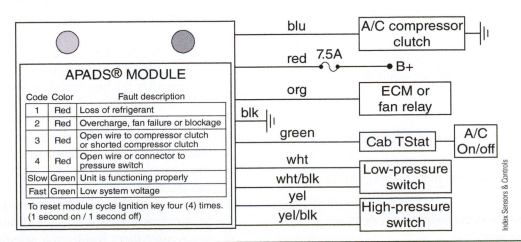

Figure 11-2 The inputs and outputs of a control module (line diagram).

John Dixon © 2014 Cengage Learning

Figure 11-3 Pressure switches used to supply the control module with operating system information.

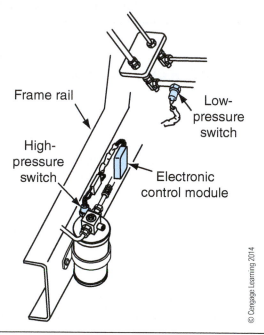

© Cengage Learning 2014

Figure 11-4 The location of pressure switches mounted on the vehicle.

The evaporator thermostat switch is the main air-conditioning system control. The TStat is wired in series with the air-conditioning on/off switch. These switches are configured to switch to battery voltage when the command for the air-conditioning compressor is requested.

OUTPUTS FOR ACPU CM-813 CONTROLLER

The CM-813 module delivers outputs to the following locations:

- A/C drive (compressor clutch drive)
- Fan (fan actuator)
- Diagnostic LEDs

The module's output signals a high side relay switch that provides battery voltage to the coil of the air-conditioning compressor when activated.

In the engine fan actuator circuit, a ground signal is provided to either an engine ECM or a fan solenoid valve to keep the engine fan off. When the high-pressure switch moves to an open state (pressure rising above 300 psi), it removes the ground from the circuit, causing the engine fan to engage.

The red and green diagnostic LEDs on the module are used to communicate system faults and system status to the technician.

APADS RULES FOR COMPRESSOR CONTROL (CM-813)

1. The air-conditioning compressor is never turned on until 15 seconds have elapsed since the ignition switch has been turned on (module power up).
2. The module initially ignores the input signal from the thermostat, turning the compressor on for 15 seconds after the 15-second engagement time has elapsed, as in rule #1. If the high- or low-pressure switches signify an out-of-bounds parameter or fault condition, the compressor is turned off.
3. The three possible inputs—TStat, HPx, and LPx—that can control the compressor operation are controlled by a module, which limits the maximum cycle rate to once every 15 seconds.
4. The evaporator thermostat (TStat) is the primary control. The air-conditioning compressor cycles at the frequency called for by the TStat, but is limited to a maximum of one cycle per 15 seconds (rule #3).
5. If the high-pressure switch (HPx) indicates a high-pressure condition, the air-conditioning compressor is allowed to stay on for an algorithmically determined variable period, limited to 10 seconds. The compressor is allowed to turn on after the high-pressure switch resets and rule #3 is satisfied.
6. When the low-pressure switch indicates a low-pressure condition, the compressor is turned off. The compressor is allowed to turn on after the low-pressure switch resets and rules #3 and #8 are satisfied.
7. If the supply voltage drops below 11.0 volts, the compressor is turned off. It is allowed to turn on again after there has been a satisfactory rise in system voltage and rule #3 is satisfied.

8. The air-conditioning system is latched off when any of the following diagnostic faults are detected:

- Low pressure
- Open clutch circuit
- Shorted clutch circuit

Engine Fan Control

Engine fan control is adaptive, dependent upon the frequency of request for operation. Once the fan has been activated, it remains on for a minimum of 45 seconds and a maximum on time of 3 minutes. When the fan is requested, the fan control output is in the "open" state.

DESCRIPTION OF DIAGNOSTIC FAULTS

Static Low Pressure. When the control module detects a low-pressure condition that would indicate a major or complete loss of refrigerant, the Static Low Pressure fault is displayed in blink code on the module.

Dynamic Low Pressure. The module detects this fault from rapid cycling of the low-pressure switch. This symptom can usually be attributed to a partial loss of the refrigerant charge. If the low-pressure switch dominates control and tries to cycle the compressor, displaying certain dynamic characteristics, the Dynamic Low Pressure fault will be displayed in blink code on the module.

High Pressure. The control module detects this fault by monitoring the activity of the high-pressure switch after the engine fan has been engaged. This fault could be caused by a faulty fan drive, debris on the condenser, restriction in the high side of the system, or by overcharging the system with refrigerant. If the high-pressure switch controls the system with a certain dynamic behavior, the High Pressure fault is displayed in blink code on the module.

Open Clutch. The module senses the current flow to the compressor clutch. If the module senses no current flow when the AC DRIVE is turned on, the Open Clutch fault is displayed in blink code on the module.

Shorted Clutch. This fault is detected by the module sensing excessive current flow to the compressor clutch. This could be caused by a short in the wire from the module to the clutch coil or a shorted clutch coil. This fault is displayed in blink code on the module.

Low Psw Open. This fault indicates a problem in the wiring to the low-pressure switch. This could be caused by an unseated connector, a broken wire in the harness, or moisture intrusion into the connectors or switch. This fault is displayed in blink code on the module.

High Psw Open. This fault indicates a problem in the wiring to the high-pressure switch. This could be caused by an unseated connector, a broken wire in the harness, or moisture intrusion into the connectors or switch. This fault is displayed in blink code on the module.

Low Voltage. This fault indicates a low voltage supply to the control module. This condition could be caused by a defective alternator, discharged batteries, or excessive electrical loads. The diagnostic fault is self-clearing, meaning the blink code is displayed only while the low-voltage condition is present **(Figure 11-5)**.

BLINK CODES

The module communicates system status and faults by a blinking red or green LED built into the module. When a fault occurs, the corresponding fault code is stored in nonvolatile memory. The module always broadcasts the appropriate blink code until it is cleared.

The module has the ability to display only one blink code at a time. The fault with the highest priority will be displayed over a lower-priority fault. The fault code table lists the faults in the order of their priority, name, description, and the number of blinks associated with that particular fault. Only the highest-priority fault is held in memory. Once the fault has been cleared, a new fault must occur to initiate a new blink code. A "blink" is approximately one quarter-second in length. Once the number of blinks has been broadcast, there is a delay time of 2 seconds before the blink code is repeated.

Clearing Blink Codes

There are two methods of clearing a blink code on the module:

- Cycling the ignition switch 1 second on, 1 second off for four consecutive cycles.
- Self-clearing, based on the persistence of the fault. If a fault condition occurs once and does not repeat, the fault code is held for an extended period of time and then cleared. If the fault is repetitive, it is always retained in memory until the problem is rectified.

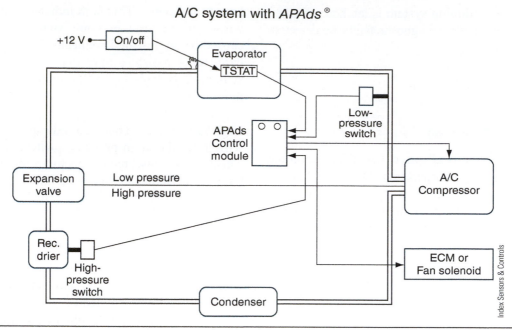

Figure 11-5 An APADs system installed on an air-conditioning system.

FAULT CODE TABLE

Priority	Fault Code Name	Fault Description	LED Blinks
1	Static Low Pressure	Full loss of charge	1
2	Dynamic Low Pressure	Partial loss of charge	1
3	High Pressure	Overcharge, blocked condenser	2
4	Open Clutch	Clutch wire open	3
5	Shorted Clutch	Shorted clutch coil	3
6	Low Psw Open	Cut wire to low psw	4
7	High Psw Open	Cut wire to high psw	4
8	Low Voltage	Low battery voltage	Rapid green
9	NO FAULTS	No faults, module functional	Red off, green slow blink

TESTING THE CM-813 MODULE

For complete testing of the CM-813 module, refer to **Figure 11-6**:

TROUBLESHOOTING

Blink Codes

Slow Green	Unit is functioning properly
Fast Green	Low system voltage
One Red Blink	Loss of refrigerant
Two Red Blinks	Fan failure or external condenser blockage
Three Red Blinks	Voltage drop out, open or shorted wire to compressor clutch
Four Red Blinks	Open wire in pressure switch circuit

Condition	Definition	Possible Cause	Action
Slow flashing green light.	Two seconds on, quarter of a second off.	Unit is functioning properly.	No action needed.
Rapid flashing green light.	Low system voltage; indicates low supply voltage condition. Voltage has dropped below 11.0 volts.	Possibly caused by a defective alternator, discharged batteries, poor ground, or an excessive electrical load.	Check voltage at the control harness by probing pins 4 (ground) and 2 (power). If voltage is less than 11.0 volts, start checking for voltage drops or a poor ground.
Ignition on, and LEDs not flashing on control module or no lights on module.	Control module not receiving battery voltage.	Fuse or circuit breaker blown.	Repair fuse or reset circuit breaker.
		Broken power or ground wire to control module.	Repair wiring.
		Defective control module.	Replace control module.
Red light flashing one blink in sequence.	Low pressure fault. This is detected by monitoring the ambient temperature and switching activity of the low pressure switch.	A partial or total loss of refrigerant.	Install service gauges and check system for leaks.
		A defective low pressure switch.	Disconnect the low side pressure switch connector and measure the resistance value of the switch. If the resistance shows 2.49 K ohms, switch is operating properly. If resistance shows less than 5 ohms, replace the switch.
		Bad Schrader-valve stem.	Verify with the set gauge. Replace if necessary.
			Note: This check must be done with the switch installed on the vehicle and the ignition in the off position and proper charge.
Red light flashing two blinks in sequence.	High-pressure fault. The system is exhibiting abnormal high-pressure activity.	Fan drive failure. Blocked air flow through condenser.	Inspect proper operation of related solenoid valve/relay and/or components.
			Remove restriction from condenser.
Red light flashing three blinks in sequence.	Open wire to A/C clutch, detected by inadequate current through the compressor clutch. If no current is sensed when the A/C drive is turned on, an open connection to the clutch coil is indicated.	A break in the wiring between the control module and the A/C compressor clutch or a break in the clutch coil.	With the use of an ohmmeter, check the resistance of the clutch coil. If the reading is less than 2.8 ohms, replace the clutch.

Condition	Definition	Possible Cause	Action
	Shorted wire or shorted A/C clutch, detected by excessive current through the compressor clutch.	This indicates either a shorted coil or shorted wiring to the clutch.	On the truck harness side of the 6-pin connector, verify the resistance between pins 1 and 4. More than 5 ohms indicates an open clutch condition, and less than 2 ohms indicates a shorted clutch condition. Next, check for a bad clutch or bad ground. If clutch resistance is greater than 5 ohms to ground, verify the ground connection before replacing the clutch. If clutch resistance is less than 2 ohms, replace the clutch. If the clutch is okay, locate the opened, frayed, or shorted wiring and repair.
	Fluctuating battery voltage.	Fluctuation greater than one volt caused by a defective voltage regulator.	Replace the alternator/ regulator unit.
Red light flashing four blinks in sequence.	Opening in wiring harness to high or low pressure switches.	Unseated connector.	Check both the module and the pressure switch connectors for loose pins.
		Break between wiring harness and pressure switch.	Check connector seals for integrity.
		Moisture intrusion into the connectors or switch.	Check pressure switch circuits for continuity. In a properly charged system, disconnect the high pressure switch and verify that the resistance between the switch's two contacts is less than 5 ohms. Disconnect the low pressure switch and verify that the resistance between the switch's two contacts is between 2.4 K and 2.6 K ohms.
		Using a non-Index pressure switch.	Install the correct pressure switch.
A/C clutch not engaging during the first 15 seconds after ignition is turned on.		Unit is functioning properly. Regardless of A/C system state at startup, the A/C clutch is disengaged for the first 15 seconds.	

Condition	Definition	Possible Cause	Action
When ignition is turned on with A/C controls set to the on position, compressor is off for 15 seconds, on for 15 seconds, then remains off indefinitely.	No voltage on the A/C ON or evaporator thermostat circuit. Six-way truck harness connector, pin 5.	Defective A/C ON switch or evaporator thermostat circuit.	With the ignition on, the A/C switch on, and the evaporator core temperature above 50°F (10°C), there should be system voltage on the truck harness connector at the A/C ON/evaporator thermostat pin. Check for bad A/C ON switch. Check for break in wiring between A/C ON switch and the evaporator thermostat. Check for break in wiring between the evaporator thermostat and the APAds module. Check for bad A/C relay.
At startup, after 15 seconds, compressor clutch doesn't engage but engine fan engages.	Control module sensing an opening in the high pressure switch circuit.	High-side pressure switch failure.	In a normally pressurized system: disconnect the high pressure switch. A good switch should measure less than 5 ohms. If not, replace the switch.
Slow flashing green, compressor on for a short period of time, off for 15 seconds in a repetitive sequence.	Indicates that the system is exhibiting abnormal high- or low-pressure activity.	Blockage in the high side of the system or in the condenser.	Repair restriction.
		Partial loss of refrigerant.	Check system for loss of refrigerant.
Slow flashing green light, compressor not engaging in defrost mode or in cold weather.		Unit is probably functioning correctly. If the ambient temperature is too low, the compressor clutch is not allowed to engage, because of low system pressure.	
Slow flashing green light, clutch is engaged, A/C not cooling.		Inoperative blower motor.	Check for proper operation of the blower motor.
		Loose or broken compressor belts.	Tighten or replace compressor belts.
Slow flashing green light or poor A/C performance.		A/C drive belt is broken, loose, or glazed.	Tighten or replace drive belt.
		Water valve left open, valve is broken, or cable is not operating properly.	Turn water valve off or replace valve or cable.
		Moisture in the system.	Check moisture indicator if equipped on drier. Replace and evacuate as necessary.
		Air ducts leaking air flow.	Repair air leakage problem.
		Loss of refrigerant charge before detected.	Check with gauge set, repair leak if necessary.

Pin number one (blue): A/C compressor clutch coil drive. An open, shorted wire, or shorted compressor clutch in this circuit causes the control module to disengage the A/C clutch and activate the red LED to blink three consecutive times. The A/C compressor will be latched off until the next ignition cycle, if open clutch is detected. (This diagnostic is not self clearing. You must cycle the ignition switch four times to clear the fault.)

Pin number two (red): Control module supply voltage, ignition switched. If voltage drops below 10.0 volts, LED will not illuminate. Low voltage signal is also detected in this circuit. This can be caused by a defective alternator, discharged batteries or excessive electrical loads. This diagnostic is self clearing, the blink code is displayed only while low voltage is present. If wire is disconnected or has an opened circuit, system will not function.

Pin number six: Not used.

Pin number five (green): Evaporator thermostat and A/C on/off input. The evaporator thermostat is main A/C system control. The t-stat is wired in series with the A/C system on/off switch. These switches are configured to switch battery voltage when A/C operation is requested. If wire is disconnected or opened, system will not function. Control module will not know there is a problem.

Pin number four (black): Control module ground. All voltage checks must be grounded at this pin. If wire is disconnected or opened, system will not function.

Six pin connector:

Pin number three (orange): Engine fan actuator circuit. A ground signal is provided in this circuit to either a engine ECM or a fan solenoid valve to keep the engine fan off. When the high pressure switch switches to an open state, (pressure rising above 300psi) it will remove the ground from this circuit causing the engine fan to engage. Depending on the control module, the fan control is adaptive, depending upon the frequency of the request for operation. Fan timing has a minimum duration of 45 seconds and a maximum of three minutes. When the fan is requested, the fan control output removes the ground going to either the ECM or the fan solenoid. If wire is disconnected or opened, engine fan runs continuously.

Pin number one (white): Low pressure switch. The switch has a 2.49k ohm resister built into the header, with a full charge of refrigerant the switch is in the open position. If the pressure switch closes due to the pressure dropping below 10psi or a loss of refrigerant, the A/C clutch will disengage. If wire is disconnected or opened, the control module will disengage the A/C clutch and will activate the red LED to blink four consecutive times.

Pin number two (white/black): Low pressure switch return. . The switch has a 2.49k ohm resister built into the header, with a full charge of refrigerant the switch is in the open position. If the pressure switch closes due to the pressure dropping below 10psi or a loss of refrigerant, the A/C clutch will disengage. If wire is disconnected or opened, the control module will disengage the A/C clutch and will activate the red LED to blink four consecutive times.

Pin number four (yellow/black): High pressure switch return. The switch has a 2.49k ohm resister built into the header, the switch is normally in the closed position. When pressure rises above 300psi the switch will activate the engine fan.. If wire is disconnected or opened, the control module will disengage the A/C clutch and will activate the red LED to blink four consecutive times.

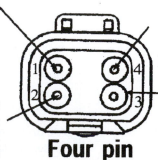

Four pin connector:

Pin number three (yellow): High pressure switch. The switch has a 2.49k ohm resister built into the header, the switch is normally in the closed position. When pressure rises above 300psi the switch will activate the engine fan.. If wire is disconnected or opened, the control module will disengage the A/C clutch and will activate the red LED to blink four consecutive times.

Figure 11-6 An ACPU control module.

To Clear Fault Codes

Clear the fault code by cycling the ignition switch four times (1 second on, 1 second off) and the green light will reappear.

ACPU CONTROL FUNCTIONS CM-820

Like the CM-813, the CM-820 can also be installed on the firewall of the truck, as in **Figure 11-7.** This figure also shows the pressure switches installed on their corresponding refrigerant lines.

A/C Start Delay. The air-conditioning compressor is not allowed to engage during the engine cranking cycle. The period of this delay begins as soon as power is supplied to the module and engine rpm is detected. This 15-second time delay period allows the engine and associated components to stabilize before the A/C compressor is allowed to operate.

Compressor Lubrication. The compressor is engaged on every ignition cycle (keeps seals and hose connections lubricated, helping to reduce system leaks).

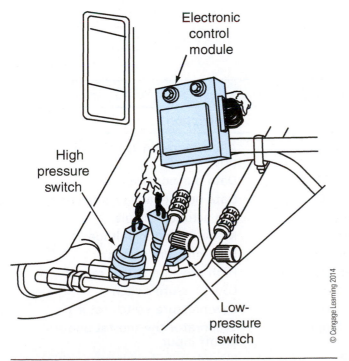

Figure 11-7 Pressure switches installed on their corresponding refrigerant lines.

The lubrication period starts after the A/C delay and when all system inputs are in a state that allows the compressor to be cycled on. The request for compressor operation lasts 15 seconds regardless of Automatic Traction Control (ATC) mode. All fan and A/C control laws govern the operation of the system during this period.

Cycle Limiting. The module will not allow the compressor clutch to cycle on or off more than once in a 15-second period.

TStat Sensor. If the TStat sensor indicates to the module that the evaporator is in a freezing condition, the module prevents the compressor from running.

High-Pressure Cutout. If the high-pressure cutout switch indicates to the module that compressor discharge pressure is too high, the module does not allow the compressor to run for an extended period of time.

Low-Pressure Cutout. If the low-pressure switch indicates to the module that compressor suction pressure is too low, the module does not allow the compressor to run.

Low-Voltage Cutout. If the module senses that the ignition voltage (V-IGN) is less than 11 volts, it does not allow the compressor to engage.

High-Voltage Cutout. If the module senses that the V-IGN is more than 16 volts, it does not allow the compressor to engage.

Diagnostic Latch Out. The compressor will be turned off when any of the following diagnostic faults are detected:

- Low pressure
- Open clutch circuit
- Shorted clutch circuit

The compressor will not cycle on again until the ignition key has been cycled off and then back on. If the fault is not present, it allows the compressor to run and store the fault code. If the fault is present, it again latches out the compressor **(Figure 11-8).**

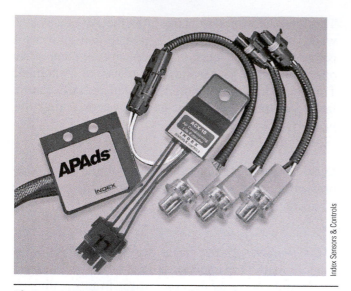

Figure 11-8 An ACPU control module.

Engine Fan Trigger

The module is capable of performing adaptive fan timing control:

- If the high-pressure switch indicates a high-pressure condition and the vehicle road speed is less than 5 mph (8 kph), the engine cooling fan is turned on for 180 seconds.

- If the high-pressure switch indicates a high-pressure condition and the vehicle road speed is 5 mph (8 kph) or greater but less than 30 mph (48 kph), the engine cooling fan is turned on for 90 seconds.

- If the high-pressure switch indicates a high-pressure condition and the vehicle road speed is 30 mph (48 kph) or greater, the engine cooling fan is turned on for 45 seconds.

- If the vehicle's parking brake is applied, the vehicle speed is 0, and the high-pressure switch detects a high-pressure condition, the engine cooling fan is requested until the parking brake is released, vehicle road speed above 0 is sensed, or the TStat circuit detects no activity for more than 2 minutes (the air-conditioning request has been turned off by the operator for more than 2 minutes).

- The fan trigger (and clutch coil) is disengaged when a low air pressure condition is broadcast over the J1587/J1708 data bus. This means that if the vehicle's air pressure is low, the air-conditioning compressor is turned off and the engine fan is locked up (as long as the fan hub assembly is designed to fail in the locked up mode).

PINOUT DEFINITION

PHYSICAL I/O TABLE				
Name	Wire Color	Wire #	Function	Description
ACDRIVE	Dk Blue	E	output	A/C compressor clutch coil drive
DATA+	Pink	A	digital I/O	Data bus positive connection
DATA-	Purple	B	digital I/O	Data bus negative connection
FAN	Orange	G	output	Fan actuator circuit
GND	Black	D	power	Module ground voltage
HPx	Yellow	Hpx A	analog input	High-pressure switch
HPx Rtn	Yellow/Blk	Hpx B	signal return	High-pressure switch return
LPx	White	Lpx B	analog input	Low-pressure switch
LPX Rtn	White/Blk	Lpx A	signal return	Low-pressure switch return
TSTAT	Green	C	digital input	Evaporator thermostat and A/C on/off input
VIGN	Red	F	power	Module supply voltage, ignition switch

INPUTS FOR ACPU CM-820 CONTROLLER

The module receives inputs from the following sensors: DATA+ and DATA- hardware interface to J1587/ J1708 vehicle data bus. These signals provide for **bidirectional** data communications (data can flow into or out of the controller).

Low-Pressure Input

The LPx is connected to a low-pressure switch. This switch is located on the suction side of the air-conditioning compressor. This "smart" device has a 2.49 K/ohm resistor installed in parallel with the electrical contact of the switch to allow diagnosing of sensor wiring and connector faults. This switch is configured to provide a low-resistance current path when it closes. The contacts of the pressure switch are open when refrigerant pressure is adequately high, above 34 psi/234 kPa **nominal** (or near that pressure with an insignificantly small variation). The contacts of the switch close when refrigerant pressure falls by approximately 10 psi (69 kPa) and open again when the pressure climbs back above 34 psi/234 kPa nominal. Low-pressure switch activity is the primary indicator of a loss of refrigerant charge and is also used to prevent compressor operation in extremely cold temperatures.

High-Pressure Input

The HPx input is connected to a high-pressure switch. This switch is located on the discharge side of the refrigerant compressor. This "smart" device has a 2.49 K/ohm resistor installed in parallel with the electrical to allow diagnosing the sensor wiring and connector faults. This switch circuit is configured to provide a low-resistance current path when it closes. The pressure switch contacts are closed when the pressure is sufficiently low (below 260 psi/1,793 kPa nominal). The contacts of the switch open when the refrigerant pressure climbs above approximately 300 psi/2,068 kPa and close back up again once the pressure goes back down to 260 psi/1,793 kPa nominal. The high-pressure switch is the primary control for the fan and is used to prevent the compressor from operating when excessive discharge pressures are present.

Evaporator Thermostat (TStat)

The TStat and the main on/off switch for the air-conditioning system are wired in series to supply battery voltage to the compressor when required. The thermostat contacts close when the temperature is greater than 38°F (3°C). The contacts of the TStat open when the evaporator temperature drops below approximately 32°F (0°C). The TStat prevents the buildup of frost on the evaporator coil by cycling the compressor off when temperatures for ice accumulation are obtained.

OUTPUTS FOR ACPU CM-820 CONTROLLER

The module produces the following output signals:

A/C DRIVE (Compressor Clutch Drive)

The module signals a high side relay switch that provides battery power to the compressor clutch coil when activated.

DATA+ and DATA-

This is the hardware interface to the J1587/J1708 vehicle data bus. These signals provide for bidirectional data communications (data can flow into or out of the controller). This is how faults are communicated to the dash panel and how the module communicates with other onboard computers.

Fan (Fan Actuator)

A low side sinking (to ground), open drain **MOSFET** is used to signal for the fan actuator circuit to activate the on-off fan drive. A fan "on" state is signified by the MOSFET being in the nonconductive state (off).

DIAGNOSTICS

Once the module detects a fault with the system, the unit broadcasts the diagnostic information over the data bus. The broadcast message appears on the dash in the following format:

EXAMPLE CODE 190 S 001 01

The first three digits of the code (190) make up the message identifier (MID). The number identifies it as an air-conditioning code. All controllers networked with the truck's data bus have a specific three-digit code identifying in what system the problem exists.

These next four digits of the code (S 001) make up the subsystem identifier (SID). This tells the technician what branch of the circuit the fault exists in.

The final two digits make up the fault code 01 indicator (FMI). There are only 14 trouble codes that are used universally.

A/C SYST REFRIG PRES HIGH	190 S 001 00	HIGH PRESSURE
A/C SYST REFRIG PRES LOW	190 S 001	LOW PRESSURE
A/C SYST LO PRES SW OPEN	190 S 005 05	LOW PRESSURE SWITCH OPEN
A/C SYST CLUTCH CIR OPEN	190 S 006 05	OPEN WIRE TO CLUTCH
A/C SYST CLUTCH CIR SHORT	190 S 006 06	SHORTED CLUTCH
A/C SYST TSTAT CIR ERRATIC	190 S 007 02	ERRATIC T-STAT
A/C SYST BAT VOLTAGE LOW	190 168 04	LOW BATTERY VOLTAGE
A/C SYST HI PRES SW OPEN	190 S 228 05	HIGH PRESSURE SWITCH OPEN
A/C SYST HI PRES SW CALIBRT	190 S 228 13	HIGH PRESSURE SWITCH LIFE HAS EXCEEDED 260,000 CYCLES

- A fault becomes active when it occurs and it remains active until it is cleared.
- A fault record can be cleared by sending a clear command with the data-bus tool, such as a Service Pro.
- Once a fault has been cleared from the module memory, it becomes active again only if the fault occurs again.
- Once a fault has been detected by the module, a fault or warning message is broadcast through the data bus and displayed on the dash (every 15 seconds) until it is cleared or becomes inactive.

High-Pressure. The condition of the high-pressure switch is monitored by the control module after the engine cooling fan has been engaged. The high-pressure fault can be caused by fan drive failure, condenser blockage, or refrigerant overcharging. If this fault is detected, the air conditioning is held off for the remainder of the ignition cycle.

High-Pressure Switch Open. This fault condition indicates that the circuit to the high-pressure switch is open. This could be caused by a break in the wiring harness, an unseated connector, or moisture intrusion into the connectors or switch. When **continuity** (circuit is closed) from the module to the switch is restored, the fault becomes inactive.

High Voltage. This fault indicates that the supply voltage on the ignition circuit feeding the module is greater than 16 volts. This can be caused by a defective alternator, voltage regulator, or surges on the supply line. The high voltage warning is canceled once the voltage of the ignition circuit returns to normal (below 16 volts).

Low Pressure. This fault is determined by monitoring the activity of the low-pressure switch. This condition can be caused by loss of refrigerant or low refrigerant pressure due to cold ambient temperature conditions. Until it is remedied, this condition latches out the air-conditioning system.

Low-Pressure Switch Open. This fault condition indicates that the circuit to the low-pressure switch is open.

This could be caused by a break in the wiring harness, an unseated connector, or moisture intrusion into the connectors or switch. When continuity from the module to the switch is restored, the fault becomes inactive.

Low Voltage. This fault indicates that the supply voltage on the ignition circuit feeding the module is lower than 11 volts. This can be caused by a defective alternator, discharged batteries, or excessive electrical loads. The low voltage warning is canceled once the voltage of the ignition circuit returns to normal (above 11 volts).

Open Clutch. This fault occurs if the module does not sense a current draw when the compressor clutch is activated. When continuity of the clutch circuit is restored, the fault becomes inactive.

Shorted Clutch. This fault occurs if the module senses excessive current flow in the compressor clutch circuit. If this fault is detected, the air conditioning will not cycle on for the remainder of the ignition cycle.

Erratic TStat. This fault occurs if the control module detects the input signal from the TStat to be rapidly switching on and off. This could be caused by a faulty A/C ON switch, TStat, or intermittent connections in the wiring or connectors.

High-Pressure Switch Calibration. This indicates that the life of the high-pressure switch has been exceeded. This may happen after the switch has exceeded 260,000 cycles.

TROUBLESHOOTING

(WHEN FAULT MESSAGE IS BROADCAST)			
Dash Message	**Definition**	**Possible Cause**	**Action**
A/C SYST REFRIG PRES HIGH 190 S 001 00	High pressure fault. The system is exhibiting abnormal high pressure activity.	Fan drive failure.	Inspect proper operation of solenoid valve/relay and/or related components.
		Blocked air flow.	Remove restriction from condenser.
		Refrigerant overcharge.	Reclaim refrigerant and charge to manufacturer's recommended specification.
		Fan circuit has not been turned on in engine control module.	Disconnect main harness to the ACPU module. If engine fan engages, replace module. If engine fan doesn't engage, check fan circuit to ECM programmed to accept ACPU input.
A/C SYST REFRIG PRES LOW 190 S 001 01	Low pressure fault. This is detected by monitoring ambient temperature and the switching activity of the low pressure switch.	Partial or total loss of refrigerant.	Check system for leaks.
		A defective low pressure switch.	Disconnect the low-side pressure switch connector and measure the resistance value of the switch. If the resistance shows 2.49 K ohms, switch is operating properly; if resistance shows less than 5 K ohms, replace switch.
			Note: This check must be done with the switch installed on the vehicle and the ignition in the off position and proper charge.
A/C SYST LO PRES SW OPEN 190 S 005 05	Open in wiring to low pressure switch. Indicates that the wiring to the low pressure switch is defective.	Unseated connector.	Check connectors at the module and the pressure switch for loose pins.
		Break between wiring harness and pressure switch.	With the use of an ohmmeter, probe the white and white/black pins and the module harness. Resistance should read 2.49 K ohms with switch installed and proper charge.
		Moisture intrusion into the connectors or switch.	Check seals and inspect wires for hole punches.
		The non use of an 8040152 pressure switch.	Install the correct pressure switch.

Dash Message	Definition	Possible Cause	Action
A/C SYST CLUTCH CIR OPEN 190 S 006 05	Open wire to A/C clutch, detected by inadequate current through the compressor clutch. If no current is sensed when the A/C drive is turned on, an open connection to the clutch coil is indicated.	A break in the wiring between the ACPU module and the A/C compressor clutch.	Check wiring with an ohmmeter for continuity.
A/C SYST CLUTCH CIR SHORT 190 S 006 06	Short wired or A/C clutch.	Detected by excessive current through the compressor clutch. This is indicative of either a shorted clutch coil or shorted wiring to the clutch.	Using an ohmmeter, check the resistance of the clutch coil. If the reading is less than 2.8 ohms, replace clutch. Locate and repair shorted or frayed wiring between the ACPU module and A/C clutch.
A/C SYST TSTAT CIR ERRATIC 190 S 007 02	Erratic evaporator thermostat. The thermostat input has been detected to be switching on and off at a rapid rate.	An erratic power signal between the A/C cab controls and ACPU module.	Check for a bad A/C switch, thermostat, or an intermittent connection in the wiring or connection between the cab controls and the ACPU module (pin C).
A/C SYST BAT VOLT LOW 190 168 04	Low system voltage. Indicates a low supply voltage condition; voltage has dropped below 11.0 volts.	Possibly caused by a defective alternator, discharged batteries, or excessive electrical loads.	Check voltage at the control module harness by probing pins D (ground) and F (power). If voltage is less than 11.0 volts, start checking for voltage drops.
A/C SYST HI PRES SW OPEN 190 S 228 05	Open in wiring to high pressure switch. Indicates that the wiring to the high pressure switch is defective.	Unseated connector.	Check connectors at the module and the pressure switch for loose pins.
		Break between wiring harness and pressure switch.	Using an ohmmeter, probe the yellow and yellow/black pins of the module harness. Resistance should read less than 5 ohms.
		Moisture intrusion into the connectors or switch.	Check seals and inspect wires for continuity.
		Not using an 8040151 pressure switch.	Install the correct pressure switch.
A/C SYST HI PRES SW CALIBRAT 190 S 228 13	Cycle life of high pressure switch has been exceeded.	High pressure switch has exceeded 260,000 cycles.	Replace with an F/L part number 8040151 pressure switch.

	(WHEN FAULT MESSAGE IS NOT BROADCASTING)		
Dash Message	**Definition**	**Possible Cause**	**Action**
No dash message	At ignition turn on, with A/C controls set to the on position, compressor is off for 15 seconds, and then remains off indefinitely.	No voltage going to A/C compressor.	Check for a break in the wiring between A/C clutch and the A/C controls.
			Check for bad A/C on/off control switch.
			Check for a bad A/C relay.
No dash message	At startup, after 15 seconds, compressor clutch doesn't engage but engine fan engages.	Defective high pressure switch.	Check switch by measuring the resistance value across the leads. If the resistance shows less than 5 ohms, switch is functioning properly. If resistance is greater than 5 ohms, replace switch.
No dash message	Compressor on for a short period of time, off for 15 seconds in a repetitive sequence.	Blockage in the high side of the system or in the condenser.	Repair restriction.
		Partial loss of refrigerant.	Check system for loss of refrigerant.
No dash message	Clutch is engaged, A/C not cooling.	Inoperative blower motor.	Check for proper operation of the blower motors.
		A/C drive belt is broken, loose, or glazed.	Tighten or replace compressor drive belt.
		Heater valve open valve is broken or control is not operating properly.	Repair problem with valve or with controller.
		Moisture in system.	Check moisture indicator on drier; replace if necessary.
		Air ducts leaking air flow.	Repair air leak problem.
		Loss of charge before detected.	Check with gauge set, repair leak if necessary.
Low air message	A/C compressor not engaging.	System air pressure is less than 60 psi.	Check for low air pressure.
No rpm message	A/C compressor not engaging.	Engine rpm is less than 300.	Check for low rpm

To Clear Fault Codes

Clear the error code by using a data bus tool such as Service Pro.

Summary

- The APAD control system is composed of an electronic control module, two pressure switches, and a conventional evaporator thermostat.

- The module receives input from the two pressure switches and the TStat, and may also receive numerous parameters read from the vehicle SAE J1587 data bus.

- The input signals are interpreted by control laws, which derive outputs to the compressor clutch coil, fan actuator circuit, and diagnostic codes.

- In the APAD system, the controller becomes the only device through which power is switched to the compressor clutch coil.

- A/C reliability is improved by actively monitoring system conditions and by controlling the compressor and the on-off fan drive.

- The CM-813 module receives inputs from the following sensors: HPx, LPx, and the TStat.

- The control module delivers outputs to the following locations: A/C drive fan, and diagnostic info to the LEDs or data bus.

- APAD and ACPU prevent rapid cycling of the compressor clutch due to high- or low-pressure conditions.

- APAD and ACPU prevent rapid cycling of the engine fan at idle.

- APAD and ACPU prevent slippage of the compressor clutch due to low voltage.

- APAD and ACPU relieve stress on the starting system by holding off the A/C until 15 seconds after startup.

- APAD and ACPU systems lubricate the A/C compressor and components year round by cycling the compressor on for 15 seconds after startup.

- APAD and ACPU display fault codes in blink format or as messages on the vehicle's dash to indicate potential A/C problems and to aid in troubleshooting.

- Fault codes may be cleared on some systems by cycling the ignition on and off four times or may be cleared using a data-bus tool like Service Pro or Prolink.

Review Questions

1. In reference to the APAD CM-813 module, what are the three inputs the module uses to control the air-conditioning compressor?

   A. Cold control switch, high-pressure switch, and low-pressure switch

   B. Refrigerant level switch, water temperature switch, and low-pressure switch

   C. High-temperature switch, low temperature switch, and high-pressure switch

   D. High-pressure switch, low-pressure switch, and evaporator thermostat

2. What are the colors of the LEDs used to display blink codes on the body of the CM-813 control module?

   A. Green, blue

   B. Red, yellow

   C. Red, green

   D. Orange, green

3. If the APAD system determines that everything is okay, what will the blink code indicate?

   A. Red LED one blink

   B. No LED lights with ignition switch on

   C. Green LED blinking rapidly

   D. Green LED blinking slowly

4. Which fault would be displayed as the highest priority?

   A. Low battery voltage

   B. Shorted clutch coil

   C. Overcharge, blocked condenser

   D. Partial loss of refrigerant charge

5. After a fault has been repaired, how is the fault erased from the module?

   A. By cycling the ignition switch four times

   B. By cycling the air-conditioning switch on and off four times

   C. By using a data-bus tool such as Service Pro

   D. Both A and B are correct, depending upon the type of module used in the system.

   E. Both A and C are correct, depending upon the type of module used in the system.

6. What could be the problem if the red LED on the APAD control module is blinking four times in sequence?

   A. Unseated connector at the high-pressure switch

   B. A broken wire in the harness to the low-pressure switch

   C. Moisture intrusion into the connector of the high-pressure switch

   D. Low-pressure switch incorrect for the application

7. What happens if the CM-820 module senses less than 11 volts from the ignition input?

   A. A/C SYST BAT VOLT LOW 190 168 04 will be displayed on the dash.

   B. The module will not allow the compressor to engage.

   C. The compressor will be cycled on and off every 15 seconds until the voltage returns to normal input voltage.

   D. Both A and C are correct.

   E. Both A and B are correct.

8. What would the driver see displayed on the dash of the truck if there was a break in the wiring between the ACPU module and the A/C compressor clutch?

   A. A/C SYST CLUTCH CIR SHORT 190 S 006 06

   B. A/C SYST TSTAT CIR ERRATIC 190 S 007 02

   C. A/C SYST CLUTCH CIR OPEN 190 S 006 05

   D. No dash message would be displayed.

9. Identify which one of the following would *not* cause A/C SYST REFRIG PRES LOW 190 S 001 01 to be displayed on the dash of the truck.

   A. Extremely low ambient temperature

   B. Partial loss of the refrigerant charge

   C. A slipping compressor clutch

   D. A thermostatic expansion valve not opening enough

   E. Total loss of the refrigerant charge

10. APAD and ACPU are capable of sending signals to control three of the following functions. Which of the following do these systems *not* control?

   A. The operation of the A/C compressor clutch

   B. The operation of the engine cooling fan

   C. The operation of the evaporator fan

   D. The communication of fault codes

12 Coach Air Conditioning

Learning Objectives

Upon completion and review of this chapter, the student should be able to:

- Determine the flow of refrigerant through a large bus air conditioning system.
- List the various components used in large bus air conditioning.
- Explain the procedures for a triple evacuation.
- Explain the procedures for a one-time evacuation procedure.
- Determine if suction and discharge pressure are acceptable.
- Perform service checks to see if refrigerant charge is correct.
- Perform refrigerant charging procedures.
- Recover the refrigerant charge for a bus air conditioning system.
- Check compressor oil and be able to adjust to correct level.
- Perform a superheat test on the system.
- Perform a low side pump down procedure.

Key Terms

check valve

filter drier

king valve

liquid line solenoid valve

maximum operating pressure

micron gauge

parcel rack

receiver

receiver tank outlet valve

semi-hermetic compressor

subcooler

vibrasorber

INTRODUCTION

Now that you have an understanding of truck air-conditioning systems, we will take a look at coach HVAC systems. Coach systems are not unlike truck systems in their principles of operation, but because of the increased heat load demands, must have a much larger refrigeration capacity. For example, a truck air conditioning system may have a refrigeration capacity of 2 to 3 tons (24,000 BTUs to 36,000 BTUs) where a coach/bus system would be in the range of 10 to 12 tons (120,000 BTUs to 144,000 BTUs).

SYSTEM OVERVIEW

In this chapter you will learn the refrigeration schematics for coach/bus HVAC systems. We will also take a look at the similarities and differences among the components, and the control systems used. And lastly we will study routine maintenance, service, and repair procedures for coach or bus air-conditioning systems.

REFRIGERATION SCHEMATIC

We will start by assuming that you have read **Chapters 3** through **7** of this text. As we proceed, follow the schematic in **Figure 12-1** while we track the refrigerant flow through the components of a large Carrier bus system.

A typical large bus system using R-134a will use the four main components of any air-conditioning system as well as those components that have been added to enhance system performance, control the system operation, and protect the system from damage.

When the vehicle operator turns on the air-conditioning system, the compressor's electromagnetic clutch will be energized, causing the compressor to begin operating. Although much larger in physical size than a truck system, it performs exactly the same function of raising the temperature and pressure of the refrigerant. The refrigerant leaves the compressor's center head through the discharge service valve as a high-temperature, high-pressure superheated vapor. Mounted directly to the center head of the compressor is a high pressure switch (HPS). This is a safety device that will terminate current flow to the compressor clutch (effectively turning off the compressor) if discharge pressure reaches an unsafe condition **(Figure 12-2)**.

From the service valve, the refrigerant enters the discharge hose, commonly referred to in the industry as a **vibrasorber**. This hose assembly isolates the refrigerant piping from engine compressor and suspension vibration that would fracture the copper piping. The refrigerant vapor now travels down the discharge line to the discharge line **check valve**. This one-way valve will allow refrigerant to flow to the condenser but will not allow liquid refrigerant to migrate back to the compressor when the compressor is in the off mode. Refer to **Figure 12-1**.

From the discharge check valve, refrigerant flows through the discharge pipe to the condenser coil. The condenser coil removes heat from the vaporous refrigerant as it passes through. Once the superheated refrigerant cools to the boiling point by removing sensible heat, it will turn to a saturated mixture (vapor/liquid mixture). Latent heat will now be removed as the refrigerant continues through the condenser. During this time, the majority of the heat is removed from the refrigerant although there is no temperature change through this section of the condenser. Once all refrigerant has changed states into a liquid, it will start to give up sensible heat again as it makes its final passes through the condenser coil. The refrigerant will leave the condenser as a high pressure subcooled liquid.

The subcooled liquid flows through the liquid line to the inlet valve of the **receiver**. The receiver inlet valve is a manual service valve **(Figure 12-3)** with a system access port designed to isolate the receiver assembly for repair. This valve should be fully back-seated during unit operation. Refrigerant flows through this valve and into the receiver. The receiver acts as a storage vessel for liquid refrigerant as evaporator demands constantly change. There are two sight glasses located on the side of the receiver body to determine the system refrigerant charge **(Figure 12-4)**.

High-pressure liquid refrigerant then flows out the **receiver tank outlet valve**. Again this is a manual service valve with system access port often referred to in the industry as a **king valve**. This valve can be used for many system maintenance procedures such as system isolation, addition and removal of refrigerant. During unit operation, this valve should be fully back-seated. As the high-temperature, high-pressure, subcooled liquid exits the receiver, it is again routed back to the condenser section to further subcool the refrigerant. This section of the condenser is called the **subcooler**. More subcooling of the refrigerant will result in less flash gas being created as the refrigerant passes through the TXV. This will improve system capacity and efficiency.

High-pressure liquid refrigerant leaves the subcooler and travels through the liquid line to the **filter drier**. The filter drier has two main functions. First it removes and retains moisture present in the refrigerant to prevent damage to the system. It also will filter out any solid particles that may have been picked up from other components and are flowing with the refrigerant. Once the refrigerant passes through the drier, it will enter the main **liquid line solenoid valve** (LLSV).

Note: Solenoid valves are either opened or closed by electromagnetism. A valve can be normally open and will close when the coil is energized or it may be normally closed and will open when the coil is energized.

The main LLSV is a normally closed valve that whenever the A/C system is running will be energized and open **(Figure 12-5)**.

When the A/C system is running, high-pressure high-temperature liquid refrigerant will travel from the

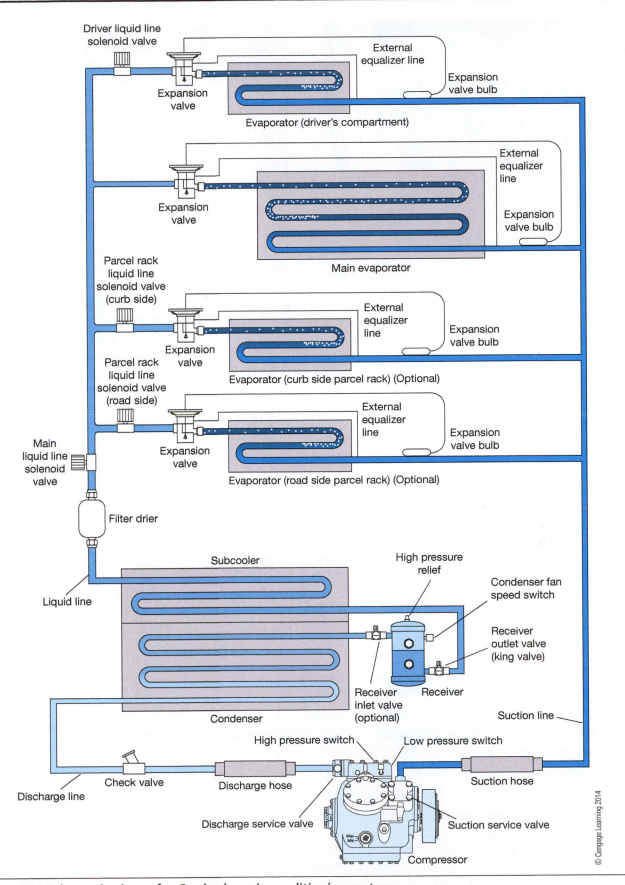

Figure 12-1 Schematic view of a Carrier bus air-conditioning system.

Figure 12-2 Compressor and electromagnetic clutch.

John Dixon © 2014 Cengage Learning

main liquid line solenoid valve to the following locations **(Figure 12-6)**:

- Main evaporator TXV
- Driver liquid line solenoid
- The two parcel rack liquid line solenoid valves (if equipped)

With the A/C turned on and the main LLSV energized and open, refrigerant will flow directly to the TXV of the main evaporator. This means that the main evaporator is **ALWAYS** in use when the system is operating.

Refrigerant will also flow through the driver liquid line solenoid. This is a normally open solenoid so the driver will also have A/C. If the driver does not want A/C, the driver LLSV will be energized and closed, effectively turning off the driver's evaporator.

Liquid refrigerant also flows from the main LLSV to the two **parcel rack** LLSVs, if equipped. Parcel

racks are evaporators that are used to provide air conditioning from above the passengers' heads. Passengers have the ability to turn the vent on or off for their own comfort. The parcel rack LLSVs are also normally open valves that will be energized to close if the vehicle operator determines that the parcel rack does not require A/C. With these valves de-energized and open, refrigerant will flow to the TXV of each parcel rack **(Figure 12-7)**.

Whenever the air conditioning is turned on, high-pressure, high-temperature subcooled liquid refrigerant will be flowing to the TXV of the main evaporator. The TXV is responsible for metering refrigerant into the evaporator coil at a rate that will maintain a constant superheat setting at the evaporator outlet. Many TXVs are equipped with a **maximum operating pressure** (MOP) setting. With this valve feature, if the suction pressure of the evaporator outlet reaches the MOP setting the valve will close, preventing further refrigerant flow until suction pressure is reduced **(Figure 12-8)**.

Once the high-pressure high-temperature subcooled liquid refrigerant passes through the TXV, the pressure of refrigerant will be reduced significantly from the high-pressure side of the air conditioning system. With the drop in refrigerant pressure, the boiling point of the refrigerant will also be reduced. The heat energy contained within the refrigerant will instantly create a latent heat change that will cause 10% to 20% of the liquid refrigerant to flash into gas. This will leave the remaining refrigerant as a saturated mixture (mixture of gas and liquid refrigerant).

The low-pressure saturated mixture leaving the TXV enters the evaporator. While the refrigerant circulates inside the evaporator, the air circulated in the passenger compartment is circulated past the

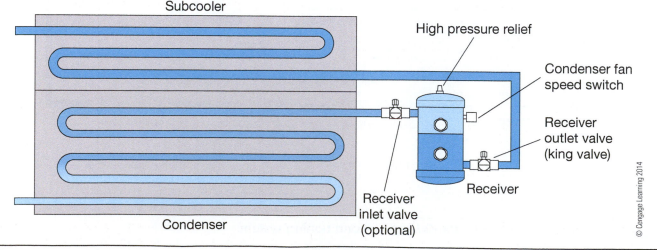

Figure 12-3 Condenser and receiver with service valves.

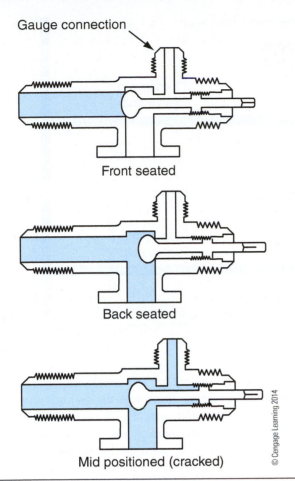

Figure 12-4 Service valves in 3 positions, front seated, back seated, and mid seated or cracked.

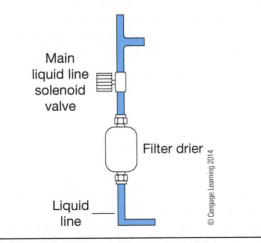

Figure 12-5 Liquid line solenoid and drier.

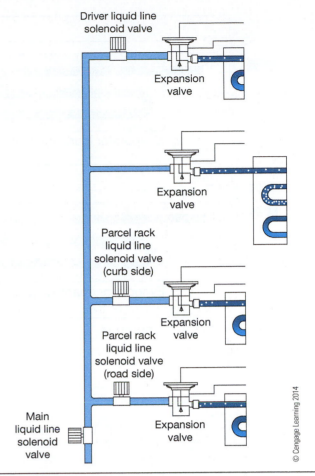

Figure 12-6 Diagram of four different paths that refrigerant can flow.

evaporator coil fins. The heat contained in the circulating air gets absorbed by the refrigerant inside the evaporator coil. This heat causes the remaining liquid to boil. The majority of the heat from the passenger compartment is removed during this latent heat exchange. Once the entire liquid refrigerant has boiled off and no more latent heat is absorbed, the refrigerant will continue to absorb sensible heat. This will cause the refrigerant to become superheated.

Some large bus air-conditioning systems are equipped with a driver's evaporator module. This module is designed to control the temperature in the driver's area and also provide defrost/defog functions of the windshield. The driver's module has a TXV and will operate in the same manner as the main evaporator.

Many large bus A/C applications are equipped with optional parcel rack evaporator modules. These units allow passengers to have individual control of the air flow in their seating area. These systems work with the main system and receive liquid refrigerant from the main liquid line. Each parcel rack evaporator has its own TXV and functions in the same manner as the evaporator.

Low-pressure low-temperature superheated vapor leaves the evaporator(s) and enters the suction line. All of the evaporators on the bus are connected to this same suction line. Refrigerant traveling through the suction line will enter the suction vibrasorber. This hose assembly isolates the vibration of the compressor

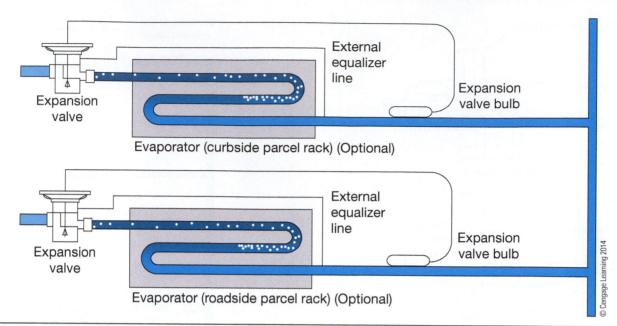

Figure 12-7 Parcel rack evaporators.

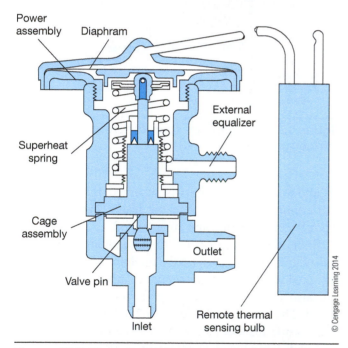

Figure 12-8 Thermostatic expansion valve.

from the vehicle engine and suspension to prevent vibration fractures of the copper refrigerant lines. The refrigerant then returns to the compressor through the suction service valve to complete the cycle **(Figure 12-9)**.

SERVICE PROCEDURES

As with any air-conditioning system, it is important that the technician understands all procedures required to service the equipment. Proper evacuation and

dehydration procedures are extremely important when it comes to servicing large bus air-conditioning systems. When performed correctly, these procedures will improve operating performance and compressor life.

The end result of improperly performing these services can be catastrophic to the entire system. An improper evacuation can result in noncondensable gases being left in the system, which will result in high discharge pressure: moisture may cause ice blockage at the expansion valve; moisture and refrigerant may react to form an acid. High discharge pressures may cause high discharge temperatures. This condition can cause the refrigerant oil to form sludge that can block the flow of refrigerant oil, starving the compressor from vital lubrication. Acid may cause copper plating of the bearing surfaces and eventual compressor failure.

A proper evacuation begins with the vacuum pump itself. The pump must be a well-maintained two-stage pump with a rated capacity of 3 to 5 cubic feet per minute (CFM). The oil used in the pump will trap some of the moisture that is removed from the refrigeration system. Moisture that gets trapped in the oil reduces the pump's overall efficiency because the oil is the sealing medium in the vacuum pump and water will alter the oil's viscosity. If moisture is left in the pump for extended periods, it may also cause damage to the seals and metal surfaces of the pump. To keep and maintain the pump in top condition, follow manufacturer's maintenance recommendations closely. The oil in the vacuum pump should be replaced after it has been used three times at the very maximum, and after a vehicle is known to have a wet or contaminated

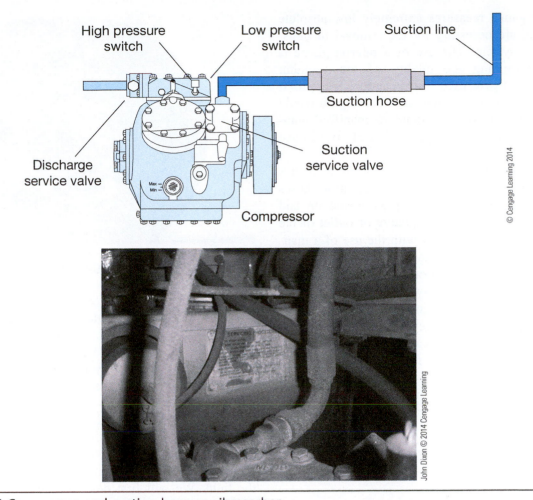

High pressure switch

Low pressure switch

Suction line

Suction hose

Discharge service valve

Suction service valve

Compressor

Figure 12-9 Compressor and suction hose or vibrasorber.

air-conditioning system. Use only oil specially formulated for the use in vacuum pumps.

To obtain the best evacuation possible, a technician should use evacuation hoses for the procedure instead of standard charging hoses like those found on a manifold gauge set. Evacuation hoses have a larger diameter and are designed especially for this purpose. Standard hoses are designed to be very flexible and to minimize leaks under high pressure, but they are not designed to prevent leaks into the hoses under vacuum. Hoses designed for the evacuation process are available from refrigeration suppliers and should be kept as short as possible.

The internal passages of the standard manifold gauge set are too small for good flow at low pressure and therefore should not be used. A large manifold gauge set made for evacuation can be used or simply attach a manifold directly to the vacuum pump.

The use of a **micron gauge** (electronic vacuum gauge) is the only real means of ensuring that a good evacuation has been achieved **(Figure 12-10)**. The compound gauge on a standard manifold is not acceptable.

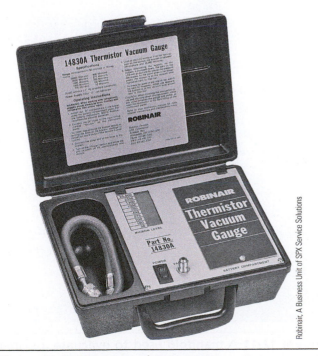

Figure 12-10 Vacuum micron gauge.

The micron gauge measures extremely low absolute pressures that are necessary for the removal of moisture from the system. The use of a micron gauge is vital in determining the quality of the vacuum; 500 microns, a figure used in evacuation, is approximately 1/50 of an inch of mercury. This figure would be impossible to read on a standard manifold compound gauge. Micron gauges are available from most refrigeration suppliers.

When performing an evacuation procedure, always use three points of the system. Evacuation hoses should be connected to the compressor suction and discharge ports and to the king valve or outlet of the receiver tank outlet valve port with the use of a manifold. As well, any electric normally closed solenoid valves must be energized and open. A simple manifold like that in **Figure 12-11** allows the technician to increase the number of evacuation points.

Be sure to leak check your evacuation setup prior to evacuation. If hoses or seals leak, proper evacuation is impossible. Repair any leaks found before beginning the evacuation procedure.

All hoses from the system should be connected to the evacuation manifold at the vacuum pump.

Figure 12-12 Compound gauge.

The vacuum pump uses 3/8-inch flare fittings and 3/8-inch hoses. Also connected to the manifold will be the micron gauge and a refrigerant recovery system. The compound gauge on the manifold gauge set will be used to monitor system pressure, not vacuum **(Figure 12-12)**. Attach the vacuum pump as shown in **Figure 12-13**.

The two basic evacuation procedures are the triple evacuation and the one-time evacuation. The triple evacuation is an extremely good procedure and may be performed with equipment that is normally found in the field. We will first describe the procedures for a triple evacuation. Before proceeding, you may choose to replace the filter drier with a section of copper tubing with the suitable fittings.

Triple Evacuation

1. Start by removing any refrigerant that remains in the system with the use of a refrigerant recovery machine.
2. Connect the vacuum hoses, vacuum pump, vacuum micron gauge, and manifold gauge set to the bus air conditioning system.
3. Back seat (close) all service valves on the system and open the vacuum pump and micron gauge valve. Energize any normally closed refrigerant solenoid valve so that it is open. Start the vacuum pump and draw the manifold and hoses into a very deep vacuum. Shut off the pump and check to see if the vacuum holds. This is how you ensure that your vacuum setup is free of leaks. You must make sure the setup holds a vacuum before you proceed. Check all hoses and seals before proceeding.
4. Mid-seat the three service valves on the bus air conditioning system. See **Figure 12-4**.

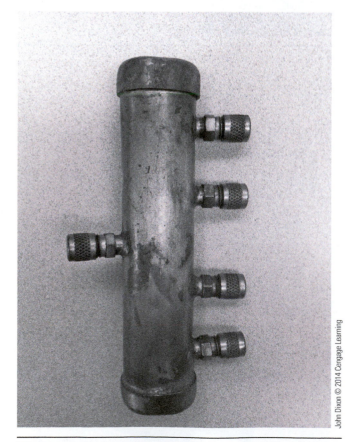

Figure 12-11 A manifold used to evacuate a system from multiple ports.

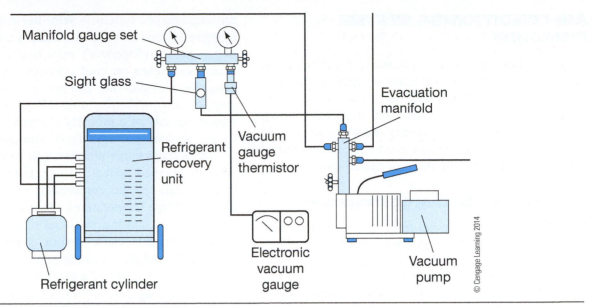

Figure 12-13 Diagram of an evacuation setup.

5. With the vacuum pump and the micron gauge open, start the vacuum pump and evacuate to a pressure below 2000 microns. Shut off the vacuum pump and wait a few minutes to be sure that the vacuum holds.

6. Close the vacuum pump valve and the valve to the micron gauge. Turn off the pump. Closing the valve to the micron gauge protects it from damage.

7. Using dry refrigerant or nitrogen, break the vacuum and bring the pressure up to roughly 2 psig. Monitor the pressure with the compound gauge.

8. Remove the refrigerant using a refrigerant recovery system. If you charged it with nitrogen, release it to the atmosphere.

9. Repeat steps 5, 6, 7, and 8 one more time.

10. After the second time you break the vacuum, replace the filter drier or replace the copper tube with a new drier. Evacuate a third time to a pressure of 500 microns. Check that the vacuum holds. If pressure rises, it may indicate that there is a leak or there is moisture remaining in the system.

11. Charge the system with refrigerant to specification per normal charging procedures.

The second evacuation method that will be described is the one-time evacuation to a deep vacuum. Although this method may seem to be less complicated, it actually may take longer in a severely contaminated system and is not as reliable as the triple evacuation.

One-Time Evacuation Procedure

1. Start by removing any refrigerant that remains in the system with the use of a refrigerant recovery machine.

2. Remove and replace the filter drier before evacuating.

3. Connect the vacuum hoses, vacuum pump, vacuum micron gauge, and manifold gauge set to the bus air conditioning system.

4. Back seat (close) all service valves on the system and open the vacuum pump and micron gauge valve. Energize any normally closed refrigerant solenoid valve so that it is open. Start the vacuum pump and draw the manifold and hoses into a very deep vacuum, Shut off the pump and check to see if the vacuum holds. This is how you ensure that your vacuum setup is free of leaks. You must make sure the setup holds a vacuum before you proceed. Check all hoses and seals before proceeding.

5. Mid-seat the three service valves on the bus air conditioning system.

6. With the vacuum pump valve and the micron gauge open, start the pump and evacuate to a pressure below 1000 microns.

7. Close the valve to the vacuum pump and turn it off. Check to see that the vacuum is holding. If pressure rises, it may indicate that there is a leak or there is moisture remaining in the system.

8. Charge the system with refrigerant to specification per normal charging procedures.

AIR-CONDITIONING SYSTEM PRESSURE

The following is a rule of thumb for technicians to determine if system pressures are acceptable. The first thing the technician must do is to take temperature readings at the condenser inlet and evaporator return air temperature. Start by properly installing a manifold gauge set and run the system in cool mode to stabilize the system. Then follow the procedure below.

Compressor Discharge Pressure

1. Record the condenser inlet air temperature with the use of a digital thermometer.
2. Add 40 to this number and total.
3. Take the temperature from #2 and convert it to a pressure using a refrigerant pressure-temperature chart.
4. Compare this calculation to the actual system discharge pressure.
5. The discharge pressure should be +/− 10%.

Compressor Suction Pressure

1. Record the evaporator return air temperature with the use of a digital thermometer.
2. Subtract 30 from this number and total.
3. Take the temperature from #2 and convert it to a pressure using a refrigerant pressure-temperature chart.
4. Compare this calculation to the actual system suction pressure.
5. The suction pressure should be +/− 5%.

Use the following chart:

Discharge Pressure		
Condenser inlet temperature	_____	°F (°C)
Plus	_____ +40°F (22.4°C)	
Equals	_____	°F (°C)
Pressure-temperature chart	_____	PSIG
Suction Pressure		
Evaporator return air temperature	_____	°F (°C)
Minus	_____ −30°F (16.8°C)	
Equals	_____	°F (°C)
Pressure-temperature chart	_____	PSIG

CHECKING REFRIGERANT CHARGE

For maximum performance, an air-conditioning system must have the correct refrigerant charge.

If a system is undercharged, you may find that the suction and discharge pressures will be lower than a properly charged system. A **semi-hermetic compressor** (compressor and drive motor share a common housing like a home refrigerator) may overheat because the compressor relies on cool refrigerant returning to the compressor to cool the compressor motor. An undercharged air conditioning unit will not have the performance capacity of a properly charged system.

If a system is overcharged, you may find that the suction and discharge pressure will be higher than in a properly charged system. In fact, the pressure may be high enough for the compressor to sustain damage. High discharge temperatures can cause compressor oil to break down, causing acids and sludge to form. These contaminants can cause severe compressor damage. An overcharged air-conditioning unit will not have the performance capacity of a properly charged system.

The following conditions must be met before checking the refrigerant charge level on a large bus system.

1. The bus must be running in the cooling mode.
2. The compressor must be fully loaded (all six cylinders in operation).
3. Discharge pressure must be at least 150 psig for an R-134a unit or 250 psig for an R-22 unit.
4. Refrigerant pressures must be stabilized. (Stabilization usually takes a couple of minutes.)
5. Check the receiver tank sight glass:
   a. If the receiver has two sight glasses, the bottom sight glass should not be empty and the top sight glass should not be full.
   b. If the receiver has only one sight glass, it should be half full.

The following conditions must be met before checking the refrigerant charge level on a small bus system.

1. The bus must be running in the cooling mode.
2. The discharge pressure must be at least 150 psig for R-134a.
3. Refrigerant pressures must be stabilized. (Stabilization usually takes a couple of minutes.)
4. Check the liquid line sight glass. You should be able to see clear glass with the occasional bubble.

Note: If your system does not have a liquid line sight glass, you must check the "Delta T" (temperature difference between return air and supply air) across the evaporator coil. With the vehicle at high idle (900 to 1100 rpm) and medium fan speed, there should be a difference of between 18°F and 20°F between the inlet and outlet of the evaporator coil.

Refrigerant Charging Procedures for Large Bus

Full Charging. When a bus air-conditioning system has been evacuated, it will require a full system refrigerant charge. A system can be fully charged using *liquid refrigerant only* through the receiver tank outlet valve (king valve).

1. Following the directions already mentioned, fully evacuate the system using either the triple or one-time evacuation method.
2. Connect a charging hose between the refrigerant cylinder and the receiver tank outlet valve.
3. Place the refrigerant cylinder on a scale and make note of the weight of the cylinder.
4. Open the liquid refrigerant hand valve on the cylinder and mid-seat the receiver tank outlet valve.
5. Allow refrigerant to flow from the tank into the unit until the correct amount of refrigerant has been dispensed. (Consult manual for correct refrigerant weight.)
6. Close the liquid refrigerant hand valve on the cylinder and back-seat the receiver tank outlet valve.
7. Run unit and check for proper charge (following the instructions in this chapter).

Note: Charging with liquid refrigerant should be done only on bus systems that have a receiver tank outlet valve and only when the unit has been properly evacuated.

Partial Charging

It may be necessary to partially charge a bus air-conditioning system when the system is low. This procedure could be performed after a refrigerant leak was repaired while the unit was in a low side pump down or when conditions prevented the complete charge from being dispensed into the unit as described above. A system should be partially charged using *refrigerant vapor only* through the compressor suction service valve with the unit running in cool mode.

1. Run the bus air conditioning and allow system pressure to stabilize.
2. Connect a charging hose from the refrigerant tank to the compressor suction service valve.
3. Place the refrigerant cylinder on a scale and make note of the weight of the cylinder.

4. Open the vapor valve on the refrigerant cylinder and mid-seat the suction service valve.
5. Charge until the correct level had been reached. (Consult manual for correct refrigerant weight.) If it is not known how much refrigerant you need to add, use the method for checking proper refrigerant level in this chapter.
6. Close the vapor valve on the refrigerant cylinder and back-seat the receiver tank outlet valve.
7. Run the system and confirm proper refrigerant charge.

WARNING *It is never acceptable to add liquid refrigerant into the suction or discharge service valves of the compressor. Extreme damage to the compressor may occur.*

REFRIGERANT RECOVERY

Note: When working with refrigerant, always adhere to local environmental laws.

Procedures for units with a receiver tank outlet valve **(Figure 12-14).**

Using a refrigerant management center, connect the high side service hose to the receiver tank outlet valve and the low side service hose to the compressor suction service valve. Mid-seat the receiver tank outlet and compressor suction valves. Any electric normally closed refrigerant solenoid valves must be energized and open. Operate the refrigerant management center as per manufacturer's instructions until all refrigerant has been removed from the system. Today's management centers not only recover the refrigerant, they also

Figure 12-14 Refrigerant management center.

recycle it by running the refrigerant through a filter and drier process, allowing the refrigerant to be reused in the unit from which it was just removed.

CHECKING COMPRESSOR OIL LEVEL

For the compressor to receive proper lubrication, the compressor crankcase oil level must be maintained correctly.

In order the properly check compressor oil, follow the procedures below.

1. Run the air-conditioning system in the cool mode.
2. Ensure that the compressor is fully loaded (all six cylinders in operation).
3. Discharge pressure must be 150 psi for R-134a and 250 psi for R-22 units.
4. Run unit until pressures are stabilized.

After 20 minutes of operation, shut off the unit and examine the compressor oil level in the compressor sight glass. The oil level should be like those shown in **Figure 12-15.** Add or remove oil as required to maintain proper levels.

Figure 12-15 Drawing of compressor sight glass with oil level.

Adding Compressor Oil

The following instructions are for adding oil to a large bus system.

1. With the system off, disconnect the wire to the compressor's electromagnetic clutch.
2. Front-seat both the compressor suction and discharge service valves, effectively isolating the compressor from the rest of the system.
3. With a refrigerant management center, remove any remaining refrigerant from the system.
4. Stop the refrigerant recovery when the system pressure is 1 to 2 psig (positive pressure).
5. Examine compressor crankcase pressure. If it rises, continue with the recovery process.
6. Once you are able to maintain a 1-2 PSIG crankcase pressure, continue.
7. Remove oil fill plug and add oil to the compressor. Watch your progress in the compressor sight glass. Once the proper level has been achieved, replace and seal the compressor fill plug.
8. Evacuate and dehydrate the compressor.
9. Front-seat manifold or management center hand valves and mid-seat the compressor suction and discharge service valves.
10. Run unit and check proper refrigerant charge as described in this chapter.

Removing Compressor Oil

The following instructions are for removing oil from a large bus system. Regular routine maintenance will require changing the compressor oil to be removed and replaced at timed intervals. It may also be necessary to remove oil from an air-conditioning system that is overcharged with oil. This oil will cause the compressor crank to slap the oil, causing it to foam. When this happens, oil will be discharged from the compressor at a much greater rate than required. The extra oil collected in the coils and hoses insulates them, reducing their ability to transfer heat. This in turn will lower system capacity. Over time, this oil will return to the compressor as a slug and can cause severe compressor damage. To safely remove compressor oil, proceed with the following steps.

1. With the system off, disconnect the wire to the compressor's electromagnetic clutch.
2. Front-seat both the compressor suction and discharge service valves, effectively isolating the compressor from the rest of the system.

3. With a refrigerant management center, remove any remaining refrigerant from the system.

4. Stop the refrigerant recovery when the system pressure is 1 to 2 psig (positive pressure).

5. Examine compressor crankcase pressure. If it rises, continue with the recovery process.

6. Once you are able to maintain a 1-2 PSIG crankcase pressure, continue.

7. Remove oil fill plug to relieve internal crankcase pressure.

8. Remove oil drain plug and allow oil to flow until the correct level is achieved by viewing the compressor sight glass. If compressor oil is being replaced, allow oil to fully drain.

9. Reinstall oil drain plug. If all oil was removed, replace oil through the fill plug opening. View your progress at the compressor sight glass. Once the proper level has been achieved, reinstall the oil fill plug.

10. Evacuate and dehydrate the compressor.

11. Front-seat manifold or management center hand valves and mid-seat the compressor suction and discharge service valves.

12. Run unit and check proper refrigerant charge as described in this chapter.

SUPERHEAT TEST PROCEDURES

When a bus air conditioning system is experiencing a complaint of insufficient cooling capacity, one of the first things a technician can be asked is what is the superheat setting of the unit. This is a very simple procedure that will be explained here.

1. Start by making certain that your manifold gauge set is calibrated and in good working order. Recalibrate as necessary.

2. Install the manifold gauges on the suction and discharge service valves.

3. With an accurate electronic thermometer, firmly attach the probe to the evaporator tail pipe near the TXV sensing bulb and insulate the probe.

4. Start up the bus and operate the system on cool for a minimum of 20 minutes. Allow unit pressures to stabilize with the compressor fully loaded (running on all six cylinders). Discharge pressure should be at least 150 psig for R-134a units and 250 psig for R-22 units.

5. Record the temperature of the evaporator tail pipe and the pressure of the refrigerant at the suction service valve.

6. Repeat step 5 a minimum of five times at 2- to 3-minute intervals, recording all information.

7. After recording the five readings, calculate the superheat for each set of readings.

8. Using a pressure-temperature chart, find the saturation temperature at the suction pressure recorded.

9. To calculate superheat, subtract the saturation temperate determined above from the temperature measured at the evaporator tail pipe.

10. Calculate the average of the five superheat calculations and compare it to the specification found in the service manual for the unit.

Superheat Checklist

Suction Pressure (taken at the suction service valve)	Thermometer Reading (at the evaporator tail pipe)	Saturated Suction Temperature (from pressure-temperature chart)	Superheat = Measured thermometer reading minus saturation temperature
PSIG	°F	°F	°F
PSIG	°F	°F	°F
PSIG	°F	°F	°F
PSIG	°F	°F	°F
PSIG	°F	°F	°F
		Total the (5) readings (divide by 5)	°F
		Average Superheat	°F

AIR-CONDITIONING TROUBLESHOOTING TIPS

The following is a guide for troubleshooting some of the most common problems of a bus air-conditioning system.

Problem	Possible Solution
Compressor iced	1. Poor evaporator air flow ■ Dirty return air filter ■ Dirty evaporator coil ■ Bad evaporator fan motor 2. Bad TXV installation ■ Location (4 or 8 o'clock) ■ Contact ■ Secured ■ Insulated 3. Excessive compressor oil 4. Superheat setting too low
Frosted evaporator	1. Poor evaporator air flow ■ Dirty return air filter ■ Dirty evaporator coil ■ Bad evaporator fan motor 2. Superheat setting too low
High head pressure	1. Poor condenser air flow ■ Dirty condenser screen ■ Dirty condenser coil ■ Bad condenser fan motor ■ Incorrect fan rotation 2. Refrigerant overcharged 3. Noncondensable in the system
High suction pressure	1. Blown head gasket 2. Broken suction reed valves 3. Stuck unloader valve
Low suction pressure	1. Poor evaporator air flow ■ Dirty return air filter ■ Dirty evaporator coil ■ Bad evaporator motor 2. Restriction in supply air duct 3. Plugged filter drier 4. Broken TXV bulb or capillary tube 5. Superheat setting too low

LOW SIDE PUMP DOWN PROCEDURES

The low side pump down allows a technician to perform any service work required that is downstream of the receiver tank outlet valve without having to first remove all the refrigerant from the system. The following are steps for performing a low side pump down.

1. Install the manifold gauge set.
2. Start the bus and allow the air-conditioning system to operate in the cool mode for 15 to 20 minutes.
3. Front-seat the receiver tank outlet valve (king valve).
4. Run the system in cool until the compressor suction pressure is 1 to 2 psig. Turn unit off.

WARNING *Do not allow compressor to run in a vacuum. Compressor damage may result.*

5. Observe suction pressure gauge. Usually the pressure will rise a small amount.
6. If the pressure climbs to high it is usually because there is refrigerant mixed with the compressor oil. It may be necessary to restart the unit and allow it to run for a few more seconds. In the event that the suction pressure stabilizes too low (vacuum), momentarily mid-seat the valve on the manifold gauge set and bleed some of the high-side discharge pressure into the suction side until the suction side has a positive pressure of 1 to 2 psig.
7. With pressure stabilized at 1 to 2 psig, front-seat the suction service valve. All refrigerant is now contained in the compressor, condenser, and receiver sections of the system. All systems downstream of the receiver tank outlet valve can be changed or serviced.
8. Once repairs or servicing has been completed, the filter drier should be replaced. Energize any normally closed refrigerant solenoid valve so that it is open. Perform an evacuation on the low side of the system through the suction service valve and the front-seated receiver tank outlet valve. Check for vacuum leaks.
9. Return the unit to service by back-seating both the suction service valve and the receiver tank outlet valve and performance test the unit to ensure repairs were done correctly.

Summary

- A typical large bus system using R-134a will use the four main components of any air-conditioning system as well as those components that have been added to enhance system performance, control the system operation and to protect the system from damage.

- The condenser coil removes heat from the vaporous refrigerant as it passes through.

- The receiver acts as a storage vessel for liquid refrigerant.

- More subcooling of the refrigerant will result in less flash gas and will improve system capacity and efficiency.

- The filter drier has two main functions: retain moisture present in the refrigerant and filter out any solid particles that may have been picked up.

- The main LLSV is a normally closed valve that whenever the A/C system is running will be energized and open.

- The main evaporator is **ALWAYS** in use when the system is operating.

- Parcel racks are evaporators that are used to provide air conditioning from above the passenger's head.

- If the suction pressure of the evaporator outlet reaches the MOP setting, the valve will close, preventing further refrigerant flow until suction pressure is reduced.

- Proper evacuation and dehydration procedures are extremely important. These procedures will improve operating performance and compressor life.

- The use of a micron gauge is the only real means of ensuring that a good evacuation has been achieved.

- There are two basic evacuation procedures: the triple evacuation and the one-time evacuation.

- For maximum performance, an air-conditioning system must have the correct refrigerant charge.

- For the compressor to receive proper lubrication, the compressor crankcase oil level must be maintained correctly.

■ When a bus air-conditioning system is experiencing a complaint of insufficient cooling capacity, one of the first things a technician can be asked is what is the superheat setting of the unit.

■ The low side pump down allows a technician to perform any service work required that is downstream of the receiver tank outlet valve without having to first remove all the refrigerant from the system.

Review Questions

1. What component effectively isolates the refrigerant tubing from the vibrations caused by the compressor, vehicle engine, and suspension?

   A. Discharge service valve

   B. TXV bulb

   C. Discharge and suction vibrasorbers

   D. Discharge check valve

2. What is the primary function of the condenser in a coach air-conditioning system?

   A. Add pressure to the high side

   B. Add heat to the refrigerant

   C. Remove air from the system

   D. Remove heat from the refrigerant

3. What is the main purpose of the discharge check valve?

   A. Trap refrigerant in the receiver

   B. Prevent liquid refrigerant from migrating from the condenser to the compressor in the off cycle

   C. Prevent liquid from accumulating in the condenser coil

   D. Prevent refrigerant from flowing into the evaporator

4. A high-pressure switch is designed to shut off the compressor when:

   A. The controller opens the switch

   B. Atmospheric pressure is unsafe

   C. Discharge pressure is unsafe

   D. Suction pressure is unsafe

5. What must be done to the receiver tank outlet valve in order to stop refrigerant flow at the receiver?

   A. Back-seat valve

   B. Front-seat valve

   C. Mid-seat valve

   D. Side-seat valve

6. The main liquid line solenoid valve will be _____ and _____ when the air conditioning is turned on.

   A. Energized, open

   B. Energized, closed

   C. De-energized, open

   D. De-energized, closed

7. The driver's liquid line solenoid valve will be _____ and _____ when the driver does not require air conditioning.

   A. Energized, open

   B. Energized, closed

   C. De-energized, open

   D. De-energized, closed

8. A TXV with a maximum operating pressure setting will _____ .

   A. Open when the controller requires it

   B. Open when discharge pressure is normal

   C. Close if suction pressure is above the valve setting

   D. Close if suction pressure is below the valve setting

9. The only way to determine if a proper evacuation has been achieved is with the use of a _____.

   A. Manifold gauge set

   B. Micron gauge

   C. Multimeter

   D. Ammeter

10. All service valves should be _____ under normal operating conditions.

    A. Frontseated

    B. Backseated

    C. Midseated

    D. Any position will work.

11. Typically, where would you expect to find the high pressure switch on a bus air-conditioning system?

    A. On the condenser

    B. On the evaporator

    C. On the compressor suction hose

    D. On the center compressor head

12. Which of the evacuation procedures described in this chapter is the most reliable method?

    A. The single-evacuation method

    B. The double-evacuation method

    C. The triple-evacuation method

    D. The four-point evacuation method

13. To accurately check the refrigerant charge on a coach air-conditioning system, what parameters must be met?

    A. If the receiver has two sight glasses, the upper sight glass should be full of refrigerant.

    B. If the receiver has one sight glass, it should be half full.

    C. If the receiver has two sight glasses, the bottom sight glass should not be empty and the top sight glass should not be full.

    D. Both A and B are correct.

    E. Both B and C are correct.

14. In order to accurately calculate the superheat of the air conditioning system, what tools and information do you need?

    A. Suction pressure

    B. Pressure-temperature chart

    C. Temperature at evaporator tail pipe

    D. All the above are required.

13 Truck-Trailer Refrigeration Equipment

Key Terms

ambient	heat of respiration	set point
auto stop/start	hot load	short cycling
box temperature	multi-temp units	thermostat
continuous run mode	precooling	
Freon	residual heat	

INTRODUCTION

For many years, only ice was used to maintain cool temperatures for the safe transport of perishable items. Due to the need to constantly replace the melted ice, the decreased cargo weight caused by the weight of the ice, and the inability to maintain a temperature much below 32°F (0°C), ice is now used only as a topping on the load to aid in maintaining humidity in the cargo space.

Fred Jones invented the first mobile refrigeration unit in the summer of 1938 to get freshly killed chickens to market without spoiling. He developed his invention in a workshop in Minneapolis, Minnesota, and produced the first Thermo King™.

Regardless of OEM, truck-trailer refrigeration equipment operates in the same manner, removing heat from where it is not wanted. In order to understand how the refrigeration components work, it is essential to understand the basics of thermodynamics discussed in **Chapter 3**.

SYSTEM OVERVIEW

Today's sophisticated refrigeration units can maintain product temperatures from −20°F to 80°F (−29°C to 27°C), regardless of the **ambient** temperature (air temperature outside the truck or trailer). This is how ice cream can get to your local supermarket when it's 100°F (38°C) outside. On the other side of the coin, this same unit can keep milk from freezing when the outside temperature is −30°F (−34°C). Most refrigeration units are powered by a diesel engine that drives a compressor. Some units also have an electric motor for quiet standby operation **(Figure 13-1)**.

The refrigeration unit maintains the desired **thermostat** set point regardless of ambient temperatures. When the cargo space temperatures are higher than the set point, the unit operates in the cool mode. When the cargo space temperatures are lower than the desired set point, the unit operates in the heat mode.

The thermostat also controls the engine speed, which in turn governs the output of the compressor. If the cargo space temperatures vary more than 2 to 3 degrees above or below set point, the unit either operates in high-speed heat or high-speed cool. Once the controlled space temperature is within plus or minus 2 degrees of the desired set point, the unit operates in low speed **(Figure 13-2)**.

Figure 13-2 A trailer refrigeration unit.

The unit operates on the principle of moving heat. During the cooling mode, the unit removes heat from the controlled space and places it where it is not objectionable (in the ambient air) **(Figure 13-3)**.

Temperature influences the deterioration rate of perishable and frozen food products more than any other factor. Depending on the product being shipped, the refrigeration unit may be required to maintain very low cargo space temperatures, even with high ambient temperatures.

For example, ice cream should be maintained at −20°F (−29°C). With ambient temperatures in the 90°F (32°C) range, the unit must work very hard. Other products, such as bananas, should be kept at 57°F (14°C) because anything less than 55°F (13°C) will damage the fruit. With ambient temperatures in the −20°F (−29°C) range, the unit must work very hard to maintain the desired heating capacity.

SYSTEM COMPONENTS

The basic components in a refrigeration unit, or "reefer" as it's called in the industry, are basically the same as those used in air conditioning, although the capacities are greatly increased.

Engine

Trailer units use a four-cylinder diesel engine to drive the large displacement compressor. The engine itself has only two speeds, high and low. As the refrigeration unit gets close to the desired temperature, the engine is placed into the low-speed setting to slow down the cooling process. The engine is also responsible for turning the fans required to move the air for the refrigeration process. One fan draws air through

Figure 13-1 The many components that make up a typical trailer refrigeration unit.

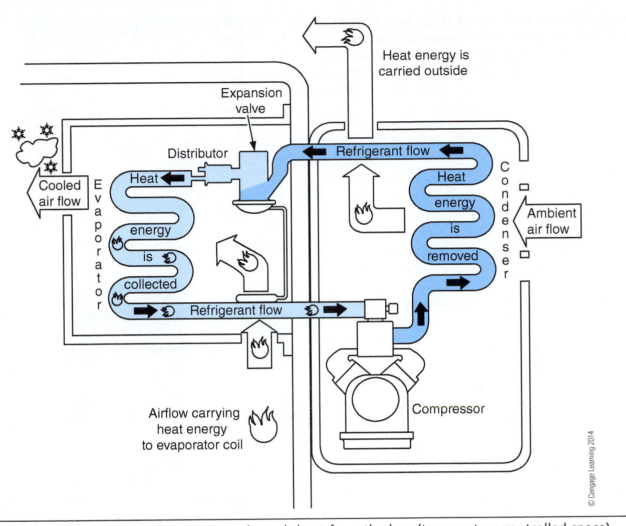

Figure 13-3 A refrigeration unit removes and expels heat from the box (temperature-controlled space).

the engine radiator and condenser section of the unit. The other fan is responsible for circulating the air through the evaporator section, as well as circulating the cold air over, around, and under the cargo. The engine drives these fans using a series of belts and pulleys.

Compressor

The compressor is responsible for pumping the refrigerant throughout the system. It may be thought of as the heart of the circulatory system. The compressor must be capable of pumping high discharge pressure and at the same time drawing a low suction pressure. Compressors range from two cylinders (found on truck units) to six cylinders (found on some trailer units).

Condenser

Just like the condenser in the air-conditioning system, this condenser is responsible for dissipating the heat (carried in the refrigerant) from the controlled

space (cargo area). The liquid refrigerant leaving the condenser then goes through a heat exchanger. The heat exchanger is not required in a basic refrigeration system, but as you will learn, enhances the efficiency of the refrigeration process.

Thermostatic Expansion Valve

The refrigerant leaving the heat exchanger then goes through a thermostatic expansion valve or TXV. The fixed orifice tubes used in air-conditioning systems are not used on larger refrigeration systems. The TXV is usually of the externally equalized type and will be discussed in detail in the next chapter. The TXV lowers the pressure of the liquid refrigerant as it enters the evaporator section.

Evaporator

The liquid refrigerant in the evaporator is boiled into vapor in the evaporator. The evaporator fan circulates the warm air of the cargo space and routes it through the evaporator coil. It is this heat that causes the liquid refrigerant in the evaporator to boil into vapor before reaching the suction side of the compressor. The outlet of the evaporator may be connected to a component known as a chute. The chute is a flexible duct that hangs from the ceiling of the trailer. The chute guides the discharge air, allowing it to circulate out its sides and out over the top of the cargo. It also directs the discharge air to the rear of the trailer so it can be drawn up under the cargo to the inlet side of the evaporator fan.

Microprocessor

The microprocessor is the brain of the entire refrigeration unit. It controls all aspects of the refrigeration and heating process as well as all stop/start features. A wiring harness connects the processor to sensors and solenoids placed on the engine, compressor, and at various points throughout the refrigeration system.

Box Temperature

Box temperature is the actual temperature inside the cargo-carrying area of the truck or trailer (van). Truck-trailer refrigeration equipment has a box temperature thermometer. It may be a dial-type, digital-type, or may be incorporated into the readout of the microprocessor.

Set Point

The **set point** is the optimum temperature for maintaining the quality of a product over the time it will be in transportation. The set point is selected by the operator with a simple thermostat dial or digital microprocessor-style touch pad.

Thermostat

The thermostat operates very much like the one that controls the heating and air-conditioning functions in most homes. You select the temperature and the equipment maintains the desired temperature or set point.

The thermostat controls the unit's mode of operation by sensing the temperature inside the controlled space and comparing it to the set point selected by the driver (operator). Setting the thermostat lower or higher than the desired temperature does not increase the speed it takes to reach set point. The unit heats or cools at the same rate (speed) regardless of the desired set point.

The thermostat may appear as a simple rotary dial type (in older units). It may also be built into the microprocessor control panel **(Figures 13-4** and **13-5).**

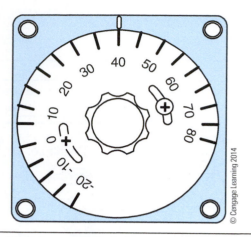

Figure 13-4 A rotary dial thermostat.

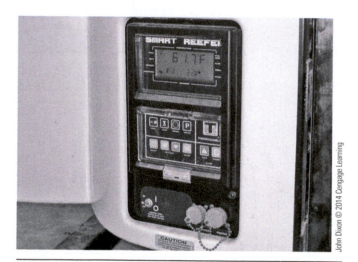

Figure 13-5 A microprocessor-controlled thermostat.

REFRIGERANT

A refrigerant is a chemical substance that has the ability to change states at low temperature and pressure. A refrigerant is used in a refrigeration system because of its low boiling point and its stability (it doesn't lose its physical properties) under high or low operation temperatures.

Refrigerant is a colorless, odorless gas that may be referred to by many different names or numbers. Many people simply refer to refrigerant as **Freon**. This would be like calling all snowmobiles a *Ski-Doo* or all personal watercraft a *Sea-Doo*. However, Freon is actually a trade name of the DuPont Chemical Corporation and does not apply to all refrigerants. It is more correct and more specific to call each refrigerant by its unique name. In modern transport refrigeration units, R-134a and R-404a are the most common refrigerants. Each refrigerant is assigned a unique color (for color-coding refrigerant containers), and each has unique characteristics.

We all know that the boiling point of water is 212°F, but R-404a boils at −51.7°F (−46.5°C) at atmospheric pressure. If you set an open container of refrigerant on a block of ice, the heat from the ice causes the refrigerant to boil like a pot of water on a stove. The refrigerant's low boiling point and its ability to change states under pressure make it an efficient heat carrier.

The refrigerant is moved through the major components of the refrigeration system by the compressor. It changes from a gas to a liquid in the condenser as it gives off heat outside the cargo space. Then it changes from a liquid to a vapor (gas) in the evaporator as it absorbs heat from the cargo space.

For many years R-12, R-502, and R-22 were used extensively in refrigeration units. These refrigerants provided good cooling performance but contain chlorine and are no longer manufactured. Today's tough environmental laws require new, environmentally friendly refrigerants. These refrigerants are chlorine-free and are known as HFCs. Refrigerant R-404a is currently used to achieve low box temperature in the −20°F (−29°C) range, while R-134a is used for applications no lower than 0°F (−17°C).

REEFER VAN CONSTRUCTION

The reefer trailer itself is constructed in much the same way as a standard van trailer, with the exception of insulation in the walls, floors, and ceiling as well as insulated rear doors. This insulation slows the rate of heat transfer, just as it does in the walls and ceilings of a house. The insulation used is usually a blown-in expanding foam type that does not absorb moisture. Insulation that absorbs moisture loses its insulating properties, also known as "R" value. Insulation is extremely important in the floor area because large amounts of heat are radiated up from an absorbent black asphalt surface. The front wall of the trailer is reinforced with a framework to mount and support the weight of the refrigeration unit. A fuel tank (anywhere from 30 to 100 U.S. gallons) is mounted to the crossmembers behind the landing gear. A conduit is used to route fuel lines and fuel pump wiring from the fuel tank through crossmembers and the upper coupler assembly.

Truck-Trailer Flooring

Today, most truck and trailer floors are fabricated from aluminum that forms ductwork under the cargo. On this type of floor, pallets are not usually required for air circulation under the product, but floor channels

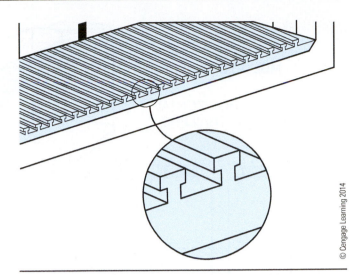

© Cengage Learning 2014

Figure 13-6 Ducted flooring does not require cargo to be loaded on skids to ensure proper air circulation.

must be kept clear of debris that could impede air flow. If a flat floor is used in a refrigerated box, the load must be placed on pallets to allow air to circulate under the load **(Figure 13-6)**.

MULTI-TEMPERATURE REFRIGERATION UNITS

Some refrigeration units have more than one compartment inside the van or truck. They may have up to three compartments, each separated by an insulated barrier wall. **Multi-temp units** are designed for deliveries to small grocery stores, convenience stores, and fast-food restaurants. The deliveries are usually a mixed load: some frozen, some fresh, and some dry cargo. These units look the same on the outside as normal refrigeration units, but each compartment has its own remotely mounted evaporator with lines connecting it to the host unit, which has its own built-in evaporator. Each compartment also has its own thermostat to control the temperature within the compartment. Thus, all compartments operate independently of each other **(Figure 13-7)**.

LOADING FACTORS

Precooling the Product

It is extremely important that products are precooled to their proper holding temperature prior to loading. If the trailer is loaded with a "**hot load**," it may take many hours to pull down to the set point. This time contributes to product damage. Truck-trailer refrigeration equipment is designed to maintain the

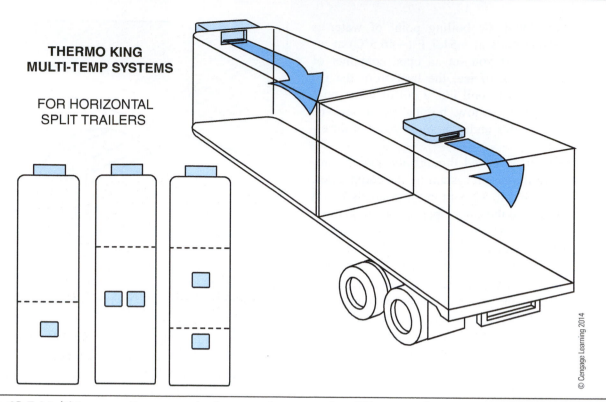

THERMO KING MULTI-TEMP SYSTEMS

FOR HORIZONTAL SPLIT TRAILERS

© Cengage Learning 2014

Figure 13-7 Multi-temp units are used when products require different refrigeration temperatures.

temperature of the product. It doesn't have the refrigeration capacity to cool a load that is warmer than the desired set point.

Precooling of the Controlled Space

Precooling of the controlled space is required before loading to remove the residual heat and humidity absorbed by the walls, floor, ceiling, and insulation. This is achieved by running the unit to the desired set point. When the unit switches into slow-speed operation, the controlled space is considered to be precooled. After the unit is precooled, the unit should be turned off before rear doors are opened for loading to prevent the refrigerated air from escaping and the warm humid air from entering the controlled space (Figure 13-8).

Air Circulation

Cool air is circulated over the top of the product, and because cold air is denser than warm air, it drops between the product to the floor, where it is drawn by the inlet side of the evaporator. There must be an air gap between the product and the floor; this is accomplished with a palletized load or a ducted aluminum floor. As the cold air flows over, through, and under the product, the cold air absorbs heat and returns to the evaporator, where the heat is removed and the cold air

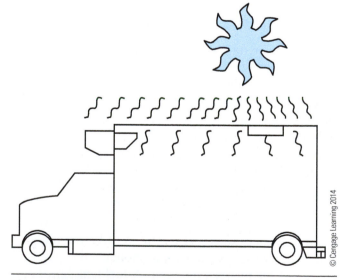

© Cengage Learning 2014

Figure 13-8 Residual heat and humidity are absorbed by the walls, floor, ceiling, and insulation.

is again circulated to absorb more heat. Constant air circulation is required to remove the heat of respiration (heat generated as living products mature or ripen) from fresh fruits and vegetables (Figure 13-9).

Pallet Positioning

Pallets must be positioned so the air can be circulated from the back of the trailer to the front. Any pallet

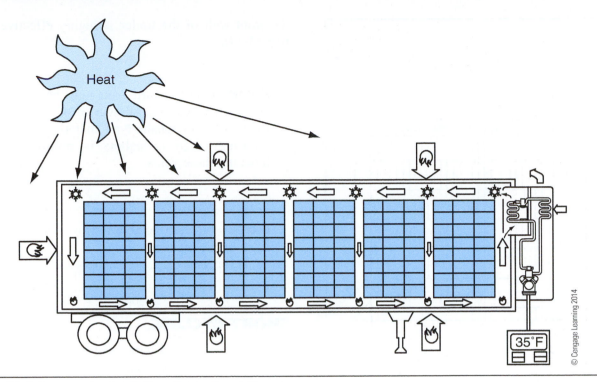

Figure 13-9 Proper air circulation throughout the cargo space.

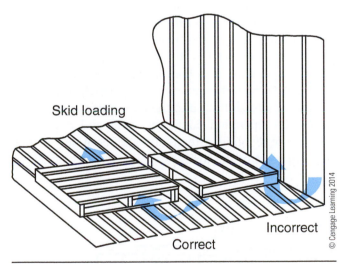

Figure 13-10 Proper skid placement does not impede air circulation underneath the cargo.

turned the wrong way impedes air flow and can seriously damage product (**Figure 13-10**).

LOADING PROCEDURES

Proper Loading

Proper air circulation is necessary to remove heat from delicate cargo. Poor air circulation is a major cause of cargo deterioration. Obstructions anywhere in the air flow path leave valuable cargo unprotected.

To prevent heat from transferring into refrigerated or frozen cargo, an air gap must be maintained between the cargo and the walls, floor, doors, and ceiling.

Side Spacing

Product must also be kept away from the side wall so as not to impede air flow around the sides of the product. Load locks should be used where necessary to prevent the load from shifting. The space between the product and wall should be consistent on both sides because air will take the path of least resistance, causing undesirable temperature variations throughout the load (**Figure 13-11**).

Roof Spacing

Ample air space must be allowed between the top of the load and the ceiling. A minimum air space of 9 inches is recommended. Cool air from the front of the trailer must freely travel across the top of the load to the rear of the refrigerated compartment. If product is stacked too high, it blocks air flow and short cycles the unit (**Figure 13-12**).

Rear Door Spacing

Air space must also be left between the cargo and the rear doors. If the cargo is stacked too close to the rear doors, air is not able to get around to the back side

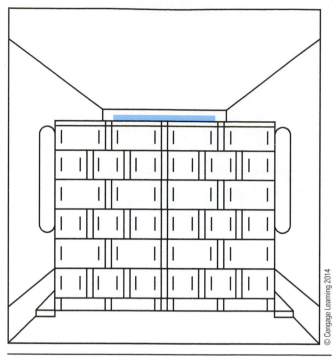

Figure 13-11 Cargo should not block air flow down the sides of the cargo space.

of the cargo and drop down to the floor, where it should be pulled under the cargo by the inlet side of the evaporator fan **(Figure 13-13).**

Front Bulkhead Spacing

A device must also be used to prevent loading of cargo too close to the unit's return air inlets. A professionally installed bulkhead may be used, but something as simple as a pallet placed upright against

the front wall of the trailer is highly effective **(Figure 13-14).**

> **Note:** Inlet or return vents must be kept free of debris that can be drawn in or restrict air flow. The cellophane-type material used for securing cargo to pallets is notorious for blocking these vents.

> **Note:** Air must be able to flow freely on all sides of the product.

SHORT CYCLING

Short cycling is a term used when a refrigeration unit alternates between heat and cool modes (or turns on or off) more often than normal.

When cargo blocks the unit's discharge air, it does not absorb the heat from the cargo space. The cold air gets routed back to the air inlet, tricking the thermostat into thinking that it has reached its set point. The thermostat then switches its mode of operation to heat mode. This warm air is then short cycled as well and gets routed back to the air inlet, which tricks the thermostat into shifting into the cool mode again. This short cycle operation continues, over and over. The unit is not cooling effectively and cool air is not reaching most of the load.

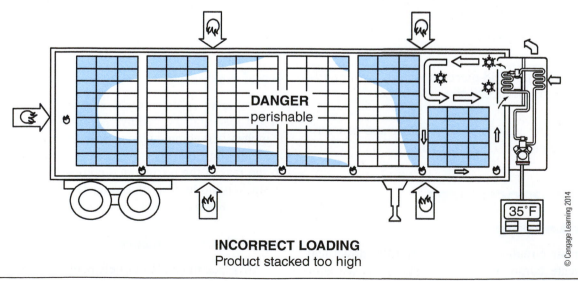

INCORRECT LOADING
Product stacked too high

Figure 13-12 Cargo should not be stacked so high that it blocks air flow from reaching the back of the cargo space.

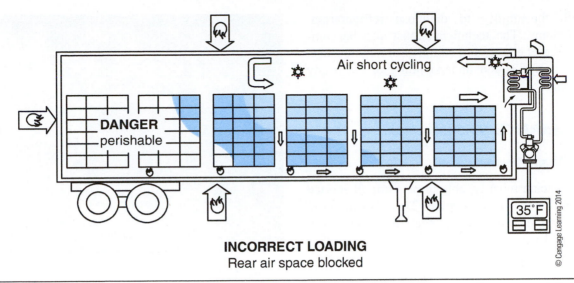

Figure 13-13 Cargo should not be positioned too close to the rear doors, blocking air flow around the backside of the cargo and back up underneath.

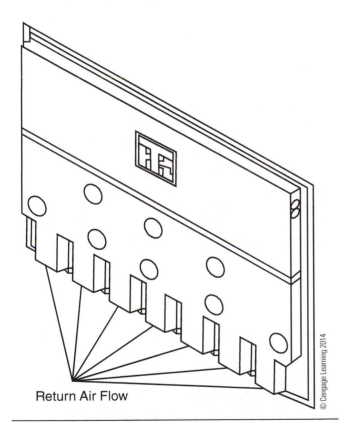

Figure 13-14 Cargo must be kept from blocking the inlet vents or air circulation will be drastically affected.

In **continuous run mode**, the unit maintains box temperature by cycling between heat and cool mode. When auto stop/start mode is selected, the unit shuts off when the box temperature reaches the desired set point, as long as a few other parameters are met. The unit must run long enough to warm the engine temperature sufficiently. It must also run long enough to charge the battery, ensuring a restart. The unit remains shut off until the box temperature rises to a predetermined restart temperature or until a maximum off time limit initiates a restart. The unit also starts if the engine block temperature falls below a predetermined temperature. Once the unit restarts, it again runs until all parameters are again met.

Manufacturers claim that the auto stop/start feature can save fuel costs by up to 80%. The actual savings can be affected by many variables, including type of product maintained, ambient temperature, frequency of door openings, and thickness and quality of box insulation.

WARNING *With the unit on-off switch in the ON position, the unit may start at any time without prior warning. Only qualified technicians should check the belts, fans, or other components that can cause severe personal injury or death.*

AUTO STOP/START

Most truck-trailer refrigeration equipment today uses an **auto stop/start** feature to save fuel. Constant run mode is still used for products that require constant air circulation.

MAINTENANCE PROCEDURES

Highly skilled technicians must maintain transport refrigeration equipment. The skill sets of the technician include specialized soldering and brazing techniques, along with the ability to diagnose refrigeration-related problems. The technician must also be competent with diesel engine and compressor maintenance, as well as have a sound understanding of electricity and electronics.

PERFORMANCE TASKS

Engine Maintenance

The technician must be able to perform all service work related to the diesel engine. This also involves routine maintenance procedures such as changing oil, oil filters, fuel filters, belt tension, as well as starting and charging systems.

Refrigeration Maintenance

Refrigeration technicians are responsible for all aspects of the refrigeration section of the unit. This includes performance testing the system for operating efficiency. The technician also performs routine maintenance, such as changing the compressor oil, changing the filter drier, and checking and adjusting refrigerant levels. Technicians must also be able to troubleshoot all the components of the refrigeration system for proper operation or problems. The technician must be able to test for refrigerant leaks and repair these problems.

Electrical System Maintenance

Electrical systems require regular maintenance to ensure proper operation of the refrigeration unit. Units

Figure 13-15 Fruit.

using auto stop/start must have starters, alternators, and batteries kept in optimum shape to ensure that the unit can restart after it has shut off. Auto stop/start puts much more wear and tear on starters and charging systems of units using this feature. The technician must be able to troubleshoot the electrical system, following schematic diagrams to ensure proper operation of the unit.

Technicians in the transport refrigeration field are usually working under the pressure of time constraints because the cargo may spoil unless the unit can be repaired quickly and effectively. Competence in all aspects of reefer repair is a must under these conditions, with the value of some cargos in the $100,000 range **(Figure 13-15).**

Summary

- Today's truck-trailer refrigeration equipment has come a long way since 1938.

- Microprocessors now control all aspects of the refrigeration and heating cycle.

- Modern advances in technology allow distributors to market fresh fruits and vegetables any time of the year.

- Loads must be precooled to the temperature to be maintained by the refrigeration unit.

- The box must also be precooled to remove residual heat and humidity.

- Correct loading is essential to product quality.

- Discharge air must be able to circulate freely around all sides of the product.

- Rising fuel costs have contributed to the technology needed for auto stop/start operation.

- The auto stop/start feature is a huge financial savings for the companies using it, as well as a way of conserving our natural resources.

Review Questions

1. The temperature inside the temperature-controlled environment is:

   A. Ambient temperature

   B. Box temperature

   C. Set point

2. *True or False:* Truck-trailer refrigeration equipment operates by bringing outside air into the box, cooling it, and then exhausting the warm, stale air to the outside.

   A. True

   B. False

3. Heat generated by a product as it ripens is:

   A. Heat of transportation

   B. Heat of perspiration

   C. Heat of dissipation

   D. Heat of respiration

4. *True or False:* Truck-trailer refrigeration equipment is designed to quickly reduce product temperature to the desired set point.

   A. True

   B. False

5. *True or False:* Setting the thermostat lower than the desired temperature increases unit output and shortens the time required to precool the box.

   A. True

   B. False

6. Important factors when loading temperature-sensitive product include:

   A. The product should be precooled to the correct temperature before loading.

   B. The box should be precooled to the correct temperature before loading.

   C. The unit should be turned off when loading.

   D. All of the above

7. Which of the following contributes to poor air circulation?

   A. Product stacked directly on flat floors

   B. Pallets without adequate openings or that are improperly oriented

   C. Plastic wrap around pallet openings

   D. All of the above

8. A refrigeration unit alternates between the heat and cool modes more often than normal. This is called:

   A. Auto stop/start

   B. Auto cycle

   C. Modulation

   D. Short cycling

9. Units are run using the auto stop/start feature when:

   A. Products require constant air circulation

   B. Optimum fuel savings are required

   C. Products don't require constant air circulation

   D. Both A and B are correct.

   E. Both B and C are correct.

10. The heat absorbed by the refrigerated compartment while it is parked empty in the sun is called:

    A. Sensible heat

    B. Latent heat

    C. Residual heat

    D. Superheat

14 Refrigeration Components

Learning Objectives

Upon completion and review of this chapter, the student should be able to:

- Describe the purpose of the compressor, condenser metering device, and evaporator.
- Explain the construction of the compressor, condenser metering device, and evaporator.
- Illustrate the operation of service valves and Schrader valves.
- Describe the purpose and construction of a vibrasorber.
- Demonstrate the operation of a thermostatic expansion valve.
- Explain the superheat setting of the TXV.
- Describe the mounting location of a sensing bulb.
- Determine the superheat setting of the TXV.
- Explain the purpose of the distributor.
- Describe the purpose of the receiver tank.
- Compare the drier materials and explain the purpose of the filter drier.
- Describe the purpose and operation of a heat exchanger.
- Explain the purpose and operation of the accumulator.
- Explain the purpose of pressure regulating devices.
- Describe the purpose and operation of the different types of refrigerant safety valves.

Key Terms

accumulator

compressor

condenser

desiccant

distributor

evaporator

filter drier

heat exchanger

moisture indicator

pressure regulating valve

receiver tank

safety valve

service valve

sight glass

suction pressure regulator

thermostatic expansion valve

vibrasorber

INTRODUCTION

Truck-trailer refrigeration equipment uses the same four basic components that we learned about in the earlier chapters of the air-conditioning section, but on a much larger scale. The four major components are the:

1. Compressor
2. Condenser
3. Metering device
4. Evaporator

No truck-trailer refrigeration equipment would work without a fifth component, the refrigerant. The refrigerant is pumped throughout the system like blood through the human body. Refrigerant is the medium that carries the heat from the box and transports it to the outside, releasing the heat through the condenser.

SYSTEM OVERVIEW

Truck-trailer refrigeration equipment consists of many different components, all working together to do their part to facilitate the refrigeration process. This chapter discusses the function of various components and how they operate in conjunction with each other to achieve an efficient, high-capacity refrigeration system. The components are listed in the order in which they would appear as the refrigerant flows when the unit is in the cooling mode.

THE COMPRESSOR

The **compressor** is the heart of all refrigeration equipment. It is responsible for pumping the refrigerant through the system so that the heat can be moved from the controlled space to the ambient air. The compressor raises the pressure of the refrigerant vapor. When the refrigerant is compressed, the temperature of the refrigerant rises, due to the fact that the heat energy of a large mass of refrigerant is concentrated together. There is no heat being applied by the compressor **(Figure 14-1)**.

The compressor must be able to create high pressure for heat transfer to take place on very hot days. It is important that the refrigerant be hotter than the temperature of the surrounding air (ambient air); otherwise, no heat transfer would take place.

The compressor must also be able to draw a very low suction pressure in order to maintain a very low box temperature. These two pressures (high discharge and low suction) help control the boiling of the

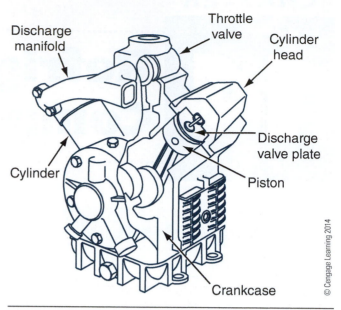

Figure 14-1 A four-cylinder compressor responsible for raising the pressure of the refrigerant vapor.

refrigerant, causing refrigerant to boil in the evaporator and condense in the condenser.

Compressor Operation

Refrigerant vapor passes through the suction throttling valve (if equipped) and enters the compressor crankcase. When the piston moves down in the cylinder, refrigerant vapor is drawn in through the suction valves. These reed-type valves are actuated by differential pressure. Refrigerant is drawn into the cylinder until the piston is at the bottom of its stroke. Once the piston begins to move up, the pressure in the cylinder will increase and the suction valve will close **(Figure 14-2)**.

During the compression stroke, refrigerant vapor is compressed at a ratio of 50 to 1. The highly compressed refrigerant vapor then passes through the discharge valve plate and enters the discharge manifold. The discharge valve plate acts like a check valve that prohibits the highly compressed vapor from moving back into the cylinder area as the piston moves down for another charge of refrigerant **(Figure 14-3)**.

SERVICE VALVES

Service valves are located at the compressor suction and discharge service ports and may also be located at various points throughout the refrigeration system for diagnostic or servicing procedures. The service valve allows the technician to install a manifold gauge set into the system for testing purposes, evacuation, recharging, or purging of the refrigeration system.

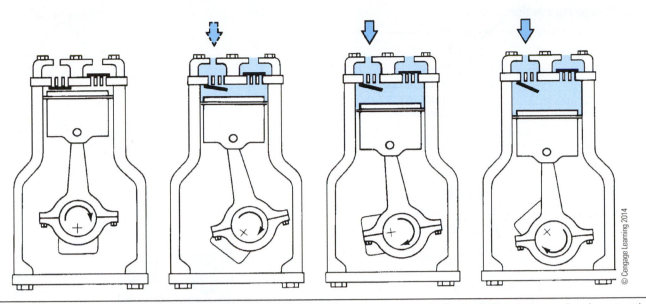

Figure 14-2 As the piston moves down in the cylinder, refrigerant vapor is forced through reed valves to the chamber above the piston.

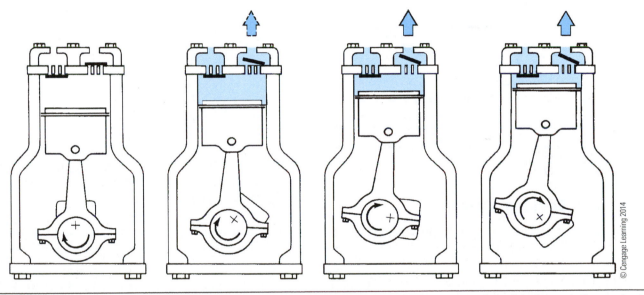

Figure 14-3 As the piston moves up the cylinder, refrigerant vapor is compressed and is forced through the discharge valve plate.

This valve not only allows access to the system for testing refrigerant pressures but can also be used to isolate the compressor from the rest of the refrigeration system. The service valve is normally left in the back-seated position, which blocks off the service port from the system. After the manifold gauge service line has been connected, the valve can be placed in the mid-seated position. This allows the refrigerant to enter the service port, allowing the gauges to read. To mid-seat the valve for gauge readings, the stem need only be turned in one-half turn (cracked) or the discharge pulses of the refrigerant leaving the compressor will cause the gauge needle to vibrate, usually knocking the gauge out of calibration **(Figure 14-4)**.

CAUTION *Care must be taken to **NEVER** front-seated the discharge service valve while the compressor is operating. Even though the high pressure cutout switch might be positioned below the valve, it would not operate fast enough to prevent major damage to the compressor and prevent possible personal injury.*

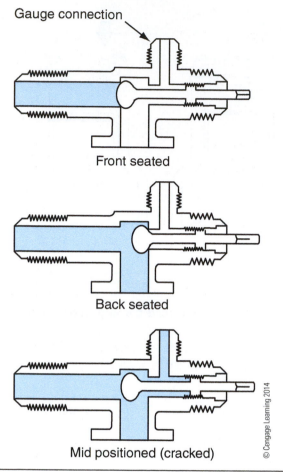

Figure 14-4 A typical service valve, shown in the three different positions that a technician may be required to use.

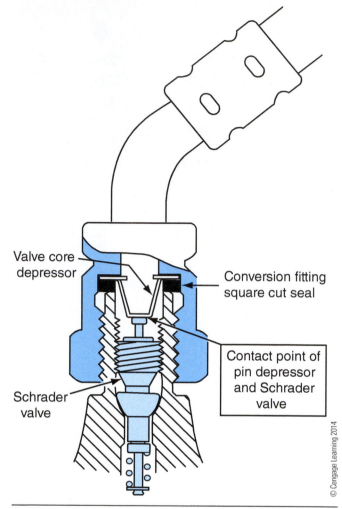

Figure 14-5 The Schrader valve is used by some manufacturers to allow the technician access to a refrigeration system.

Schrader Service Valves

This type of service valve only provides access to the system, but does not allow the service technician to isolate the compressor from the system. The service hose must be fitted with a depressor in order to push the end of the stem inward, opening the valve. This type of valve is virtually identical to the type of valve used in pneumatic tires. This is not a common service valve in truck-trailer refrigeration but is occasionally used. Care must be exercised with this type of valve because it may not always be seated properly when the hose depressor is removed from the fitting, in turn allowing the charge of refrigerant to leak from the system (**Figure 14-5**).

Vibrasorbers

Vibrasorbers are flexible refrigerant suction and discharge lines positioned at the compressor. They are designed to prevent engine and compressor vibration from being transmitted to the copper piping of the refrigeration system. There are two types of vibrasorbers. A typical discharge vibrasorber is made up of a bellows-shaped stainless steel center and a covering of braided stainless steel wire. A typical suction vibrasorber is made up of reinforced fabric-covered hoses, often using replaceable mechanical fittings (**Figures 14-6** and **14-7**).

When replacing discharge vibrasorbers that may be either soldered or brazed into position, great care must be taken not to overheat the fitting. Overheating can loosen either the braided covering or the metal bellows, causing the new vibrasorber to leak. When soldering or brazing a new vibrasorber into position, a heat sink should be used to draw heat away from the end caps, preventing overheating.

THE CONDENSER

The **condenser** is located outside the controlled space. Its function is to release the heat from the controlled space to the outside air.

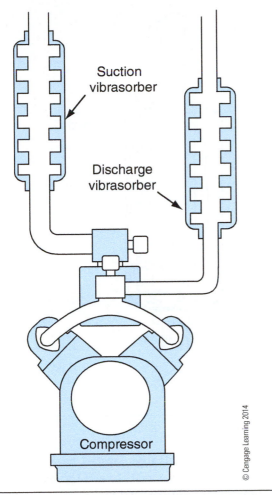

Figure 14-6 The vibrasorbers are placed close to the compressor's suction and discharge ports.

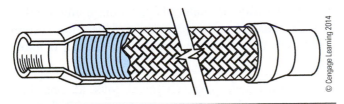

Figure 14-7 A typical discharge vibrasorber is made up of a bellows-shaped stainless steel center.

The condenser relies on two key principles of heat transfer:

- Heat energy always moves from hotter to colder.
- Heat energy transfers quickly between objects having a large temperature difference.

The condenser consists of copper tubing running through sharp aluminum fins. The copper tubing contains the refrigerant, while the aluminum fins increase the surface area of the condenser. Because the fins are attached firmly to the copper tubing,

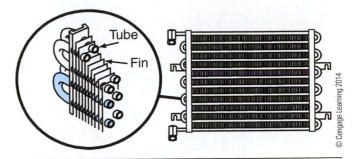

Figure 14-8 The condenser consists of copper tubing running through sharp aluminum fins.

conduction transfers the heat of the refrigerant to the fins for maximum exposure of heat to the cool ambient air.

As the high-pressure, superheated refrigerant passes through the condenser, heat is removed from the refrigerant and is transferred to the cooler ambient air. This cooling of the refrigerant causes it to change states and condense into a high-pressure liquid by the time it reaches the end of the condenser **(Figure 14-8).**

THE RECEIVER TANK

The **receiver tank** acts as a storage tank for liquid refrigerant leaving the condenser section. The tank's assembly usually contains one or two sight glasses to provide an indication of the refrigerant level. The **sight glass** is a clear glass window with a small white ball. The ball, known as a pith ball, is located within the sight glass. The ball floats within the window to indicate the refrigerant level. Only when certain testing procedures are followed can the sight glass method of refrigerant level testing be considered reasonably accurate.

Most receiver tanks use service valves to add or remove refrigerant or to trap refrigerant in the receiver so that service work may be performed on other parts of the refrigeration system. If the receiver tank outlet valve is front seated while the unit is running, all refrigerant will be trapped between the compressor discharge and the receiver tank. This allows the service technician to service any components downstream of the receiver tank outlet valve to the compressor suction service valve. This, of course, assumes that the receiver tank outlet valve does not have an internal leak.

The receiver tank may also be equipped with a bypass check valve and a hot gas line. During the heating process, the higher pressure gas leaving the compressor is pumped into the receiver through the bypass check valve to push out the liquid refrigerant trapped in the

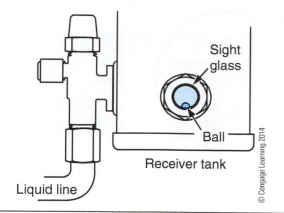

Figure 14-9 The sight glass contains a ball to indicate refrigerant level under certain conditions.

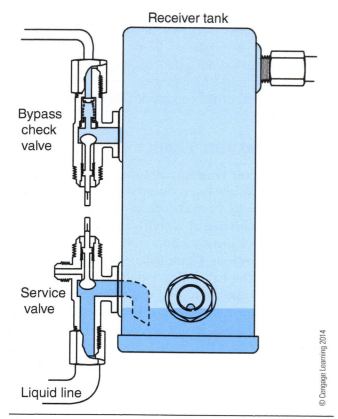

Figure 14-10 A typical receiver tank, including service valve and bypass check valve.

receiver tank. This excess refrigerant is required for the heating process **(Figures 14-9 and 14-10).**

FILTER DRIER

The **filter drier** is an extremely important component in the refrigeration system and serves two main functions. It is designed to work as a filter and a drier. The filter drier is located in the liquid line between the receiver outlet and the thermostatic expansion valve.

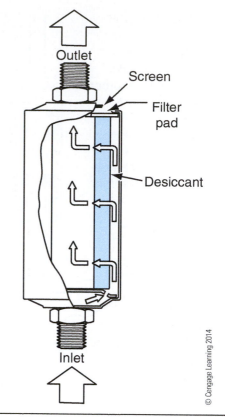

Figure 14-11 A filter drier.

A filter drier may also be located between the compressor and condenser or between the evaporator and the compressor **(Figure 14-11).**

Drier Materials

Three different materials are commonly used to absorb moisture and acid in the refrigeration system. These chemicals can be formed into a bead or block form.

1. Silica gel: This chemical was at one time commonly used as a drying agent. It does not have the drying capacity of some of the new chemicals and as a result is not used as much today.

2. Activated alumina: This material has a greater ability to absorb acid than the other **desiccant** types but lacks the moisture-absorbing ability. The acids formed in the refrigeration system when moisture enters, or when a hermetic compressor fails, are hydrochloric and hydrofluoric. These acids are extremely corrosive and must be kept at a safe level. Hermetic compressors are electrically driven compressors that have the electric motor sealed with the compressor. This is the style of compressor used in your home refrigerator or freezer. Some newer

truck-trailer refrigeration units may use a hermetic or semi-hermetic compressor and use the diesel engine to generate their power supply. They may also use them as standby systems when the vehicle is parked with a product onboard and can be plugged into a power source.

3. Molecular sieve: This chemical has the greatest moisture absorbing characteristics of the three but lacks the acid retention of activated alumina.

Drier manufacturers must take into consideration the end use of the drier and use these chemicals accordingly.

Liquid Line Installation

The liquid line drier works during normal unit operation. This is when the drier is located between the receiver tank outlet and the thermostatic expansion valve. In this position, the drier removes moisture from the refrigerant to aid in preventing the thermostatic expansion valve from freezing up or corrosion from developing.

Drier design and service interval recommendations have changed over the years. Early R-12 units used a fabric/neoprene suction line vibrasorber that allowed moisture to infiltrate the system. Large-capacity driers with 12-month service intervals were used with these systems. With the new refrigerants, improved vibrasorber, and high performance driers, service intervals have increased to 24 months, or whenever major refrigeration service is performed.

When using R-134a, the level of moisture in the system must be maintained below 6 to 10 ppm (parts per million), virtually nothing. If this level is not maintained, the heat and pressure of the refrigeration system cause the moisture to form both hydrochloric and hydrofluoric acid. Both of these are extremely detrimental to the refrigeration system.

To get a better idea of the quantity of moisture permissible in a refrigeration system, we will look at the following example. If one drop of water was added to 3 pounds of R-134a, it would be equal to 40 ppm, or more than twice the maximum allowable limit.

A drier's ability to retain moisture changes with fluctuations in temperature. The colder it gets, the more moisture a drier retains. When a drier is manufactured, the fittings are sealed to prevent moisture from entering. These sealing caps should only be removed when the drier is being installed into the system, just before evacuation. If these caps are removed on a humid day, the drier absorbs its full quantity of moisture in approximately 15 minutes. The term *hygroscopic* may describe a drier as well as compressor oil. This means that it will readily absorb moisture.

Filtration

The drier is also designed to act as a filter to remove dirt and sludge from compressor oil breakdown, as well as flux from soldering repairs and wear metal particles from the compressor. The filter pads normally remove solids of 5 microns and larger, helping to reduce compressor wear and plugged thermostatic expansion valve screens.

Vapor Line Installation or Low Side Installation

When the drier is located between the evaporator outlet and the compressor suction side, it may be referred to as a scrubber. Scrubbers are designed to remove all particles 5 microns or larger from the refrigerant vapor that is returning to the compressor. This is particularly important if a compressor has failed at one time or another. These driers may be installed and changed at regular intervals, or may be installed and run for a specified amount of time to scrub the refrigeration system after a catastrophic compressor failure.

MOISTURE INDICATORS

Moisture can collect in a refrigeration system to the point that the drier can no longer absorb any more of it. If the unit is equipped with a **moisture indicator**, the technician has a visual indication of moisture accumulation before problems develop. To provide this indication, moisture indicators usually contain a chemical disc designed to change color as moisture passes by the window. A moisture reference color chart is on the face of the indicator. For example, with one manufacturer, the color blue may represent a dry system and the color pink may represent a wet system. If the indicator color represents a wet system, the refrigeration system must be serviced **(Figure 14-12).**

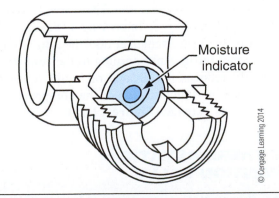

© Cengage Learning 2014

Figure 14-12 Moisture indicator center pellet changes color to indicate moisture in the refrigeration system.

Some manufacturers don't use a dry-eye (moisture indicator), relying instead on changing the drier at regular intervals in their routine maintenance programs. They hope this maintenance provides a better opportunity to control moisture levels than strictly relying on the dry-eye. The dry-eye's ability to indicate moisture depends on the liquid line temperature because refrigerant can absorb more moisture when it is cold, thereby preventing any problems until the refrigerant passes the thermostatic expansion valve, where the refrigerant drops in both temperature and pressure. The moisture then immediately changes to ice, causing a restriction in the thermostatic expansion valve.

HEAT EXCHANGER

The **heat exchanger** is located in the liquid line between the receiver and the TXV. Also, the outlet of the evaporator and the suction line of the compressor are routed through it.

This component performs two important functions that enable the refrigeration system to operate more efficiently. It subcools the liquid refrigerant before it reaches the TXV, and it also evaporates any liquid refrigeration that might get to the outlet of the evaporator.

Heat Exchanger Operation

As the cold vapor passes over the hot liquid line, the hot liquid is further subcooled before it enters the TXV. This subcooling of the refrigerant is important because when the pressure of the refrigerant is reduced by the TXV, it starts to boil immediately. This boiling is called *flash gas* and can boil as much as 30% of the liquid refrigerant, leaving only 70% of the refrigerant to do its job in the evaporator. The heat that causes the refrigerant to boil is the heat of the liquid refrigerant before it enters the TXV. Therefore, subcooling the liquid before it reaches the TXV greatly reduces the amount of flash gas, improving efficiency. The other part of the heat exchanger adds heat to any low-pressure liquid refrigerant returning to the compressor. This greatly reduces the chance of liquid refrigerant getting back to the suction side of the compressor and causing serious damage **(Figure 14-13)**.

THE THERMOSTATIC EXPANSION VALVE

The **thermostatic expansion valve**, or TXV, is the division between the high pressure and the low pressure sides of the refrigeration system. The TXV

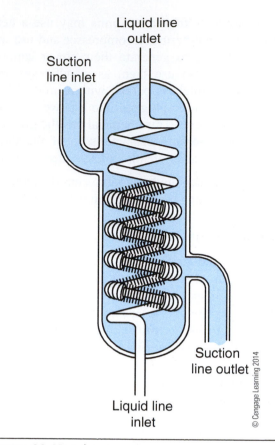

Figure 14-13 A heat exchanger.

modulates the flow of liquid refrigerant to the evaporator. The TXV senses the temperature and/or pressure of the evaporator outlet and adjusts the inlet to match it. The restriction helps maintain high pressure in the condenser so that heat transfer will take place. This gives the superheated refrigerant time to boil and condense.

The low pressure developed at the inlet of the evaporator lowers the boiling point and temperature of the refrigerant. The boiling point of the refrigerant in the evaporator can be reduced so much that refrigerant in the evaporator boils at temperatures below −20°F.

The thermostatic expansion valve continuously adjusts the inlet flow of refrigerant to the changing evaporator outlet temperature and pressure. If too much refrigerant is allowed into the evaporator, the evaporator won't be able to absorb enough heat to make it all boil before it leaves the evaporator. If the amount of refrigerant is excessive, the boiling point may never be reached, resulting in lost efficiency and, more likely, compressor damage.

If too little liquid refrigerant is metered into the evaporator, the refrigerant all boils off long before it reaches the end of the evaporator. Because most of the heat transfer takes place while the refrigerant is boiling, any point in the evaporator after the refrigerant has boiled off is lost efficiency.

The expansion valve is designed to sense the evaporator temperature and pressure and, in turn, control the refrigerant flow to maintain a fully active evaporator under all load conditions.

Operation

The operation of the valves is determined by their fundamental pressures:

1. Sensing bulb pressure applied to one side of the diaphragm tries to open the valve against spring pressure. The bulb is usually charged with the same refrigerant used by the system in which it is installed.
2. Evaporator outlet or compressor suction pressure applied to the opposite side of the diaphragm helps to make the valve responsive to compressor suction pressure.
3. Spring pressure, which is applied to the needle assembly and diaphragm on the evaporator side, constantly tries to close the valve.

If the sensing bulb pressure were removed, the expansion valve would be closed by the spring pressure and equalizer line pressure. It is very important that the sensing bulb accurately senses evaporator outlet temperature. The sensing bulb is attached to the tail pipe (outlet side) of the evaporator so that it monitors the temperature of the vapor leaving the evaporator. It is wrapped with a sticky cork tape to insulate it from the surrounding air, ensuring that the bulb makes an accurate reading.

As the temperature of the vapor in the evaporator rises, the sensing element temperature also rises, increasing the pressure of the liquid or vapor in the sensing bulb. This pressure is transmitted through the capillary tube, where it is applied to the diaphragm, which in turn opens the needle valve, allowing more refrigerant to enter the evaporator. As all this occurs, evaporator pressure rises, in turn causing the valve to close slightly to maintain equilibrium.

If at this time the valve does not feed enough refrigerant, the evaporator pressure falls. The sensing bulb temperature then increases due to the warmer vapor leaving the evaporator and the valve again opens further, allowing more refrigerant to enter the evaporator until the three pressures are again in balance.

As the load on the evaporator increases (trailer door opened), the liquid refrigerant evaporates at a faster rate, thereby increasing the evaporator pressure. The increase in pressure causes an increase in temperature

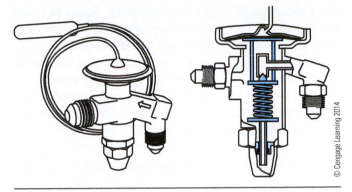

Figure 14-14 A thermostatic expansion valve.

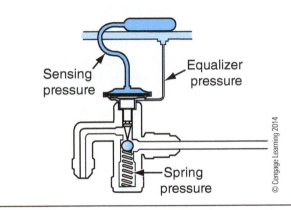

Figure 14-15 An externally equalized thermostatic expansion valve uses an equalizing line.

in the evaporator, so the increasing sensing bulb pressure on the top side of the diaphragm is equal to the pressure of the evaporator on the lower side of the diaphragm. The two pressures tend to cancel out each other as the valve adjusts to changes in load.

The pressure that is applied to the lower side of the diaphragm can be taken from the inlet side of the evaporator (internally equalized) or, on larger evaporators, it can be taken from the outlet of the evaporator (externally equalized). An externally equalized TXV uses an equalizer line (**Figures 14-14** and **14-15**).

The Equalizer Line

The equalizer line carries evaporator outlet pressure to the expansion valve diaphragm. In the power head, equalizer pressure works with spring pressure to close the expansion valve and reduce the flow of refrigerant.

Valve Superheat

For optimum operation of the evaporator, it would be desirable to have the last bit of liquid refrigerant evaporate (or change state) just as it leaves the evaporator. Due to sudden changes of demand or load on the evaporator, it might be possible for liquid to return to the compressor, thereby causing possible compressor damage. To prevent this, the TXV is designed to allow all of the refrigerant to evaporate far enough back in the evaporator so that the refrigerant vapor can absorb enough heat to be 8 to 12 degrees superheated by the time it reaches the evaporator outlet. This helps prevent liquid refrigerant from entering the compressor.

Overview of Determining Superheat

To determine the superheat setting of a valve, it is necessary to measure the evaporator outlet temperature and the evaporator pressure, preferably at or near the outlet of the evaporator. With the aid of a pressure-temperature chart, look up the pressure of the refrigerant and then determine the temperature. (Pressure-temperature charts can be found in **Chapter 5** of this text as **Figures 5-12** and **5-13**.) Take the temperature from the chart and subtract it from the temperature measured at the evaporator outlet. Here again are the four easy steps:

- Step 1. Determine the suction pressure with an accurate gauge at the compressor suction service valve.
- Step 2. Using the refrigeration pressure-temperature charts, determine the saturation temperature at the observed suction pressure.
- Step 3. Measure the temperature of the suction gas at the evaporator outlet.
- Step 4. Subtract the saturation temperature read from the chart in Step 2 from the temperature measured in Step 3. The difference between the two is the superheat of the suction gas returning to the compressor.

Most manufacturers recommend a superheat setting of between 8°F and 12°F to provide a balance of protection and efficiency.

Example

An operating refrigeration unit is tested and the technician determines that the unit is operating with R-134a in the system. The suction pressure is recorded at 31 psig and the temperature recorded at the evaporator outlet is 45°F. What is the superheat setting of the TXV?

Step 1. Suction pressure = 31 psig

Step 2. Converted suction pressure to temperature = 36°F

Step 3. Suction temperature = 45°F

Step 4. 45°F − 36°F = 9°F

The superheat setting of the valve is 9°F.

Sensing Element Charges

Normally, the bulb or element is a liquid/vapor charge of the same refrigerant as the system it is to be used on. This can sometimes be a problem because as the temperature of the sensing element rises, a corresponding opening of the valve takes place. This might cause an overloading or liquid slugging of the compressor under some conditions.

To prevent this, some manufacturers use a TXV with a vapor-charged or cross-charge sensing element, which controls the maximum opening point of the TXV. As an example, if evaporator pressure rises above a safe limit (35 to 50 psi), the TXV closes until the compressor is able to lower evaporator pressure, at which time the TXV returns to a normal operating position. Carrier Transicold refers to this built-in safety factor as "MOP," which means maximum operating pressure. This feature eliminates the need for a suction throttling valve, which will be discussed later in this chapter.

Sensing Bulb Location

The mounting of the TXV sensing bulb is of particular importance because this could cause the valve to incorrectly control the flow of refrigerant to the evaporator. First, it must have a good mechanical contact with the evaporator outlet and be insulated so it will not be affected by the surrounding air temperature. Second, it must be located on the line in a position such that it can monitor actual vapor or line temperature. If it is mounted in the 6 o'clock position, the oil in the line might provide insulation between the sensing element and the actual vapor temperature. If it is mounted in the 12 o'clock position, the sensing bulb charge would be in direct contact with the line, again possibly causing an incorrect sensing of vapor temperature. Follow, where possible, the manufacturer's specifications with respect to location of the sensing bulb.

DISTRIBUTOR TUBE

To be sure each pass in the evaporator receives an equal amount of refrigerant, a **distributor** tube and header system are connected between the TXV outlet and the evaporator inlet. As refrigerant flows from the expansion valve, the distributor divides the flow into several routes through the evaporator coil for greater efficiency. The distributor is equipped with a passage so that during the heating and defrost cycle, hot gas is pumped into the evaporator circuit, bypassing the TXV.

EVAPORATOR

The **evaporator** receives the boiling refrigerant as it leaves the distributor. As the refrigerant boils, it absorbs heat from the controlled space through the external fins, which cool the air (or absorb the heat from the air) as it passes through them. The refrigerant boils because the pressure of the refrigerant is significantly lowered by the TXV. If the air contains any moisture, it soon condenses on the fins, freezing and restricting the air flow, thus reducing the evaporator coil efficiency. This necessitates some method of defrosting the coil, which will be discussed later.

Evaporator Construction

Most evaporators are constructed of copper tubing that has been swedged internally to aluminum fins. A tight metal-to-metal contact must be made for maximum heat transfer. The tubing configuration and the number of tubing passes usually determine the BTU output of the evaporator coil **(Figure 14-16)**.

ACCUMULATOR

The function of the **accumulator** is to separate liquid refrigerant from vaporous refrigerant before it enters the compressor.

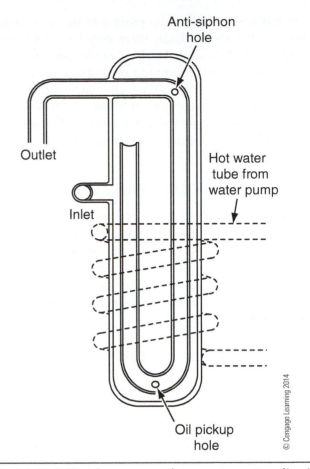

Figure 14-17 The accumulator separates liquid refrigerant from vaporous refrigerant.

When a refrigeration system is operating intermittently or when it is used as a heat pump, it is possible for large quantities of liquid refrigerant to pass through the suction line and enter the compressor. This can cause a number of problems, including broken pistons, damaged connecting rods, broken valves, blown head gaskets, and damaged compressor bearings. The accumulator normally has the capacity to hold the entire refrigerant charge to prevent the above from happening **(Figure 14-17)**.

Operation

When refrigerant vapor and liquid enter the accumulator, the liquid refrigerant falls to the bottom and the vapor is returned through the ''U'' shaped tube to the compressor. As the vapor passes through the ''U'' tube, it picks up a small amount of liquid refrigerant and oil through the metering hole in the bottom of the tube, returning it to the compressor for cooling and lubrication. To prevent the compressor from drawing too much liquid refrigerant and oil through the metering hole, an anti-siphon hole is placed in the top section of the ''U'' tube.

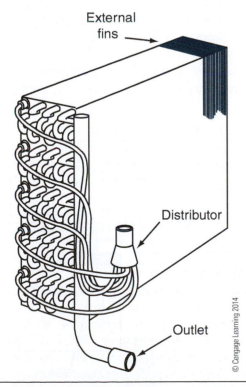

Figure 14-16 Refrigerant in the evaporator absorbs heat from the controlled space.

To aid in the evaporation process of the refrigerant trapped in the accumulator, there may be a device to heat the shell of the unit. The heat can be supplied by a coil of tubing wrapped around the outside of the accumulator with engine coolant circulated through it, very much like a heater core. It could also be in close proximity to the engine muffler or use an electrical strip heater that is thermostatically controlled.

PRESSURE REGULATING DEVICES

The pressure regulating valve may serve two purposes in a refrigeration system. It may be used to regulate evaporator pressure to prevent the evaporator from freezing. Or it may be used to regulate the compressor suction or crankcase pressure, to prevent overloading the engine during the heat/defrost cycle or upon startup (Figure 14-18).

Evaporator Pressure Regulator

Evaporator pressure regulators are primarily designed to control evaporator pressure regardless of compressor suction pressure. The pressure setting of the valve is that which is equal to $30°F-32°F$ (pressure depends upon type of refrigerant) inside the evaporator coil. With normal mechanical heat loss, you could expect to find the evaporator fin temperature to be between $40°F$ and $50°F$, well above freezing. This regulator, due to its ability to totally stop refrigerant flow, requires an oil bypass line from the base of the evaporator to the compressor suction. This prevents a total loss of compressor oil under some operation conditions. This type of valve is not used in very many applications because many cargos require the evaporator to reach very low pressures in order to obtain low box temperatures.

Suction Pressure Regulator

Suction pressure regulators are designed to limit the crankcase suction pressure during the heat/defrost cycle or on startup. If the compressor suction pressure is too high, the drive engine will be overloaded, increasing fuel consumption and engine wear (Figure 14-19).

Operation

During startup, when evaporator and crankcase pressure is high, the valve is closed. When the compressor has lowered the internal suction pressure of the compressor below the set point of the valve, it begins to open, allowing the compressor to lower the evaporator pressure. When evaporator pressure has been lowered to the valve setting, the valve completely opens, allowing the compressor to lower evaporator pressure still more. This low pressure is required for the refrigeration system to maintain low box temperatures. During the heat/defrost cycle, when high pressure vapor is pumped from the compressor through the distributor tube to the evaporator, the suction pressure becomes quite high. The force of the high-pressure refrigerant collapses the bellows against spring pressure, which in turn draws the piston over to the left, blocking the inlet flow of refrigerant. The restriction caused by the suction pressure regulator provides the restriction needed for the compressor to pump against during the heat/defrost cycle. This causes the compressor to pump the high-pressure

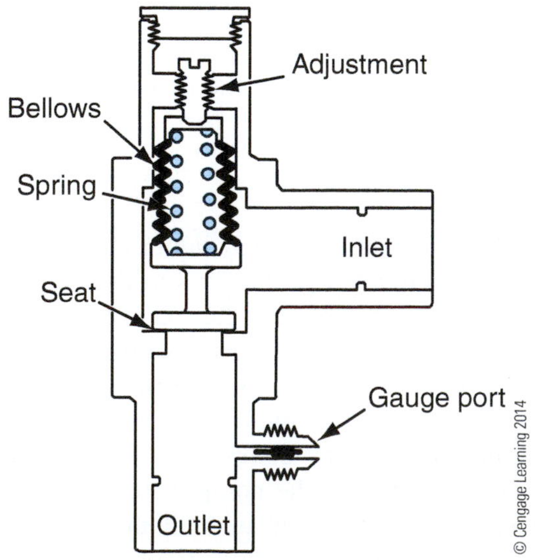

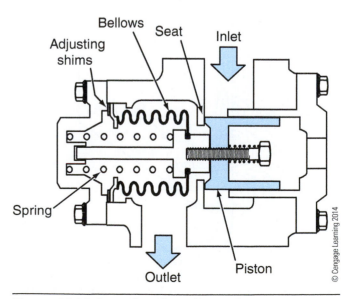

Figure 14-18 A suction pressure regulator.

Figure 14-19 A suction throttling valve.

(temperature) refrigerant into the evaporator for the heating/defrosting process. Suction pressure regulators do not totally restrict refrigerant flow, so they do not require an oil bypass line. These valves are adjustable by increasing or decreasing spring pressure.

SAFETY VALVES

Most refrigeration units using more than 1 pound of refrigerant are equipped with some type of pressure-relieving **safety valve**. Their purpose is to prevent possible explosions by safely relieving high refrigeration pressure in the event of fire, coil blockage, or overheating of the unit.

Two types of safety valves are currently used. One type is made of a spring-loaded piston in which excessive refrigerant pressure working on one side of the piston overcomes spring force at a predetermined pressure, allowing the refrigerant to escape through a relief port passage. This type of safety valve may have a slight refrigerant leak after it has relieved excessive pressure, but should reseal itself. This valve is much like the type used on the hot water heater in your house **(Figure 14-20)**.

The second style of safety valve is a fusible metal plug. These valves sense temperature only, and are often designed to release at about 200°F to 220°F (109°C), which is about 415 to 450 psi. The core material of this valve is designed to melt away, allowing the refrigerant to escape. Once the valve has released the pressure, it is necessary to replace the valve. These valves are one-time applications. This valve is of the same type found in acetylene cylinders used for welding **(Figure 14-21)**.

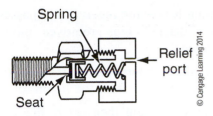

Figure 14-20 Safety valve, spring-loaded piston type.

Figure 14-21 Safety valve, fusible metal plug type.

PERFORMANCE TASKS

The service tasks the technician is required to make to the refrigeration system include regularly scheduled changes of the filter drier and changes of the compressor oil. The rest of the components require no regularly scheduled maintenance. The refrigeration system should be performance tested to ensure that all the components in the system are performing their intended jobs. These routine checks are extremely important because they can allow the technician to catch problems in the refrigeration system before a costly perishable load is lost.

Summary

- In this chapter, we have discussed the four main components of a mechanical refrigeration system that must be used for the refrigeration process to take place.

- There are many other components of the refrigeration system. These components are not absolutely necessary for the refrigeration process but improve the operating efficiency and capacity of the system.

- Starting from the compressor, superheated refrigerant passes through the discharge service valve, entering the discharge vibrasorber.

- The vibrasorber isolates the refrigeration system from vibration caused by the engine and compressor.

- From the vibrasorber, the refrigerant enters the condenser, where it gives up heat to the ambient air.

- Refrigerant cools in the condenser and condenses from a gas to a liquid.

- The liquid refrigerant leaves the condenser and enters the receiver, where it is stored until it is needed.

- Refrigerant leaves the receiver and passes through the filter drier, which removes moisture and contaminants.

- The refrigerant enters the heat exchanger, which further removes heat from the liquid refrigerant.

- The liquid refrigerant then enters the TXV and is metered into the distributor and then the evaporator.

- The TXV balances the inlet flow of refrigerant to the outlet pressure and temperature so that all the refrigerant has time to change states from a liquid to a gas before it exits the evaporator.

- Refrigerant exits the evaporator and enters the accumulator (if equipped).

- The accumulator separates the liquid refrigerant from the vapor in order to prevent liquid refrigerant from entering the compressor.

- Exiting the accumulator, the refrigerant flows through the suction line, through the suction vibrasorber, and on to the suction service valve, then into the suction pressure regulator (if equipped).

- The regulator controls the load placed on the drive engine or electric motor.

- The refrigerant flows out of the suction pressure regulator and into the suction side of the compressor.

- The refrigerant is then compressed and starts its journey through the system again.

Review Questions

1. The compressor:

   A. Pumps the refrigerant through the system

   B. Separates liquid refrigerant from vaporous refrigerant

   C. Acts as a storage tank for liquid refrigerant

   D. Releases heat from the controlled space to the outside air

2. Service valves allow the technician to install a manifold gauge set for testing purposes and _____ of the refrigeration system.

   A. evacuation

   B. recharging

   C. purging

   D. All of the above

3. The function of the condenser is:

   A. To release the heat from the controlled space to the outside air

   B. To store liquid refrigerant until it is needed

   C. To filter the refrigerant and remove dirt and sludge

   D. To absorb heat from the controlled space

4. The expansion valve:

   A. Creates low pressure in the condenser

   B. Creates high pressure in the evaporator

   C. Controls the pressure and boiling point in the condenser and evaporator

   D. All of the above

5. Which components divide the system between the low side and high side?

   A. The compressor and expansion valve

   B. The expansion valve and condenser

   C. The condenser and evaporator

   D. All of the above

6. What is the purpose of a heat exchanger?

    A. To remove the heat from the controlled space

    B. To enable the refrigeration system to operate more efficiently

    C. To facilitate the condensing of the superheated refrigerant in the condenser section

    D. To cool the vaporous refrigerant returning to the compressor

7. In regard to the thermostatic expansion valve, which force applied to the diaphragm causes the valve to open, allowing refrigerant to pass?

    A. Spring pressure

    B. Evaporator outlet pressure

    C. Compressor suction pressure

    D. Sensing bulb pressure

8. The purpose of the evaporator is:

    A. To allow refrigerant to change states from gas to liquid before reaching the TXV

    B. To aid in the cooling process of the drive engine

    C. To allow the refrigerant to absorb heat from the controlled space

    D. To dry up any moisture that may condense on the fins of the evaporator

9. What is the function of the accumulator?

    A. To enable the refrigeration system to operate more efficiently

    B. To allow the refrigerant to absorb heat from the controlled space

    C. To create high pressure in the evaporator

    D. To separate liquid refrigerant from vaporous refrigerant before it enters the compressor

10. Why do some truck-trailer refrigeration units use a suction pressure regulator?

    A. To boost suction pressure, thereby increasing compressor output

    B. To limit suction pressure, protecting the drive engine from being overloaded

    C. To create high pressure in the evaporator

    D. To control the pressure and boiling point in the condenser

    E. Both B and C are correct.

15 Refrigerant Flow Control

Learning Objectives

Upon completion and review of this chapter, the student should be able to:

- Describe the purpose of refrigerant cycle control valves.
- Explain the operation of the three-way valve in the cool cycle.
- Describe the operation of the three-way valve in the heat cycle.
- Explain the purpose of the pilot solenoid.
- Describe the purpose of the condenser pressure bypass valve.
- Explain the purpose of a check valve.
- Explain the cool/heat/defrost cycle of a three-way valve refrigeration system.
- Explain the cool/heat/defrost cycle of a solenoid control refrigeration system.
- Explain the cool/heat/defrost cycle of a four-way reversing valve refrigeration system.

Key Terms

check valve

condenser pressure bypass valve

cool cycle

defrost cycle

four-way valve
(reversing system)

heat cycle

pilot solenoid

quench valve

solenoid control valve

three-way valve

INTRODUCTION

Mobile refrigeration technicians must possess a solid understanding of refrigeration equipment produced by many different manufacturers. Although all of the systems perform the same functions of heating, cooling, and defrosting, they all do it in different ways. Therefore, in order for technicians to work safely and efficiently, they must possess the aforementioned skills.

SYSTEM OVERVIEW

In this chapter, we will discuss many different components and how they work to route refrigerant flow for the heat, cool, and defrost operating modes. As you work your way through the refrigeration schematics, you will discover how manufacturers use these additional components to provide an efficient operating unit over a wide range of ambient and box temperatures. You will also see how these efficiencies increase fuel economy, an extremely important topic for fleet managers everywhere. We will take a look at three different ways in which manufacturers control the flow of refrigerant throughout the system to perform the functions of cooling, heating, and defrosting. The systems we will look at are the three-way valve, the **solenoid control valve**, and the four-way valve.

REFRIGERANT CYCLE CONTROL VALVES

These valves are used to control the unit's mode of operation from **cool cycle** to **heat cycle** or **defrost cycle**. This is done by directing the flow of superheated refrigerant through the condenser (cool mode) or directly to the evaporator (heat/defrost mode). The heat of the compressed refrigerant developed by the compressor is normally given off as it passes through the condenser circuit, allowing the refrigerant to

condense and circulate as in a normal cool cycle. If heat/defrost is required, one of several systems may be used: a three-way valve, solenoid controlled valves, liquid and vapor line solenoids, or a four-way valve. These valves direct the refrigerant from the compressor directly to the evaporator. The evaporator can then give up the refrigerant's heat to the cargo space or use the heat solely for defrosting the evaporator coil and drip pan. In order to defrost only, either the evaporator fan must stop turning or the output of the fan must be blocked so that it does not distribute air through the cargo space.

Three-Way Valve

The **three-way valve** is located in the discharge line just downstream of the discharge vibrasorber **(Figure 15-1)**.

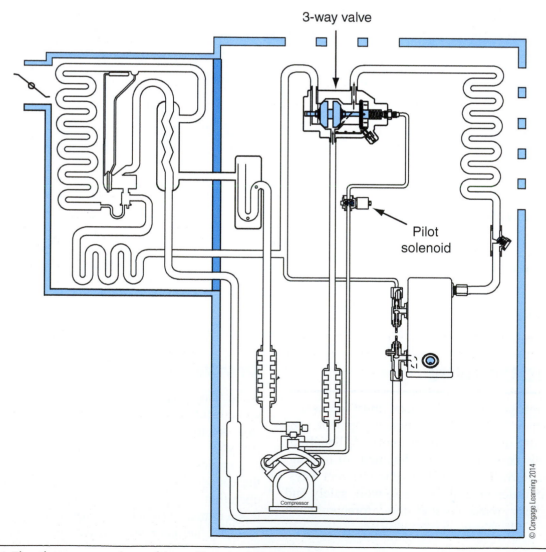

Figure 15-1 The three-way valve refrigeration system.

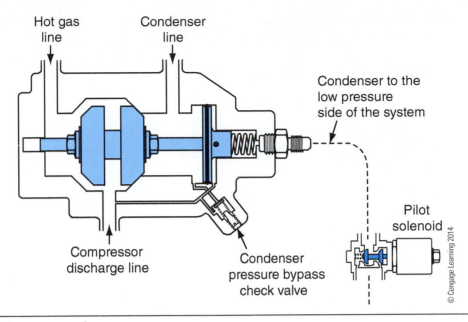

Figure 15-2 The three-way valve.

In cool operation, the valve directs the flow of refrigerant into the condenser section. In the condenser, the superheated gas is cooled by the ambient air and condenses to a high-pressure liquid before it is routed through to the TXV and evaporator. In the heat/defrost mode, the hot gas from the compressor flows through the three-way valve and straight into the evaporator section. The three-way valve is controlled by another valve called the pilot solenoid **(Figure 15-2)**.

Pilot Solenoid

The **pilot solenoid** is an electrically activated valve and is energized by the microprocessor (thermostat on older units) only during the heat or defrost mode **(Figure 15-3)**.

Three-Way Valve Operation

The three-way valve operates on the principles of spring pressure and differential refrigerant pressure. When the unit is not operating, the large spring holds the spool valve tightly against the backside of the three-way valve (cool mode) **(Figure 15-4)**.

When the unit is operating in the cool mode, the seat of the spool valve blocks the flow of refrigerant to the evaporator side of the valve, leaving the condenser side of the valve open for the refrigerant to flow into the condenser. In the cool mode, the pilot solenoid is de-energized, blocking the three-way valve end cap from the suction side of the compressor. This allows high-pressure, hot gas from the compressor discharge line to flow through the bleed passage into the end cap

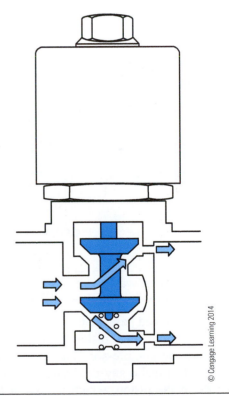

Figure 15-3 The pilot solenoid electrically controls the flow of refrigerant for the heat/defrost mode of operation.

section of the valve. This pressure is now equal on both sides of the spool valve piston. Spring pressure holds the spool valve all the way to the left, blocking refrigerant flow to the evaporator **(Figure 15-5)**.

In the heat/defrost mode, the pilot solenoid is energized and is open. This allows the refrigerant

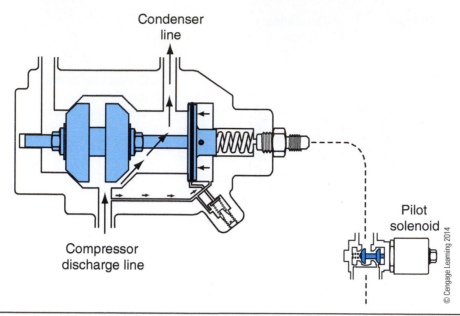

Figure 15-4 The three-way valve, directing the flow of refrigerant to the compressor to the condenser.

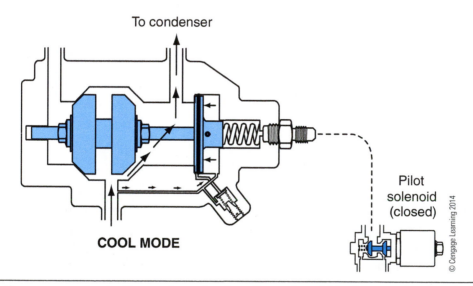

Figure 15-5 The three-way valve in the cool mode; the pilot solenoid is de-energized and closed.

pressure from the bleed passage that has built up on the right side of the spool valve piston to be drawn into the suction side of the compressor. The bleed passage is much too small to make any difference in pressure on the right side of the piston, which is now at the same low pressure as the suction line. Discharge pressure acting on the left side of the spool valve piston easily overcomes spring pressure, causing the piston to shift over to the right. In this position, the spool valve blocks the flow of refrigerant to the condenser and opens the path through to the evaporator side **(Figure 15-6).**

If the processor then requests that the unit go back into the cool mode, the electrical path feeding the pilot solenoid is broken and the valve closes. This allows the discharge pressure to again build up in the end cap

section of the valve (via the bleed passage) and pressurize the right side of the spool valve piston. It is important to note that the pressure is actually equal on both sides of the spool valve piston. The valve then shifts by spring pressure to the left again, blocking the evaporator side and opening up the condenser side.

CONDENSER PRESSURE BYPASS VALVE

The end cap of the three-way valve may also contain a **condenser pressure bypass valve**. The function of this valve is to ensure that condenser pressure does not exceed discharge pressure. The condenser pressure helps to hold the spool valve piston

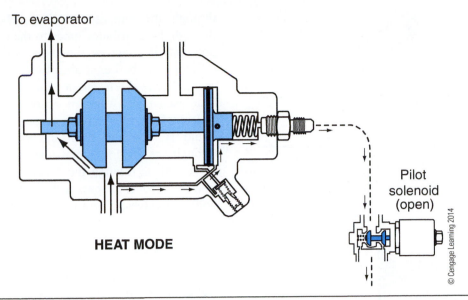

To evaporator

HEAT MODE

Pilot
solenoid
(open)

© Cengage Learning 2014

Figure 15-6 The three-way valve in the heat/defrost mode; the pilot solenoid is energized and open.

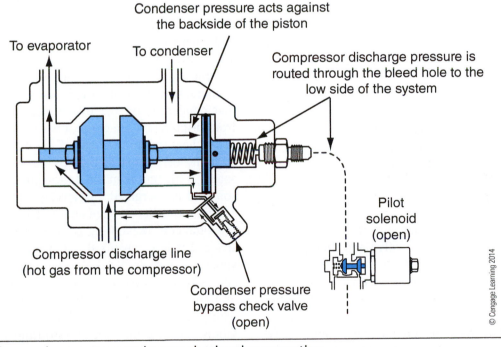

Condenser pressure acts against
the backside of the piston

To evaporator   To condenser

Compressor discharge pressure is
routed through the bleed hole to the
low side of the system

Compressor discharge line
(hot gas from the compressor)

Pilot
solenoid
(open)

Condenser pressure
bypass check valve
(open)

© Cengage Learning 2014

Figure 15-7 The condenser pressure bypass check valve operation.

in the heat position, but if this pressure exceeds discharge pressure, the valve has trouble shifting back into the cool position. The condenser pressure bypass check valve allows the higher pressure of the condenser to bleed into the discharge line until the condenser pressure drops to the level of the discharge pressure **(Figure 15-7)**.

CHECK VALVES

Check valves are used by most manufacturers of refrigeration equipment. Check valves are used in the

refrigeration system to allow refrigerant flow in one direction and stop its flow in the opposite direction. There are two different styles of check valves used in the industry, serviceable and non-serviceable.

The serviceable check valve incorporates a removable cap so that the service technician can replace the seal and spring assembly. The bodies of these valves are usually manufactured from brass. The cap uses a copper sealing washer that must be replaced whenever the cap is removed **(Figure 15-8)**.

The non-serviceable in-line check valve is used as part of an effort to minimize potential refrigerant leaks

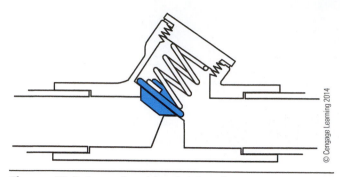

Figure 15-8 A serviceable check valve.

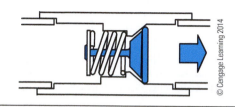

Figure 15-9 A non-serviceable in-line check valve.

(the fewer refrigerant seals, the fewer refrigerant leaks) **(Figure 15-9)**.

REFRIGERANT FLOW FOR THREE-WAY VALVE SYSTEMS (THERMO KING UNITS)

Cool Cycle

Starting at the compressor, the refrigerant leaves through the discharge service valve and then goes on to the discharge vibrasorber **(Figure 15-10)**. The refrigerant then flows to the three-way valve and is directed through to the condenser as the evaporator side of the valve is sealed by the spool valve. As the refrigerant flows through the condenser, it changes states from a superheated gas to a subcooled liquid. The refrigerant then flows through the condenser check valve. This valve prevents the flow of refrigerant back through into the condenser during the heat/defrost mode. The liquid refrigerant is then stored in the receiver tank until it is required by the TXV. The bypass check valve will not allow the refrigerant to pass, leaving only the receiver outlet valve (king valve) for the refrigerant to exit the receiver tank. The liquid refrigerant now flows through the liquid line and on through the filter drier. The refrigerant then flows through the heat exchanger, where it gives up further heat to the cold suction line. The liquid then passes

through the expansion valve, where it is metered through the distributor tubes to the evaporator coil. The pressure drop across the expansion valve causes the refrigerant to boil as it absorbs heat from the controlled space. The cold refrigerant vapor then passes through the heat exchanger, where it absorbs more heat. The refrigerant then goes into the accumulator tank, where any liquid is separated from the vapor before it passes through the suction vibrasorber to the suction service valve and on through the suction throttling valve. And last, the refrigerant passes through the compressor suction reed valves, where the cycle starts again.

Heat Cycle

The high-pressure, superheated refrigerant vapor leaves the compressor through the discharge service valve, and then moves through the discharge vibrasorber to the three-way valve **(Figure 15-11)**. With the pilot solenoid energized, the three-way valve is shifted to the heat position, blocking off the condenser and opening a passage to the drip tray and evaporator. The refrigerant is then routed through the drip pan, melting the ice prior to entering the distributor and evaporator. (The drip pan is used to catch the water that condenses on the evaporator coil, preventing it from dripping on the cargo, possibly causing damage.) In the heat mode, the heat of the evaporator is blown over the cargo space, thereby heating the controlled space. If the unit is in the defrost mode, a damper door solenoid is energized, effectively blocking off air circulation to the controlled space. This melts the ice and dries the coil, preventing unnecessarily heating up the space that is meant to stay cold or frozen. The cool, vaporous refrigerant, along with some liquid refrigerant, is then passed through the heat exchanger to the accumulator, where the liquid is separated and the vapor returns to the compressor through the suction vibrasorber and the suction throttling valve.

The accumulator tank may be heated externally to aid in vaporizing the liquid refrigerant. This is done by wrapping the accumulator with a coil and circulating hot engine coolant through it.

A hot gas line is also teed off from the drip pan heater and is routed to the bypass check valve into the receiver tank. This hot, high-pressure refrigerant pressurizes the trapped refrigerant left in the receiver from the cooling cycle. The higher-pressure refrigerant in the receiver tank closes off the condenser check valve, blocking off the condenser. The refrigerant has only one possible exit from the receiver: the receiver tank outlet valve. The liquid refrigerant then is passed

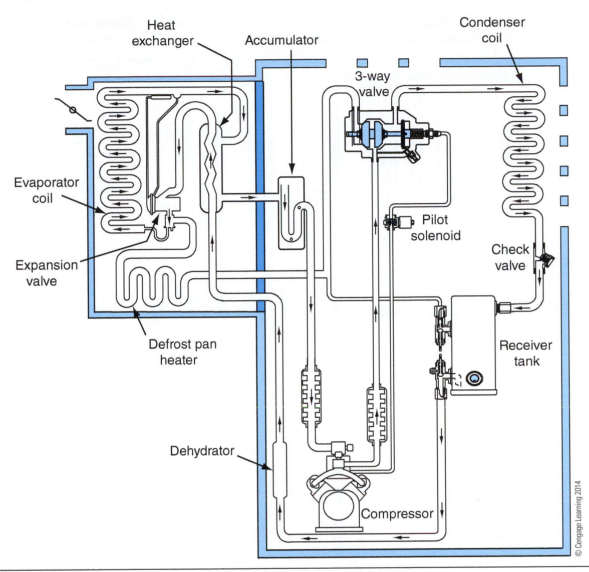

Figure 15-10 The three-way valve system operating in the cool mode.

through the drier heat exchanger and then is passed through a notch in the seat or an internal orifice in the body of the thermostatic expansion valve. The refrigerant then mixes with the hot vapor from the compressor in the distributor. This refrigerant that was once trapped in the receiver tank can now be used in the heat/defrost cycle to allow the compressor to achieve higher pressure and thereby more heat.

Defrost Cycle

The flow of refrigerant when the unit is in the defrost cycle is identical to the heat cycle. The only exception is that during the defrost cycle, air is not circulated through the loaded area. Closing of the damper door traps heat in the evaporator compartment. This heat in turn melts ice that has built up on the coil

and the water drips into the defrost pan. The defrost pan guides the water to a pair of drain tubes. These tubes exit the evaporator section and extend down the front wall of the trailer. From there, the water flows onto the ground at the front of the trailer.

When trailer box temperatures become very low, the water produced by the melting ice freezes in the drain pan. For this reason, the unit uses a defrost pan heater. Hot gas runs through this coil in the heat/defrost mode to keep the water above the freezing point so it can flow to the ground outside.

Most units are kept from entering the defrost mode until the evaporator temperature reaches a point below approximately 45°F (7.2°C). The defrost cycle terminates when the evaporator coil temperature reaches approximately 55°F (12.8°C). At this temperature, there can be no ice left on the coil to impede air flow.

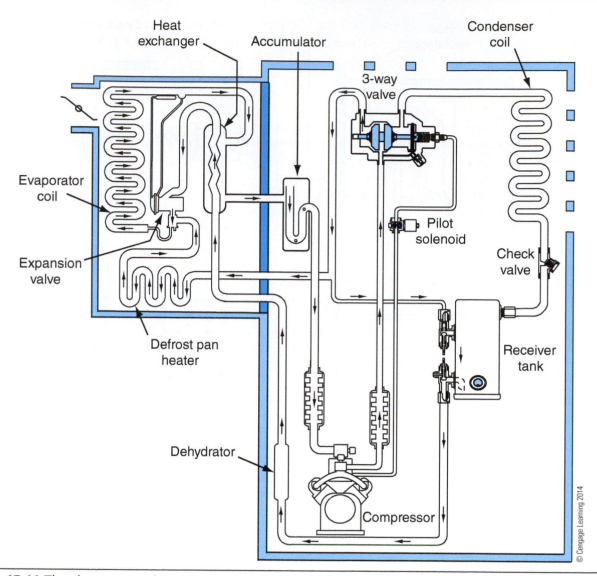

Figure 15-11 The three-way valve system operating in the heat mode.

Note: It is harmful to the cargo to have warm air circulated through temperature-sensitive products that are required to be kept frozen or refrigerated. For this reason, if a problem is detected with the defrost door (not closing), it must be repaired immediately because the unit will not come out of the heat/defrost cycle until the whole trailer has warmed up far above the set point.

SOLENOID CONTROL SYSTEM (CARRIER)

Unlike a single three-way control valve, as used by some manufacturers, the solenoid control system uses three or four liquid and vapor line solenoids to control the flow of refrigerant for the heat, cool, and defrost cycles. One of the liquid line solenoids, called SV1, is situated between the condenser and the receiver tank. SV1 is a normally open solenoid. Another liquid line solenoid, called SV2, is in line between the receiver tank and the TXV. This solenoid is normally closed. The unit also has one or two vapor line solenoids called SV3, and if equipped, SV4. Both of these valves are also normally closed.

Operation of the Solenoid Control System (Carrier)

In the cool mode, SV3 and SV4 are both closed, SV1 is open, and SV2 is energized and open. The refrigerant vapor is pumped through the condenser, where it changes state, to the receiver, drier, TXV, evaporator, and back to the compressor.

In the heat/defrost mode, SV1 is energized to close, SV2 is energized and open (unless pressure rises too high), and SV3 is energized to open. If the unit is equipped with two vapor line solenoids, SV4 will be energized and open initially during the heat cycle. After 60 seconds, SV3 will be energized and open as long as system pressure requirements are met. With SV1 closed, the refrigerant still enters the condenser, where it continues to condense until the condenser is filled with liquid refrigerant. After the condenser is full, the refrigerant is pumped through SV3 and SV4 and is directed to the evaporator drip pan and evaporator coil, and then returns to the compressor.

Cooling Cycle

In the cooling cycle, starting at the compressor discharge service valve, the superheated refrigerant flows through the discharge line to the discharge vibrasorber **(Figure 15-12)**. The refrigerant then passes through the discharge check valve. This valve prevents liquid refrigerant trapped in the condenser from migrating back to the compressor in the off cycle.

(If liquid refrigerant were able to get back and fill the cylinders of the compressor, the compressor could sustain damage upon startup.)

The refrigerant flows up to solenoid valves 3 and 4, called SV3 and SV4. In the cool mode, both of these valves are de-energized and closed, blocking refrigerant passage. The superheated refrigerant now flows to the condenser coil, where it gives up its heat to the ambient air through its large finned surface. Liquid refrigerant leaving the condenser enters the condenser pressure control solenoid, called SV1. This solenoid is normally open and is de-energized whenever the unit runs in the cool mode. Once the liquid refrigerant leaves SV1, it is then routed through to the receiver tank, where the excess liquid refrigerant is stored for the cooling process. The liquid refrigerant cannot go through the bypass check valve, which is closed to refrigerant passing in this direction.

From the receiver, the liquid refrigerant flows out the liquid line to the shutoff valve (known as the king valve). From the king valve, the refrigerant is passed through the subcooler, which is another portion of the condenser. As the component's name implies, it further subcools the liquid refrigerant, resulting in less

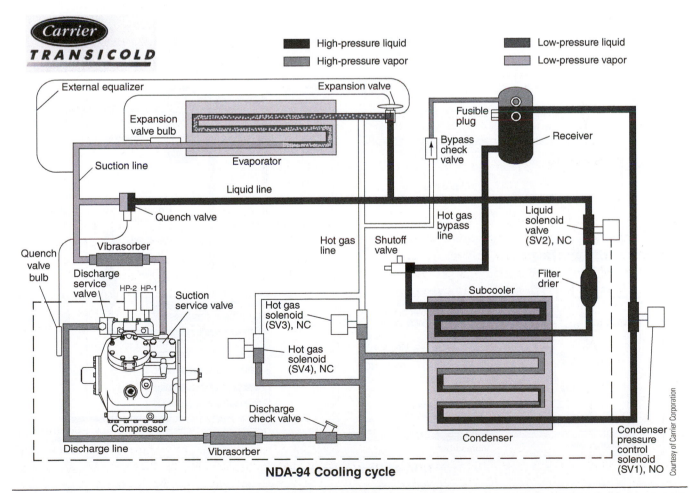

Figure 15-12 A solenoid-controlled system operating in the cool mode.

flash gas being created as the refrigerant is metered through the TXV. With less flash gas created, the unit operates with more capacity and efficiency because there is more liquid to evaporate and absorb heat from the controlled space. Upon exiting the subcooler, the refrigerant passes through the filter drier and on to SV2. This liquid line solenoid is a normally closed valve that is energized and opens during the cooling mode. The liquid refrigerant passes through SV2 to the TXV. The TXV meters refrigerant into the evaporator to maintain a constant superheat of the refrigerant at the evaporator outlet. The restriction caused by the TXV creates a pressure drop in the refrigerant. The low-pressure liquid refrigerant then enters the distributor and flows through the evaporator. The pressure drop across the expansion valve causes the liquid refrigerant to boil. It is the heat from the controlled space that causes the refrigerant to boil as it absorbs heat through the evaporator fins.

Teed into the same liquid line that feeds the TXV is a component not used on the three-way valve system, called the **quench valve**. The quench valve is a small TXV-style valve that monitors the discharge temperature of the refrigerant leaving the compressor. It does this using a sensing bulb and a capillary tube, just like the TXV. The sensing bulb is secured to the discharge line at the compressor. If the discharge temperature exceeds safe limits, the quench valve opens to allow a small amount of liquid refrigerant into the suction line. The liquid refrigerant flashes because of the pressure drop and cools the compressor. This flashing refrigerant lowers the discharge temperature in the compressor resulting from extremely high discharge pressure. High temperatures can cause breakdown of the compressor oil, in turn causing system damage.

The low-pressure, low-temperature superheated refrigerant leaves the evaporator and enters the suction line, and then travels through the suction vibrasorber. From the suction vibrasorber, the refrigerant vapor completes the cycle as it returns to the compressor suction service valve.

Heating Cycle

In the heat/defrost cycle, starting at the compressor discharge service valve, the high-pressure superheated refrigerant passes through the discharge line to the discharge vibrasorber. The refrigerant then passes through the discharge check valve **(Figure 15-13)**.

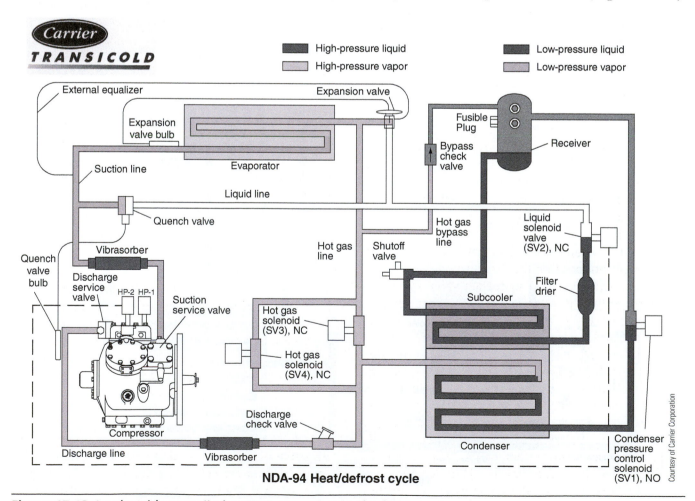

NDA-94 Heat/defrost cycle

Figure 15-13 A solenoid-controlled system operating in the heat mode.

This valve prevents liquid refrigerant trapped in the condenser from migrating back to the compressor in the off cycle. (If liquid refrigerant were able to get back and fill the cylinders of the compressor, the compressor could sustain damage upon startup.) The refrigerant then flows to solenoid control valves 3 and 4, called SV3 and SV4. These valves are normally closed; when the unit is initially placed in the heat mode, SV4 is energized and open, allowing refrigerant to pass. Sixty seconds after SV4 is energized, if the temperature and pressure requirements are met, the microprocessor opens SV3 as well. Hot gas then passes through SV3 and SV4 and flows directly into the evaporator.

Refrigerant also flows into the condenser, but is not able to get past SV1 at the end of the condenser. SV1 is a normally open valve that is energized and closed in the heat/defrost mode. Before the heat/defrost cycle can become effective, the condenser section must first be packed to capacity with liquid refrigerant. (Hot gas will still condense in the condenser.) Once the condenser is full, the hot refrigerant passes through the hot gas line toward the evaporator.

The hot gas line also tees off to feed the bypass check valve. Because the pressure of the refrigerant is higher, it flows through the check valve and pressurizes the excess refrigerant trapped in the receiver tank from the cooling cycle. The liquid refrigerant in the receiver now flows through the subcooler, filter drier, and SV2. SV2 is a normally closed valve that is controlled by the head pressure control switch (HP2) when in heat/defrost mode. SV2 is cycled open and closed by the compressor head pressure switch (HP2) as it allows the liquid stored in the receiver to migrate to the evaporator. Because so much refrigerant must be packed into the condenser, the excess refrigerant trapped in the receiver is critically needed for the heat/defrost cycle.

Teed into the same liquid line that feeds the TXV is a component not used on the three-way valve system, called the quench valve. The quench valve is a small TXV-style valve that monitors the discharge temperature of the refrigerant leaving the compressor. It does this using a sensing bulb and a capillary tube, just like the TXV. The sensing bulb is secured to the discharge line at the compressor. If the discharge temperature exceeds safe limits, the quench valve opens to allow a small amount of liquid refrigerant into the suction line. The liquid refrigerant flashes because of the pressure drop and cools the compressor. This lowers the discharge temperature in the compressor resulting from extremely high discharge pressure. High temperatures can cause breakdown of the compressor oil, in turn causing system damage.

The superheated refrigerant that does get past SV3 and SV4 then enters the evaporator through the hot gas line. This hot refrigerant can then transfer its heat to the evaporator coil. The evaporator fan that blows air through the coil then circulates the warm air throughout the cargo space. The transfer of heat energy from the refrigerant, as well as the suction of the compressor, causes the pressure to drop, with a corresponding drop in the boiling temperature of the refrigerant. Because of the drop in pressure, the refrigerant does not condense.

The low-pressure, low-temperature superheated refrigerant leaves the evaporator and enters the suction line, and then travels through the suction vibrasorber. From the suction vibrasorber, the refrigerant vapor completes the cycle as it returns to the compressor suction service valve.

The heat that is transferred from the refrigerant through to the evaporator fins is produced by the compressor as it compresses the refrigerant. This is known as the heat of compression.

Defrost Cycle

The flow of refrigerant when the unit is in the defrost cycle is identical to that of the heat cycle. The defrost mode differs from the heat mode in that air is not circulated over the cargo space. This is accomplished by stopping the evaporator fan from turning in the defrost mode. This can be done in one of two ways. If the evaporator fan is electrically powered, it has the power turned off during the defrost mode. If the fan is belt driven, an electromagnetic clutch system can be used.

It is harmful to have warm air circulate through temperature-sensitive cargos that require refrigeration or need to be kept frozen.

FOUR-WAY VALVE (TRANE/ARCTIC TRAVELER)

Four-way valve systems, sometimes called reversing systems, are still used by some manufacturers. The four-way valve system uses a valve capable of changing the direction of refrigerant flow through the components. This valve is located close to the compressor in the high- and low-pressure vapor lines.

Four-Way Valve Operation

This valve completely changes the direction of refrigerant flow in the heat/defrost mode by turning the evaporator into the condenser and the condenser into the evaporator. By changing the direction of refrigerant flow of the hot gas, heating or cooling can be developed in the cargo area. This system requires two TXV valves, due to the changing of the normal condenser to an evaporator during the heat/defrost cycle.

Cool Cycle

The high-pressure superheated vapor leaves the compressor through the discharge service valve, then goes through the discharge vibrasorber to the four-way valve. The pilot solenoid (cool) is energized, opening a passage to the suction side of the compressor. This differential pressure causes the valve to shift to the left, opening the passage to the condenser side of the four-way valve. **Figure 15-14** shows more detail about the four-way reversing valve. When the valve is in the cool mode, the pilot solenoid opens a passage between port B and the suction line. The difference in pressure causes the high discharge pressure to push the spool over to the left.

The refrigerant then enters the condenser, where it gives up its heat to the cooler ambient air. The liquid refrigerant then passes through the drier to the heat exchanger, and then to the liquid line check valve. It then passes through the evaporator expansion valve and distributor tubes, where it is metered into the evaporator coil. The pressure drop across the expansion valve causes the liquid refrigerant to boil as it absorbs heat from the cargo space while it passes through the evaporator. The cold refrigerant vapor then passes through the heat exchanger and then to the four-way valve before it enters the accumulator tank. The refrigerant then flows back to the suction side of the compressor.

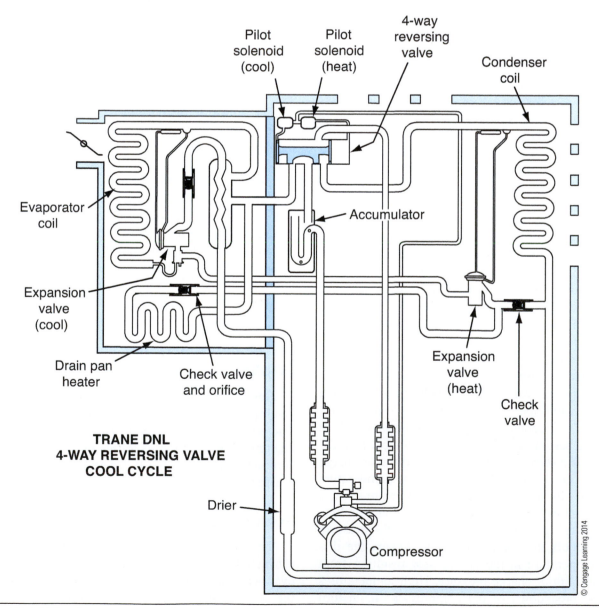

Figure 15-14 The four-way reversing valve and the pilot solenoid in the cool mode (detailed view).

During the cool mode, to prevent the refrigerant from passing through the drip tray and four-way valve, a check valve and orifice are located between the drip tray and the distributor. A second check valve, which is teed in this line and known as the check valve (heat), is forced closed due to the high pressure in the condenser circuit and the low pressure on the evaporator circuit.

Heat Cycle

During the heat/defrost cycle, the four-way reversing valve is shifted to change the operation of the evaporator and condenser. As stated previously, this requires two TXVs and some additional check valves. The condenser now becomes the evaporator, absorbing heat from the ambient air. In cold environments, it may be necessary to use shutters to hold in the engine heat in order to boil the refrigerant into a vapor. The high-pressure superheated refrigerant leaves the compressor through the discharge service valve, then moves through the discharge vibrasorber to the four-way valve. The pilot solenoid (heat) is energized, opening a passage to the suction side of the compressor. This differential pressure causes the valve to shift to the right, opening the passage to the evaporator. **Figure 15-15** shows the four-way reversing valve in more detail. When the valve is in the heat mode, the pilot solenoid opens a passage between port A and the suction line. The difference in pressure causes the high discharge pressure to push the spool over to the right.

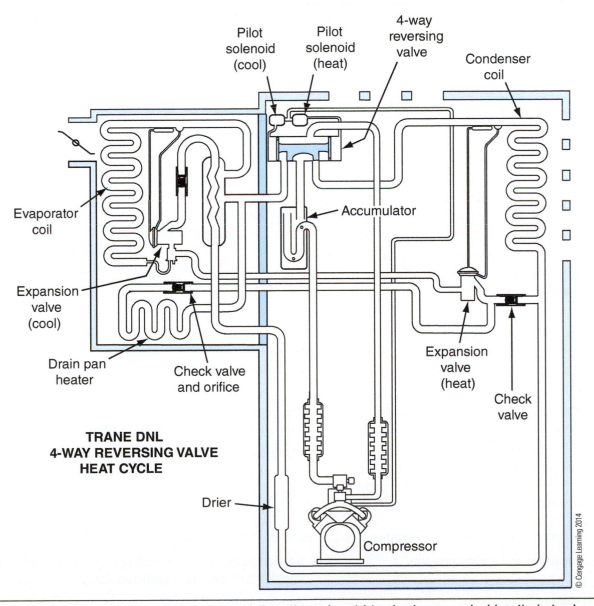

Figure 15-15 The four-way reversing valve and the pilot solenoid in the heat mode (detailed view).

The four-way valve then directs the flow of refrigerant through to the heat exchanger and then on into the evaporator (which is now acting like a condenser), where it gives up heat to the air passing through the evaporator coil. Some of the refrigerant vapor passes through the drip pan, the check valve, and the orifice, entering the condenser as a low-pressure saturated mixture (vapor and liquid together). The liquid refrigerant leaving the evaporator is prevented from returning through the heat exchanger and drier by a check valve in the cool line. The liquid refrigerant then passes through the expansion valve (heat) at the inlet of the condenser, then through the check valve heat and then to the condenser, where it changes states

by absorbing heat from the ambient air. The low-pressure vapor then returns to the four-way valve, which directs it through the accumulator and then back to the suction side of the compressor.

Four-Way Valve Defrost Cycle

Like the other systems previously mentioned, air flow through the evaporator section must be stopped. This can be accomplished by stopping the fan motor or completely blocking air flow with a defrost damper door. It is harmful to have warm air circulated through temperature-sensitive cargos that require refrigeration or need to be kept frozen.

Summary

- Refrigerant cycle control valves are used to control the refrigeration unit's mode of operation from cool to heat/defrost mode. This is done by directing the flow of superheated refrigerant.

- When the three-way valve is in the cool cycle, it directs the flow of refrigerant into the condenser section.

- When the three-way valve is in the heat cycle, it directs the flow of refrigerant into the evaporator section.

- The function of the condenser pressure bypass valve is to ensure that condenser pressure does not exceed discharge pressure, enabling the three-way valve to shift from the heat cycle to the cool cycle.

- Check valves are used in the refrigeration system to allow refrigerant flow in one direction and to stop it in the opposite direction.

- The solenoid control system uses three or four liquid and vapor line solenoids to control the flow of refrigerant for the heat, cool, and defrost cycles.

- The four-way valve completely changes the direction of refrigerant flow in the heat/defrost mode by turning the evaporator into the condenser and the condenser into the evaporator.

- The four-way valve system requires two TXV valves due to the changing of the normal condenser to an evaporator during the heat/defrost cycle.

- The flow of refrigerant in the defrost cycle is identical to that of the unit's heating cycle, regardless of manufacturer. All units also use some means of stopping the air flow through the cargo space in the defrost mode.

Review Questions

1. With a refrigeration unit that uses a three-way valve, in the cool mode of operation, which of the following statements is true?

   A. The pilot solenoid is energized and the three-way valve is shifted to the left (evaporator sealed).

   B. The pilot solenoid is energized and the three-way valve is shifted to the right (condenser sealed).

   C. The pilot solenoid is de-energized and the three-way valve is shifted to the left (evaporator sealed).

   D. The pilot solenoid is de-energized and the three-way valve is shifted to the right (condenser sealed).

2. There are two forces working together to try to close the TXV. What are these forces?

   A. Equalizer and spring pressure

   B. Suction bulb and equalizer pressure

   C. Suction bulb, equalizer, and spring pressure all work together to close the expansion valve

   D. Sensing bulb and spring pressure

3. The accumulator tank is especially important during the heat mode on a three-way valve system because:

    A. Superheated refrigerant pushed from the evaporator will damage the compressor.

    B. Some liquid refrigerant is pushed from the evaporator as the unit switches from cool to heat.

    C. Refrigerant oil may be pushed from the evaporator as the unit switches from cool to heat.

    D. Both A and C are correct.

    E. Both B and C are correct.

4. To improve the performance of the heat/defrost mode, which component allows hot gas to enter the receiver tank?

    A. Condenser pressure bypass

    B. Accumulator bypass valve

    C. Condenser check valve

    D. Bypass check valve

5. When the solenoid control system is in the cooling mode, the coil state and valve position for SV3 and SV4 are _____.

    A. energized and open

    B. energized and closed

    C. de-energized and closed

    D. de-energized and open

6. What is the purpose of the discharge check valve on a solenoid control system?

    A. To prevent liquid refrigerant from migrating from the condenser to the compressor in the off cycle

    B. To trap refrigerant in the receiver

    C. To prevent liquid refrigerant from flowing from the receiver to the hot gas line

    D. To prevent refrigerant from flowing into the evaporator

7. When the solenoid control system is in the cool mode, the coil state and valve position for SV1 are _____.

    A. energized and open

    B. energized and closed

    C. de-energized and closed

    D. de-energized and open

8. In what position must the manual liquid line shutoff or (king valve) be to stop the flow of refrigerant on a unit with a solenoid control system?

    A. Back seated

    B. Front seated

    C. Mid seated

    D. Removed

9. Which factor alone will cause a decrease in the amount of flash gas formed when the liquid refrigerant passes through the TXV of a solenoid control system?

    A. More subcooling

    B. Less subcooling

    C. Increased discharge pressure

    D. Smaller condenser

10. When the solenoid control system is in the cool mode, the coil state and valve position for SV2 are _____.

    A. energized and open

    B. energized and closed

    C. de-energized and closed

    D. de-energized and open

11. The purpose of the TXV is to meter refrigerant into the evaporator in order to maintain a constant _____ of the refrigerant at the evaporator outlet.

    A. superheat   C. subcooling
    B. pressure   D. flow

12. If the discharge temperature exceeds safe limits on a solenoid control refrigeration system, the _____ opens to allow a small amount of liquid refrigerant into the suction line.

    A. TXV   C. fusible plug
    B. SV3   D. quench valve

13. The state of the refrigerant returning to the compressor through the suction vibrasorber is a low-pressure, low-temperature _____.

    A. subcooled liquid   C. superheated vapor
    B. subcooled vapor   D. superheated liquid

14. When the solenoid control system is in the heat/defrost mode, the coil state and valve position for SV4 are _____.

    A. energized and open   C. controlled by HP2
    B. energized and closed   D. de-energized and open

15. When the solenoid control system is in the heat/defrost mode, the coil state and valve position for SV3 are _____.

    A. energized and open   C. controlled by HP2
    B. energized and closed   D. controlled by the microprocessor

16. When the solenoid control system is in the heat/defrost mode, the coil state and valve position for SV1 are _____.

    A. energized and open   C. controlled by HP2
    B. energized and closed   D. de-energized and open

17. When the solenoid control system is in the heat/defrost mode, the coil state and valve position for SV2 are _____.

    A. controlled by HP-1   C. controlled by HP2
    B. energized and closed   D. de-energized and open

18. When the solenoid control system is in the heat/defrost mode, the hot gas line delivers high-temperature, high-pressure _____ to the evaporator.

    A. superheated liquid   C. subcooled vapor
    B. superheated vapor   D. subcooled liquid

19. What is the only difference between the heat and defrost modes of operation?

    A. Air is not circulated over the load in defrost mode.   C. SV4 is energized in heat mode but not in defrost mode.
    B. Defrost mode can be used only when ambient temperature is greater than box temperature.   D. SV4 is energized in defrost mode but not in heat mode.

16 Truck-Trailer Refrigeration Electrical Components

Learning Objectives

Upon completion and review of this chapter, the student should be able to:

- Describe the construction of storage batteries.
- Work safely around storage batteries.
- Explain two different ways in which batteries may be shipped from the manufacturer.
- Describe the three different types of batteries used today and how maintenance is performed on them.
- Explain how batteries are rated.
- Describe storage procedures for batteries.
- Explain testing procedures to ensure that the battery can function at its rated performance level.
- Describe how a battery should be recharged and list the steps involved in jump-starting a unit with a low battery.
- List the steps involved in removing and replacing a battery.
- Explain the function of the charging system.
- List the major components of a typical alternator and the function of those parts.
- Explain the function of a voltage regulator.
- Describe the steps involved in performing an alternator output test.
- List the steps involved in removing and replacing an alternator.
- Describe the two styles of starters used in the transport refrigeration industry.
- Explain the purpose of an overrunning clutch.
- Describe the steps involved in performing a starter test.
- Explain the purpose of the various truck-trailer refrigeration safety switches.

Key Terms

Battery Council International (BCI)

cold cranking amps

conventional starter motor

deep cycle

dry charged

fast charging

field circuit

gassing

gear reduction starter motor

hydrometer

load test

open circuit voltage test

overrunning clutch

rectifier

refractometer

reserve capacity

slip rings

slow charging

specific gravity

wet charged

INTRODUCTION

Technicians working on transport refrigeration equipment must have a solid understanding of how electrical systems work. This chapter does not deal directly with teaching the technician basic electricity, but assumes the student has a basic grasp of how electrical systems operate. For more information on the basics of electricity, please refer to Sean Bennett's *Heavy Duty Truck Systems, Fifth Edition* (Clifton Park: Delmar Cengage Learning, 2011). One of the most common problems encountered by today's transport refrigeration technicians is the complaint that "the reefer won't start." This chapter takes a systematic approach to diagnosing common no-start complaints, as well as looking at the various safety switches that protect the unit and may also prevent the unit from starting.

SYSTEM OVERVIEW

This chapter gives the technician a firm understanding of how batteries are constructed, how they work, and the general maintenance required for the different types of batteries. The technician will also learn how to test a battery to determine if the battery is serviceable or must be replaced.

When the battery is the problem, the technician should not stop there, but should determine why the battery failed. Was it old age or is there a problem with the charging circuit? In this chapter, the technician will learn how to test the alternator's output to determine if the problem is in the charging circuit. If both the battery and charging circuit check out (it may be necessary to jump-start the unit to verify the charging circuit), the starting motor itself should be tested to ensure that it is not trying to draw too many amps from the battery.

When all of the above have been tested and confirmed to be in good condition (or remedied), and the unit engine will still not start, the problem may lie in one of the unit's protective safety switches. These switches generally interrupt power to the engine's fuel solenoid. If power is not delivered to this solenoid, the fuel supply to the engine's injection pump is effectively shut off and the engine stops running. If all else fails, there may be mechanical problems internal to the engine itself.

STORAGE BATTERIES

Transport refrigeration units use 12-volt direct current (DC) automotive-type storage batteries like the battery in most light trucks. The storage battery is the energy source for the refrigeration system. It is important to remember that a battery does not store electricity but rather stores a series of chemicals, and through a chemical process, electricity is produced. Basically, there are two types of lead in an acid mixture, which react to produce an electrical pressure called voltage. This electrochemical reaction changes chemical energy into electrical energy and is the basis for all automotive-style batteries.

Battery Construction

Automotive batteries consist of diluted sulphuric acid electrolyte as well as positive and negative electrodes, in the form of plates. These plates are made of lead or lead-derived materials; consequently, these batteries are often referred to as lead-acid batteries.

Figure 16-1 illustrates the components that make up a typical battery. Batteries are separated into

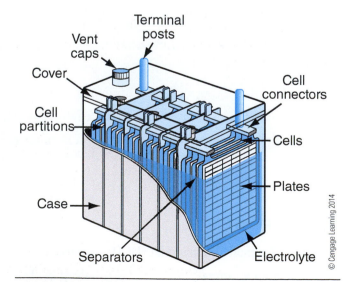

Figure 16-1 A battery cross-section with components labeled.

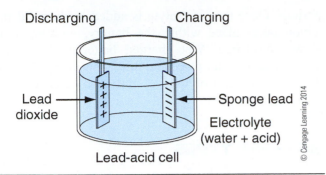

Figure 16-2 A diagram of a single battery cell.

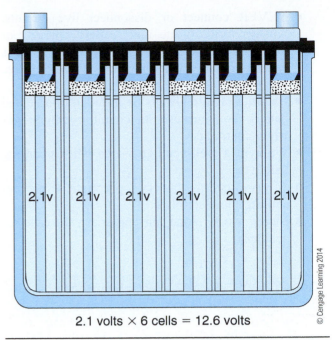

2.1 volts × 6 cells = 12.6 volts

Figure 16-3 A diagram of six battery cells connected together in series to form a 12-volt battery.

several cells (six cells for a 12 V battery) and in each cell there are several battery elements, all bathed in electrolyte solution.

Cell Operation

When two dissimilar metals are placed in an acid bath, an electrical potential is produced across the poles. **Figure 16-2** illustrates basic battery cell construction. The cell produces voltage due to a chemical reaction between the plates and the electrolyte. The positive plate is usually made of a reddish-brown material called lead dioxide (PbO_2). The negative plate is made of a grayish material called spongy lead (Pb). Electrolyte is a solution of water and sulphuric acid. When all three are combined, a cell element is formed.

Cell Voltage

Each of the six cell elements in a 12 V battery produces approximately 2.1 volts, regardless of the size or number of the plates. The cells are connected in series, which produces a total voltage of 12.6 V.

Figure 16-3 illustrates how cells are connected in series to produce a 12 V battery.

BATTERY SAFETY

When working around batteries, technicians should exercise caution to avoid personal injury and to protect the safety of others in the immediate area.

Figure 16-4 illustrates a typical warning label affixed to the case of a battery. Batteries contain sulphuric acid and generate hydrogen gas that is highly flammable. Always comply with the following safety tips when working with batteries:

■ Keep flames or sparks away from batteries. Make sure there is no smoking around batteries.
■ Always wear eye protection and rubber gloves to protect yourself from chemical burns when handling batteries.

Figure 16-4 The warning label found on a battery.

- NEVER connect or disconnect live circuits. Always turn off the unit, battery charger, or tester when attaching or removing leads. (Sparks can be produced when making or breaking live circuits.)
- Batteries should always be installed in a vented battery box because they emit hydrogen gas when charging.
- Work in a well-ventilated area when charging batteries.
- Always keep the battery top upright to prevent spilling the electrolyte.
- Never work alone on batteries, in case of accident.

CAUTION *Battery electrolyte contains sulphuric acid, which can cause severe personal injury (burns) and damage to clothing and equipment. If electrolyte is accidentally spilled or splashed on a person's body or clothing, it must be immediately neutralized by washing with a solution of baking soda and water. The solution should be 0.25 pounds baking soda to 1 quart water (115 grams baking soda to 1 liter of water).*

Electrolyte splashed into the eyes is extremely hazardous. The eyes should immediately be held open and flushed with cool clean water for about 5 minutes; then seek medical treatment at once.

BATTERIES

Automotive-style batteries may be shipped from the manufacturer in two different ways: **wet charged** and **dry charged**.

Dry Charged Batteries

Dry charged batteries are delivered to the distributor with no electrolyte in the cells. The internal cell components are electrically charged, washed, and assembled into the case. While the batteries are in this state, they require no special storage and do not require periodic charging. Dry charged batteries are activated by adding electrolyte just before the battery is put into service.

Wet Charged Batteries

Wet charged batteries are delivered to the distributor with their full supply of electrolyte in each cell. These batteries require special care during storage

periods. Once the electrolyte is added, a chemical reaction takes place within the battery, causing it to slowly discharge. Wet charged batteries should be stored in a cool, dry place and fully slow charged every 30 days. The electrolyte level must also be checked at charging time.

BATTERY TYPES

Batteries may be further categorized according to the amount of maintenance required to keep them at optimum charge level. Batteries can be classified as conventional, low maintenance, and maintenance-free. Conventional batteries can be purchased as either wet or dry charged, whereas low maintenance and maintenance-free batteries are usually wet charged because they retain their charge during storage much better than conventional batteries. The major difference between the three types of batteries from a servicing point of view is how often the electrolyte level, terminals, and cables must be checked and maintained. Each of the three types uses different materials for the positive and negative plate construction. This causes different operation and service characteristics.

Conventional Batteries

Conventional batteries require more maintenance than low maintenance or maintenance-free batteries. This is due to the chemical composition of the plates, which causes the water in the electrolyte to turn into gas; therefore, the water must be replenished. When adding water, use only distilled water because minerals and chemicals commonly found in drinking water react with the plate materials and shorten battery life. The water level should be no higher than an eighth of an inch (3.2 mm) below the bottom of the vent well. To avoid permanent damage, make sure the electrolyte level never drops below the top of the plates. In addition, avoid overfilling because some of electrolyte will be washed away, leaving a weaker solution. Refer to **Figure 16-5** for correct electrolyte level. Gasification can cause the electrolyte to condense on the top of the battery case and cause the battery to discharge by providing a current path between the positive and negative terminals. This also causes corrosion of the battery terminals, cables, and cable ends. Conventional batteries are more easily overcharged than low maintenance and maintenance-free batteries. They also discharge more quickly when stored in a wet charged state. Conventional batteries perform well in **deep cycle** applications. Deep cycling occurs when a battery goes from fully charged

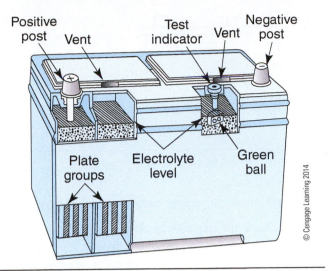

Figure 16-5 Filling the battery to the correct level of electrolyte, making sure the plates are below the surface of the electrolyte.

down to a low state of charge and is then fully recharged during system operation.

Low Maintenance Batteries

Low maintenance batteries are manufactured to require less maintenance than conventional batteries. This is because there is less electrolyte gasification, so the level of the electrolyte is reduced much more slowly than with conventional batteries. Corrosion of the battery terminals, cables, and cable ends is also reduced. Low maintenance batteries are more resilient to overcharging than conventional batteries, but have a shorter life in deep cycle applications.

Maintenance-Free Batteries

Maintenance-free batteries are designed to not require electrolyte replenishment under normal operating conditions. Battery terminals, cables, and cable ends require almost no maintenance because of greatly reduced gasification and water loss. Maintenance-free batteries retain their charge longer than conventional or low maintenance batteries when stored in a wet charged state. Maintenance-free batteries generally require higher voltage regulator settings than low maintenance or conventional batteries.

BATTERY RATINGS

Storage batteries can be selected for their application by three different ratings. They are **cold cranking amps** rating, the **reserve capacity** rating, and the **BCI (Battery Council International)** dimensional group number.

Cold Cranking Amps

Batteries are performance rated by their cold cranking amps (CCA) designation. This is the load in amperes that a battery can sustain for 30 seconds at 0°F (−17.8°C) and not fall below 1.2 V per cell, or 7.2 V on a 12 V battery. The CCA rating indicates how much power a battery can deliver in extremely cold conditions.

The battery's main function is to start the engine, so it is imperative that the battery has sufficient capacity to accomplish this task. **Figure 16-6** illustrates how a battery's capacity drops as ambient temperature drops. The engine also requires more power to turn and start as the ambient temperature drops as well.

Reserve Capacity

The reserve capacity rating indicates the number of minutes a new fully charged battery at 80°F (26.7°C) can sustain a load of 25 amperes before the battery voltage drops to 1.75 V per cell or 10.5 V for a 12 V battery. This is a rating devised by the automotive industry to indicate how long a battery can provide enough power to keep the

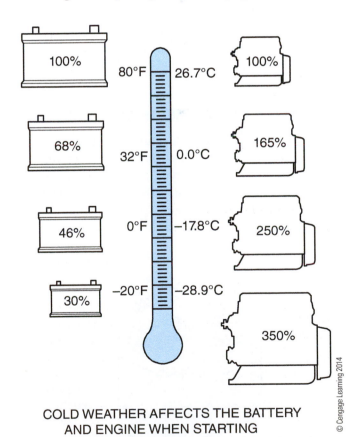

COLD WEATHER AFFECTS THE BATTERY
AND ENGINE WHEN STARTING

Figure 16-6 A diagram illustrating how the battery's capacity is severely debilitated as the ambient temperature is reduced and resistance is caused by the engine as it becomes harder to turn in decreasing temperatures.

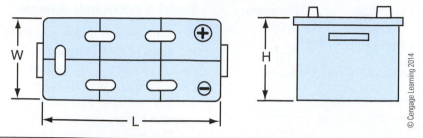

Figure 16-7 A diagram of the battery's dimensions used to configure its BCI number.

ignition, head and tail lights, windshield wipers, and heater operating if the charging system fails.

Battery Council International (BCI) Group Dimensional Number

The Battery Council International (BCI) number indicates a battery's physical dimensions. As an example, a group 31 series battery is always 13 inches long (33 cm), 6.8 inches wide (17.3 cm), and 9.4 inches high (23.9 cm). **Figure 16-7** illustrates the BCI dimensional measurements. This rating has nothing to do with the battery's performance capacity, just the physical dimensions. A battery's performance characteristic is determined by the internal components, such as the number of plates per cell, not its physical size.

BATTERY MAINTENANCE

As previously stated, conventional batteries require more maintenance than low maintenance or maintenance-free batteries. However, all batteries should be inspected periodically. **Figure 16-8** illustrates a battery requiring maintenance.

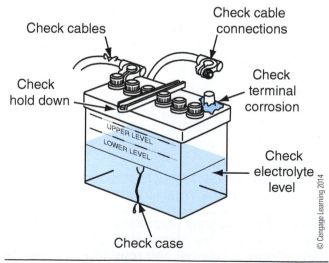

Figure 16-8 The areas a technician should look at when servicing a battery for routine maintenance.

Below are some general battery inspection procedures:

1. Check for loose or broken terminal posts.
2. Check for a cracked or broken battery case. Ensure sides of battery case are straight and not bulged out. This condition would indicate the battery is frozen or has been frozen.
3. If the case is ruptured or the terminal is broken, replace the battery immediately.
4. Inspect the case for dirt, moisture, and corrosion.
5. The battery case and terminals should be cleaned with a baking soda and water solution. If necessary, use a heavy brush on the battery terminals. Make sure the soda solution is not allowed to mix with the electrolyte in the battery.
6. Battery cables should be checked for cleanliness and tightness. A loose or dirty connection increases resistance to current flow. This condition can cause poor charging and system operation. Replace worn or frayed cables; keep all connections clean and tight.
7. Check the battery electrolyte level. Add water if the electrolyte level is below the top of the plates. Do not overfill, making sure to keep the level below the vent cap openings.

Note: Even maintenance-free batteries should be checked monthly, especially if the battery is in severe service conditions such as extreme heat.

Battery Storage

Wet charged batteries slowly self-discharge while in storage. The rate at which the battery discharges depends upon the type of battery and the storage conditions to which it is subjected. All batteries should be stored in a clean, cool, and dry environment and tested at regular intervals. The best storage temperatures are between 50°F and 60°F (10°C to 16°C).

Truth or Urban Legend

''Never leave batteries on the ground or on a concrete floor because all the power will leak out.''

This one is an urban legend because it does not matter what batteries sit on. This myth probably originated from a time when batteries were shipped in porous wooden cases. There are some truths behind this myth, though! All batteries do self-discharge over time when they are not being charged. If dust and dirt build up on the battery tops, sulphuric acid carbonizes the grime into an electrical conductor, acting like a short circuit across the terminals and quickly draining the power. Cold temperatures also reduce available power from a battery. And thermal gradients can reduce the life of a large battery. This can occur when the air temperature around a battery is much warmer than the surface it is sitting on.

....

Note: The battery's state of charge should be checked every 30–45 days and it should be charged whenever the capacity of the battery has been reduced to 75%. Battery testing and charging will be discussed later in this chapter.

....

BATTERY TESTING

There are many ways of testing a battery's state of charge and its ability to perform, but the most common methods are the **open circuit voltage test** and the **hydrometer** test for state of charge, and the **load test** for the battery's performance rating.

Hydrometer Testing

The hydrometer can be used to test a battery's state of charge. This is accomplished by measuring the **specific gravity** (SG) of the electrolyte solution. Water has an SG of 1.000, whereas sulphuric acid has an SG of 1.835. A fully charged battery contains approximately 65% water and 35% sulphuric acid. When the specific gravity of a fully charged battery is tested, the SG should be 1.265 at 80°F (26.7°C). The SG of the battery's cells varies with the temperature to which the battery is subjected. When a SG test is performed on a

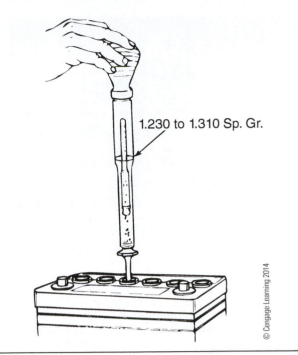

1.230 to 1.310 Sp. Gr.

© Cengage Learning 2014

Figure 16-9 The correct method of accurately reading a hydrometer.

battery using a hydrometer, the temperature must be compensated for in order to make an accurate determination about the condition of the battery. In order to do this, 0.004 must be added to the actual reading for every 10°F (5.5°C) above 80°F (26.7°C). And 0.004 must be subtracted from the SG reading for every 10°F (5.5°C) below 80°F (26.7°C).

Figure 16-9 illustrates the proper use of a hydrometer to obtain accurate data. The specific gravity of the electrolyte solution can also be measured with a **refractometer**. With a refractometer, there is no need to correct for temperature. This instrument is much like the one used to test antifreeze strength.

Figure 16-10 illustrates the proper use of a refractometer. Follow the list of instructions when using a refractometer:

1. Wear proper eye protection.
2. Remove battery caps.
3. Starting at one end of the battery, extract a drop of electrolyte and place it on the refractometer lens, then close the prism.
4. Holding the refractometer up to the light, take a reading on the chart inside the view finder.
5. Record the reading and do the same on the remaining five cells.

The following table compares the battery's state of charge with the SG reading. A battery should be recharged when its capacity drops to 75%.

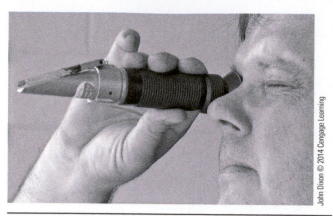

John Dixon © 2014 Cengage Learning

Figure 16-10 The correct method of using a refractometer to get a specific gravity reading of the electrolyte.

Electrolyte Specific Gravity	% of Charge
1.260	100
1.230	75
1.200	50
1.100	0

....

Note: A difference of more than 0.050 SG between individual cells indicates a defective battery. The unequal consumption of electrolyte is usually due to an internal defect or deterioration from extended use. A battery found to be in this condition should be replaced.

....

The SG measurement of the battery can also be used to determine at what temperature the battery will freeze. The following table indicates the temperature at which the battery will freeze according to its SG level and state of charge.

Electrolyte Specific Gravity	Freeze Point
1260	−71.3°F (−57.38C)
1250	−62°F (−52.3°C)
1230	−16°F (−26.6°C)
1200	0°F (−17.8°C)
1100	19°F (−7.3°C)

Open Circuit Voltage Test

The open circuit voltage test is another method of determining a battery's state of charge. This test is performed when the battery is at rest and is not being charged or discharged. The battery contains a surface charge if the unit has been running. This surface charge must be removed from the battery before testing can be performed. On a reefer, the glow plug circuit can be energized for 1 minute; then allow 2 minutes for the battery to recover before proceeding. Connect the positive probe of a voltmeter to the battery positive terminal and the negative probe to the battery negative terminal.

Meter Reading	Battery Condition
12.66 V	full charged
12.48 V	75% charged
12.30 V	50% charged
11.76 V	0 charge

The battery should be recharged if the open circuit voltage test indicates that the battery voltage is less than 12.4 V.

Load Test

A load test is a true measurement of a battery's ability to perform. There are two ways of load testing a battery, one with a load tester and the other using the reefer unit's engine.

Before a load test is performed, the battery must be fully charged and at room temperature. Always follow the instructions of the instrument manufacturer before performing a load test.

Generally a load test is performed as follows, but as mentioned earlier, follow the manufacturer's instructions.

Using a Commercial Battery Load Tester

1. Turn the load knob on the load tester to the NO LOAD position and connect the terminals to the battery.
2. Turn the load knob until 50% of the CCA rating of the battery is being drawn.
3. Hold the load for 15 seconds while monitoring the battery voltage. The voltage should not go below 9.6 V at a temperature of 70°F (21°C).

Testing Truck Batteries

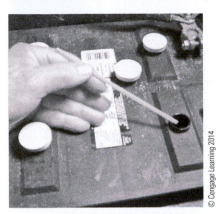

P4-1 Dip the refractometer probe into the battery cell, wetting just the tip of the probe. Throughout this procedure, remember that battery electrolyte is corrosive.

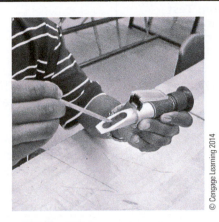

P4-2 Deposit a drop of electrolyte onto the refractometer read-lens as shown. Close the refractive lid.

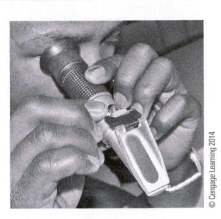

P4-3 Raise the refractometer view scope to your eye and point the refractive window toward a light source, preferably natural. The shaded area in the view finder correlates to a specific gravity reading.

P4-4 Connect a digital AVR to test a battery by connecting the polarized clamps as shown. Then connect amp pickup lead with its arrow pointing in the direction of current flow.

P4-5 A digital AVR with inductive pickup connected to a battery bank ready for a load test.

P4-6 Turn the load test knob CW as shown to load the battery. Typically, you will load to ½ CCA for 15 seconds and observe the voltmeter reading, which should not drop below 9.6 volts.

4. Replace battery if voltage falls below the minimum specification. Consult the following chart to compensate for temperature:

Battery Temperature	Minimum Test Voltage
70°F (21°C)	9.6 V
60° (15.5°C)	9.5 V
50° (10°C)	9.4 V
40° (4.4°C)	9.3 V
30° (−1.1°C)	9.1 V
20° (−6.6°C)	8.9 V
10° (−12.2°C)	8.7 V
0° (−17.7°C)	8.5 V

Using the Reefer Unit's Engine

1. Connect a voltmeter to the terminals of the battery. (Be sure the test leads are on the battery posts, and not the cable terminals.)
2. Disconnect the engine run solenoid at the injection pump to prevent the engine from starting.
3. Crank the engine for approximately 30 seconds while monitoring the battery voltage.
4. If the battery voltage drops below 9.0 V, the battery has a low capacity and must be replaced.

BATTERY CHARGING

A battery charger can be used when tests indicate that the battery is in a low state of charge. When a battery discharges, current flows from the positive terminal through the path of the circuit, and back to the negative terminal. Current within the battery flows from the negative to the positive terminal. **Figure 16-11** illustrates the direction of current flow as a battery discharges. In order to recharge the battery, the direction

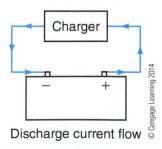

Discharge current flow

© Cengage Learning 2014

Figure 16-11 The direction of current flow as a battery discharges.

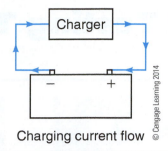

Charging current flow

© Cengage Learning 2014

Figure 16-12 The direction of current flow required to charge a battery.

of current flow must be reversed. **Figure 16-12** illustrates the direction of current flow required to recharge the battery. This action will restore the chemicals to their active state.

Batteries can be charged at one of two different rates, those being **fast charging** or **slow charging**. When a battery is charged, it should be recharged at the same rate as it was discharged. A battery should also be at a temperature of between 60°F and 80°F (15.5°C to 26.6°C).

Slow Charging

The slow charging method should be used when a battery has been discharged slowly over time. This could possibly happen because a battery has been stored for a period of time and not periodically charged or because there is a small short that slowly drains the battery. The rate at which a slow charge should typically be performed is between 6 and 8 amps over an 8- to 10-hour period. This being said, in a severe case where a battery has stood in a discharged state for a long period of time, the battery may require up to 3 days of charging at a slow rate to bring it back up to its full rate of charge.

Fast Charging

The fast rate of charge delivers a high charging rate to the battery for a short period of time. This method of charging should be used when a battery has been discharged quickly, such as in the case of a starting failure where the operator of the reefer has cranked the engine until the battery can no longer provide enough energy to turn the engine over. In this condition, the battery requires fast charging. A 12 V battery would typically be fast charged at a rate of 40 amps for up to 2 hours.

When fast charging batteries, it is important to make sure the electrolyte solution does not get too hot. If the temperature of the electrolyte reaches 125°F (51.7°C) or if it **gasses** (bubbles) violently, reduce the charge rate to slow charge, 6 to 10 amps. Overheating causes the plates in the battery to warp or buckle and can cause a short circuit within a cell. If the electrolyte gasses violently, it can strip active material from the surface of the plates, thereby reducing the battery's capacity.

> **CAUTION** *Always follow safety practices when recharging batteries. Connect the cable from the battery charger to the battery first, before plugging the charger in and turning it on. Keep sparks and flames away from the battery and NEVER smoke around a battery. When using a battery charger, always read the manufacturer's instructions before using the equipment.*

JUMP-STARTING A UNIT

Sometimes it may be necessary to get the unit started when there is insufficient charge in the battery and no time to properly recharge the battery. In these cases, the unit may be successfully started by connecting the stalled battery with a known good battery.

Figure 16-13 illustrates the proper cable placement to jump-start a unit. To avoid personal injury and damage to the equipment, follow proper safety procedures when jump-starting a unit. Be sure the voltage of the jump-starting battery is the same as the stalled battery. Make sure the battery used for jump-starting is not running another piece of equipment at the time. Cover both batteries with a damp cloth. Follow the proper sequence for connecting the jump-starting cables from the stalled to the jump-starting batteries:

1. Connect one end of the positive jumper cable to the positive (+) post of the stalled battery.

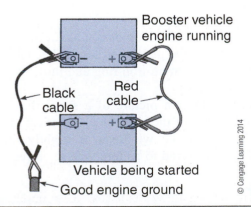

Figure 16-13 The correct method of placing the battery jumper cables when jump-starting or boosting a low battery.

2. Connect the other end of the jumper cable to the positive (+) post of the jump-starting battery.
3. Connect one end of the negative jumper cable to the negative (−) post of the jumper battery.
4. Connect the other end of the negative jumper cable to a good chassis ground connection on the stalled unit, away from the discharged battery. Do not connect the jumper cable to the negative (−) post of the stalled battery.
5. Start the stalled unit.
6. Disconnect the jumper cables in exactly the reverse order as that used to connect the jumper cables.

Battery Removal and Installation

When removing or installing a battery, always follow battery safety procedures. Check the orientation of the positive battery and install the new battery in the same manner.

1. Remove the ground (negative) cable first before the positive cable.
2. Inspect the battery mounting tray and hold down assembly; clean and replace it as necessary. This can be performed with a mixture of baking soda and water.
3. Check the condition of the battery cables and terminal ends. Replace them if insulation is worn or frayed or if corrosion is a problem.
4. Make sure that the new battery is fully charged before it is installed into the unit.
5. Use a battery lifting strap to lower the battery into the battery tray.
6. Install the battery hold down clamps evenly and securely. Do not overtighten. A battery must be mounted securely. Vibration can loosen connections, crack the case, and loosen internal components.
7. Clean battery terminals and cable clamps before installation. Install the positive battery cable first before the negative terminal. Note that this is the reverse of the sequence of removal. This prevents shorting the battery while installing the positive cable if the wrench comes into contact with a chassis ground. Once the cable ends are installed, the terminals should be coated with a dielectric grease to prevent corrosion.

CHARGING SYSTEMS

The battery is responsible for providing the energy to start the engine of the refrigeration unit.

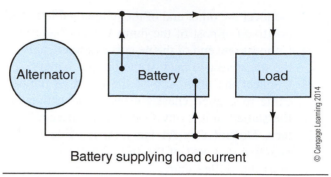

Battery supplying load current

Figure 16-14 The current flow of a charging system when the alternator is not producing voltage.

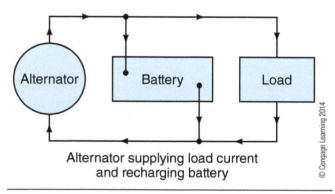

Alternator supplying load current
and recharging battery

Figure 16-15 The current flow when the alternator is charging the battery as well as supplying the power for any electrical load on the unit.

Figure 16-14 illustrates the circuit of a battery delivering power to a load. Once the battery has performed this task, the energy it has expended must be replenished to ensure that the battery has enough power to again restart the unit.

Figure 16-15 illustrates the circuit for a charging system to replenish the battery as well as supply current to the load. The charging system converts mechanical energy into electrical energy when the engine is running. It is the job of the alternator to supply the current to recharge the battery and provide the power supply to operate electrical components used to control the refrigeration, heating, and defrosting functions. During peak operation, the battery may be required in addition to the alternator to provide current for system loads. **Figure 16-16** illustrates the circuit when the battery and alternator supply load current.

The rate at which the alternator charges the battery is controlled by a voltage regulator.

ALTERNATOR COMPONENTS

The alternator can be divided up into individual components that must all perform certain tasks in order for the charging system to operate effectively.

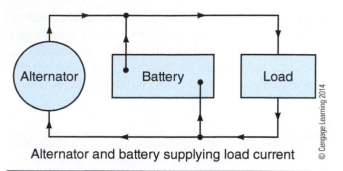

Alternator and battery supplying load current

Figure 16-16 A charging system when the unit is running at peak electrical capacity; in this case, both the alternator and the battery supply current to the load.

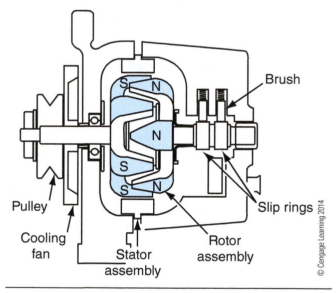

Figure 16-17 An alternator with components labeled (cross-section).

Figure 16-17 shows a cutaway of a typical alternator.

Stator

The stator is a wire loop or conductor that remains stationary within the alternator case. The stator is an assembly made from an iron ring and three individual groups of windings.

Figure 16-18 is a picture of a stator that is contained in the case of the alternator. Electrical impulses occur in each of the windings at slightly different times. This is referred to as three-phase current.

Rotor

A rotating magnet spins inside the stator and is known as a rotor. The rotor consists of a wire coil contained between two interlocking iron sections

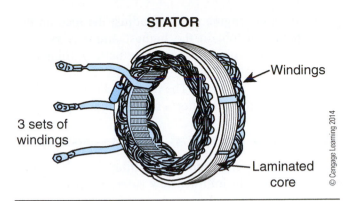

STATOR

Windings

3 sets of windings

Laminated core

Figure 16-18 A stator, consisting of an iron ring and three groups of coils or windings.

mounted on a rotating shaft. Each lobe on the rotor is a magnetic pole. This provides several movements of north and south poles for each revolution of the shaft, causing high output at low rpm.

Figure 16-19 pictures a typical rotor assembly. As previously stated, the rotor is a magnet, but not of the permanent type. Direct current is applied to the coil through brushes and **slip rings**. The iron lobes of the rotor become magnetized as current is applied. Because external current must be supplied to start the alternator charging, it is considered to be externally excited. The circuit that provides the current to set up the magnetic field in the rotor is called the **field circuit**.

Rectifier Diodes

Alternators produce alternating current (AC), which must be changed into direct current (DC) before

it can become useful in recharging the battery and supplying power to the various electrical components on the refrigeration unit. This is accomplished by a diode assembly called a **rectifier**.

Figure 16-20 pictures the rectifier diodes, which convert the current from AC to DC. The three coils located in the stator are connected to six diodes, called rectifier diodes. These diodes are usually located within the frame of the alternator.

Field Diode

The field diode, sometimes referred to as the diode trio, is used in some applications to separate the field circuit from the regulator circuit. The field diode also supplies current to the field coil from the stator after initial excitation.

Voltage Regulator

The output of the alternator must be controlled or it produces too much voltage. The voltage regulator is responsible for control of the alternator output. The voltage regulator controls two important tasks. First, it allows current from the battery to excite the rotor field coils. And secondly, it keeps the charging level within a safe set of values during charging. The regulator controls alternator output by increasing or decreasing the strength of the rotor's magnetic field. It does this by controlling the amount of current from the battery to the rotor's field coil. A regulator actually opens and closes the circuit to the field coil, and is capable of

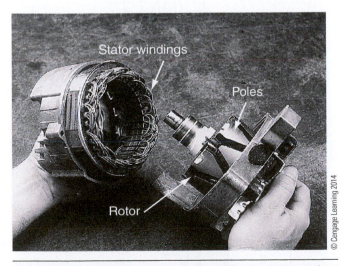

Figure 16-19 The rotor assembly is a rotating nonpermanent magnet inside the alternator housing; battery power must be fed to the slip rings to energize the rotor's magnetic field.

Figure 16-20 The rectifier diodes, used to convert the alternating current produced by the alternator into DC current (required to charge the battery and run the electrical loads of the unit).

doing this several times per second during normal operation. If the field circuit is open, the rotor is demagnetized and no stator current is generated.

A voltage regulator is made up of resistors, transistors, diodes, and thermistors, all enclosed in a sealed unit. In modern-day alternators, the regulator is contained within the alternator case.

Alternator Output Test

An alternator output test checks an alternator's ability to deliver its rated output of voltage and current. This test should be performed any time an overcharging or undercharging condition is suspected. The output current and voltage should meet the specifications of the alternator or there may be a problem with the alternator or regulator.

Figure 16-21 shows a typical tester used for testing the capacity of batteries and alternators.

> **CAUTION** *Always follow the manufacturer's instructions for testing alternator output.*

The following is a typical procedure for performance testing the alternator's output:

1. Make sure the knob controlling the load is backed off all the way.
2. Ensure that the volt and amp meters are zeroed and adjust if necessary.
3. Connect the tester load leads to the battery terminals: red to positive, black to negative.
4. Connect the clamp-on amps pickup around the battery ground (–) cable.
5. Start the unit up and place it in high speed.

6. With the engine running, adjust the load on the battery to obtain the highest ammeter reading possible without causing the battery voltage to drop below 12 V.
7. Read the ammeter. The reading should be within 10% of the alternator's rated output capacity.

Alternator Removal and Installation

When removing and reinstalling an alternator, technicians should follow a few safety measures.

1. The battery must be disconnected when installing or removing an alternator.
2. The negative battery cable should be removed first.
3. Alternator polarity must be identified before connecting the alternator to the battery. If polarity is incorrect, the rectifier diodes will be ruined.
4. Make sure that the alternator pulley and the drive pulley are in alignment to ensure maximum alternator belt life.
5. The alternator belt should wrap around the pulley by a minimum of 100° to prevent belt slippage, belt and pulley wear, and damage to the front alternator bearing due to overheating. See **Figure 16-22** for proper alternator belt alignment.
6. Check the belt tension and verify by consulting the maintenance manual for the correct belt

Figure 16-21 A tester used to load test batteries, starters, and alternators.

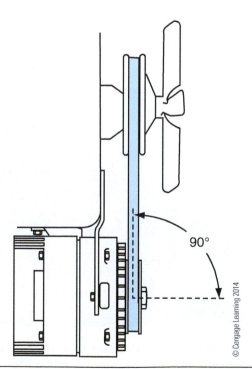

Figure 16-22 Proper alternator pulley to drive pulley alignment, critical to ensure long belt life.

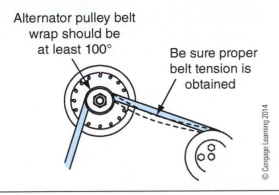

Figure 16-23 The amount of belt contact required to prevent the belt from slipping; as the alternator output increases, so does the alternator's resistance to turn. Slippage is prevented by a balance between proper belt pulley contact, along with recommended belt tension.

   tension value. See **Figure 16-23** for correct belt installation.

   7. If charging the battery on the unit, disconnect the positive cable to prevent possible damage to the alternator or regulator.

STARTERS

   Getting the reefer unit started is possibly the most important function of the electrical system. The starting system performs this function by changing electrical energy from the battery into mechanical energy in the starter motor. The starter motor can then transfer this mechanical energy through gears to the engine flywheel, thus turning the engine's crankshaft. As the crankshaft turns, the pistons draw the air while injectors deliver fuel into the cylinders, which is then compressed and ignited to start the engine. Most engines require a cranking speed of around 200 rpm to start. Starter motors are rated by power output in kilowatts; the greater the output, the greater the cranking power.

Starter Motor Types

   There are two types of starters used by the manufacturers of transport refrigeration equipment. The first type (used on older model units) uses a **conventional starting motor** and the other uses a **gear reduction starter motor.**

Conventional Starter Motors

   Figure 16-24 illustrates a conventional starter motor. The pinion gear on the conventional starter is on the same shaft as the motor armature and thus rotates at the same speed. Conventional starter motors use an electric solenoid to pull the plunger in, which in turn pulls the lever, causing the pinion gear to mesh

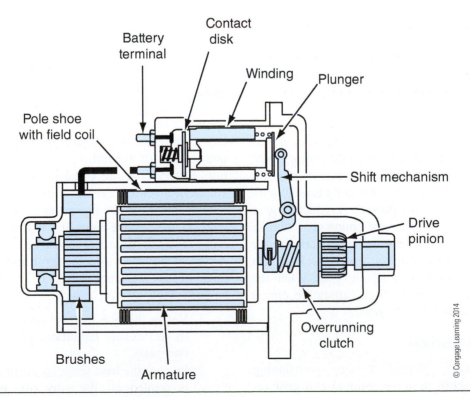

Figure 16-24 Conventional starter motor (cross-section).

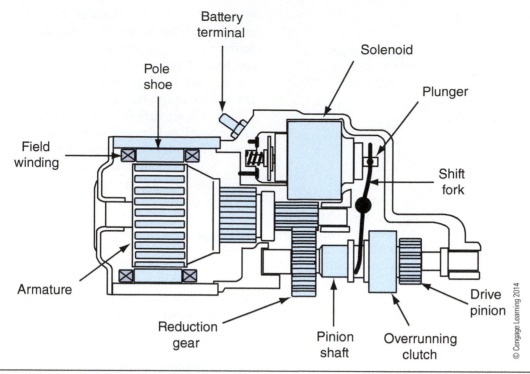

Figure 16-25 A gear reduction starter motor (cross-section).

with the ring gear. When the plunger bottoms in the solenoid body a magnetic switch closes, delivering full battery power through to the starter motor. When the engine starts, an **overrunning clutch** causes the pinion gear to disengage from the ring gear.

Gear Reduction Starter Motors

Figure 16-25 illustrates a gear reduction starter motor. The gear reduction starter uses a smaller high-speed motor and a set of reduction gears. This type of starter is smaller and lighter in weight than a conventional starter and also operates at higher speeds. The gearing ratio of the reduction gears turns the pinion gear at one-fourth to one-third the speed of the motor. Even at one-fourth the speed of the motor, the pinion gear still rotates faster than a conventional starter motor and with much greater torque (cranking power). A reduction gear is mounted on the same shaft as the pinion gear. Unlike a conventional starter, the solenoid plunger acts directly on the pinion gear (not through a drive lever) to push the pinion gear in to mesh with the ring gear. When the engine starts, an overrunning clutch causes the pinion gear to disengage from the ring gear.

Overrunning Clutch

A one-way clutch, referred to as an overrunning clutch, is used on both the conventional and gear reduction starters. This clutch prevents damage to the

starter motor once the engine has been started. It does this by disengaging the pinion gear from its drive shaft with the use of spring-loaded wedged rollers.

Figure 16-26 illustrates the components of an overrunning clutch. Without the use of an overrunning clutch, the starter would be destroyed by the speed of the flywheel. Starters were never designed to operate at these speeds.

STARTER TESTING

A starter amp draw test is a common way of ensuring that a starter is in good condition. This test provides a quick check of the entire starting system. This test also tests battery cranking voltage. The following is an example of a starter draw test, but always follow the recommendations of the equipment manufacturer. Always refer to the service manual for the actual amp draw range at which the starter should be expected to perform. A typical starter on a reefer application would be in the range of 150–300 amps at 12.5 V. Testing should always be performed with the engine at operating temperature.

1. Verify that the battery used for the draw test is in serviceable condition. Charge the battery if necessary.
2. Prepare the load tester by ensuring the load knob is rotated all the way off, counterclockwise. Check the meters and adjust to zero if necessary.

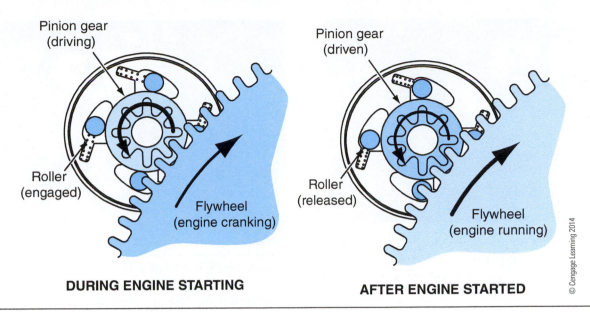

DURING ENGINE STARTING   **AFTER ENGINE STARTED**

© Cengage Learning 2014

Figure 16-26 A diagram of an overturning clutch, which prevents damage to the starter as the engine starts up.

3. Connect the leads of the load tester to the battery, red on positive and black on negative.

....

Note: The battery voltage should be at least 12.2 V (50% charged). If not, recharge the battery.

....

4. Adjust the ammeter to zero using the adjustment control knob.
5. Connect the amp inductive pickup clamp around the negative battery cable or cables.
6. Remove the wire from the run solenoid at the injection pump to prevent the engine from starting.
7. Crank the engine over while watching the ammeter and voltmeter. Cranking speed should be normal. Current draw should not exceed the maximum specification. Cranking voltage should be at or above the minimum specification.
8. Remove the leads from the tester and prepare the engine to run.

Test Results

If the starter draw current is higher than specified and cranking speed is low, this usually indicates a problem with the starter itself. However, slow cranking speed can also be an indication of engine problems. A low cranking speed with low current draw, but high cranking voltage, usually indicates high resistance in the starter circuit. Keep in mind that the battery must be fully charged and all terminal connections tight to ensure accurate results.

REFRIGERATION UNIT SAFETY SWITCHES

Manufacturers of transport refrigeration equipment install protective switches throughout the unit to protect people from any personal injury as well as to protect the unit from physical damage caused by running a unit under abnormal conditions. Safety switches, as the name implies, are designed to act in the event of low engine oil pressure, high engine temperature, and high compressor discharge pressure, and may also be used on some units for low compressor oil pressure. These switches are all part of the electrical system and shut down the engine if a problem occurs in any of the circuits. In addition to a problem with the starting or charging system, a unit may not start because parameters have not been met to satisfy one of the unit's safety switches. The following switches are some of the more common switches that could be expected to be found on a reefer unit.

Low Engine Oil Pressure Safety Switch

The low engine oil pressure switch is used to protect the diesel engine from internal damage in the event that oil pressure is lost.

Figure 16-27 shows a low oil pressure switch installed in the oil filter housing. Catastrophic engine damage will result if it is permitted to run with insufficient oil pressure.

The switch is installed in an oil galley supplied with pressurized oil from the oil pump. If the oil pressure drops below a preset point, about 15 psi, the engine shuts down.

Figure 16-27 The oil pressure safety switch located in the oil filter housing; this switch stops the engine from running if engine oil pressure is lost, thereby preventing complete engine failure.

High Engine Coolant Temperature

This switch is normally located in the cylinder head, where it monitors the coolant temperature. If the temperature rises above a preset point, the bimetal switch contacts within the switch open, causing the engine to shut down. On new style units, this switch may also be an electronic switch that can also be used as a sensor by the microprocessor to indicate coolant temperature for the display panel.

Figure 16-28 shows the location of the coolant temperature sensing switch.

Figure 16-28 A water temperature safety switch located in thermostat housing. (It could also be located in the engine head.) This switch stops the engine from running if engine coolant temperatures become too high.

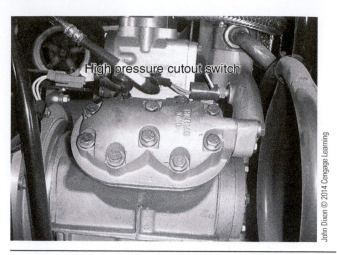

Figure 16-29 The high pressure cutout switch, which stops the engine from running in the event of high refrigerant operating pressure.

High Compressor Discharge Pressure

This switch may be installed in the compressor cylinder head, discharge manifold, or in the line between the compressor and condenser.

Figure 16-29 shows the high pressure cutout switch installed in the compressor discharge manifold. The switch shuts the engine down if the compressor discharge pressure exceeds a predetermined level, protecting the compressor and other components within the refrigeration system from excessive pressure.

Low Compressor Oil Pressure Switch

This switch is often an option. Again, it is part of the safety circuit. In the event that a compressor loses oil pressure, the unit shuts down. This is good insurance, if you consider the price of the safety switch compared to the price of a new or rebuilt compressor.

PERFORMANCE TASKS

The charging system should be inspected by the technician at regular preventive maintenance (PM) service times. The first thing that should be done is a visual inspection of the charging system components. A number of problems that would reduce charging system performance can be identified and corrected.

The battery should be tested for proper electrolyte level and state of charge. When the battery is fully charged, specific gravity should be between 1.25 and 1.27 at 80°F (26.7°C).

1. Check the battery terminals and cables. Make sure they are clean and tight.

2. Inspect the alternator drive belt for proper tension and belt condition; replace if cracked, frayed, or glazed.

3. Inspect the alternator, paying special attention to wiring connections and/or damaged wires. Tighten any loose connections.

Summary

- Transport refrigeration equipment uses 12 V automotive-type storage batteries.

- Batteries work on the principle that when two dissimilar metals are placed in an acid bath, an electrical potential is produced across the poles.

- Each cell of a battery produces approximately 2.1 V.

- 12 V batteries use six cells connected in series.

- Always follow safety precautions when working around batteries.

- Batteries may be shipped from the manufacturer as wet charged or dry charged.

- Batteries may be classified by the amount of maintenance required. The three basic types are conventional, low maintenance, and maintenance-free.

- Batteries may be rated by their physical size (BCI), by their cold cranking amp rating, or by their reserve capacity.

- Batteries should be stored in a clean, dry environment between 50°F and 60°F (10°C to 16°C).

- Batteries can be tested for their condition by an open circuit voltage test, by a measurement of the electrolyte's specific gravity, or by a load test.

- If a battery is discharged quickly, it should be charged fast. If the battery is discharged slowly, it should be charged slowly.

- Jump-starting a dead battery is performed when the unit's main battery is in a low state of charge when the unit must run. The jumped battery will require maintenance or replacement as soon as possible.

- The charging system is responsible for recharging the battery once the unit is started and also for providing current for system loads.

- The rectifier converts alternating current into direct current.

- The voltage regulator controls the alternator output voltage within a safe limit, as specified by the equipment manufacturer.

- An alternator output test confirms how many amps the alternator can produce when it is fully loaded, in comparison to the alternator's rated output.

- Starters for refrigeration units fall into two different categories, conventional starter motors and gear reduction starter motors.

- An overrunning clutch protects the starter from excessive engine rpm by disengaging the pinion gear from the armature of the starter.

- Starters are tested by measuring the number of amps they draw when starting the engine, in comparison to the starter's rated amp draw.

- Safety switches protect people from any personal injury as well as protect the unit from physical damage caused by running a unit under abnormal conditions.

Review Questions

1. What are the dimensions of a group 31 storage battery?

   A. 13 inches long, 6.8 inches wide, and 9.4 inches high

   B. 12 inches long, 10 inches wide, and 9 inches high

   C. 11 inches long, 11 inches wide, and 11 inches high

   D. 10 inches long, 10 inches wide, and 9 inches high

2. What type of battery will require that its water level be topped up more often during normal use?

   A. Maintenance-free batteries

   B. Low maintenance batteries

   C. Conventional batteries

   D. Dry charged batteries

3. When jump-starting a unit with a known good battery, what is the last connection to be made before starting the stalled unit?

A. Connect one end of the positive jumper cable to the positive (+) post of the stalled battery.

B. Connect one end of the negative jumper cable to the negative (−) post of the jumper battery.

C. Connect the other end of the jumper cable to the positive (+) post of the jump-starting battery.

D. Connect the other end of the negative jumper cable to a good chassis-ground connection on the stalled unit, away from the discharged battery.

4. Which characteristic about a battery can the cold cranking amp rating tell you?

A. It indicates the number of minutes a fully charged battery at 80°F (26.6°C) can sustain a load of 25 amps before the battery voltage drops to 1.75 V per cell or 10.5 V for a 12 V battery.

B. It indicates how much power a fully charged battery can deliver when ambient temperatures become extremely cold.

C. It indicates the load in amperes that a battery can sustain for 30 seconds at 0°F (−17.8°C) and not fall below 1.2 V per cell, or 7.2 V on a 12 V battery.

D. Both statements A and B are correct.

E. Both statements B and C are correct.

5. How should you charge a battery that has lost it charge over a long period of time?

A. The battery should be fast charged at a rate of 40 amps for up to 2 hours.

B. The battery should be slow charged at 6–8 amps for 8 to 10 hours.

6. When using the refrigeration unit's engine to load test a battery, after 30 seconds of cranking, the battery voltage should not go below _____ volts or the battery is defective.

A. 11

B. 10

C. 9

D. 8

7. How many volts are produced in each cell of a battery?

A. 2.1

B. 6.0

C. 9.6

D. 12.0

8. A battery's reserve capacity in measured in:

A. Amperes

B. Amp-hours

C. Watts

D. Minutes

9. What is the state of charge of a battery that has a specific gravity of 1.190 at 80°F (26.7°C)?

A. Fully charged

B. About three-fourths charged

C. About one-half charged

D. Completely discharged

10. A battery being load tested with a commercial load tester is discharged at one-half its CCA rating for _____ seconds.

A. 5

B. 10

C. 15

D. 20

11. Which component provides the power to operate electrical components used to control the refrigeration functions when the engine is running during normal system operation?

    A. Voltage regulator

    B. Battery

    C. Stator

    D. Alternator

12. The solenoid on a gear reduction starter:

    A. Pulls a drive lever to mesh the gears

    B. Pushes the pinion gear into mesh with the ring gear

    C. Is held in place by the pull-in coil

    D. Disengages the pinion gear from the starter armature

13. When the engine is started, the pinion gear is disconnected from the starter armature by the use of a

    A. Magnetic switch

    B. Plunger

    C. Overrunning clutch

    D. Switch return spring

14. When performing a starter draw test, if low current draw is detected, the most likely cause is:

    A. High resistance

    B. A bad starter

    C. A discharged battery

    D. A short in the starter

15. When a starter draw test is performed, if higher than specified current draw is detected, the most likely cause is:

    A. A discharged battery

    B. High resistance

    C. Battery terminal corrosion

    D. Engine problems or a defective starter

Truck and Trailer Refrigeration Maintenance

Learning Objectives

Upon completion and review of this chapter, the student should be able to:

- List the various components that must be checked and maintained on a routine PM service of a refrigeration unit.
- Describe all items that must be checked while servicing a diesel engine.
- Accurately check and adjust engine coolant strength.
- Describe refrigeration maintenance.
- List the steps required to perform a compressor pump down.
- List the steps required to perform a low side pump down.
- Describe the procedures used to replace a filter drier.
- Explain evacuation techniques.
- Explain a pre-trip inspection.
- Describe leak testing procedures.
- Describe soldering and brazing techniques.
- List structural items that should be checked during a PM.

Key Terms

bleeding

compressor pump down

flux

heat sink

low side pump down

magnehelic gauge

pre-trip

preventive maintenance (PM) service

INTRODUCTION

Proper maintenance of the refrigeration unit is extremely important in order to ensure the reliability and performance of the unit. Loads carried in the truck or trailer can be very expensive, reaching values of

$100,000 or more. Making sure the reefer unit (refrigeration unit) is reliable is a big concern for the shipper and the receiver. Often shipping receivers have a temperature tracking device hidden in with the cargo so that they are able to establish whether the product was shipped at the correct temperature. If the receiver is not satisfied with the results, the receiver may even refuse to accept the load, resulting in lost profit for shippers and possibly insurance negotiations for the shipping company.

SYSTEM OVERVIEW

In this chapter, the technician will learn the procedures that are necessary to perform a complete **preventive maintenance (PM) service** on a reefer unit, including engine service, fuel system service, compressor service, drive belt service, refrigeration system checks, glow plug system checks, and structural maintenance checks.

Only a unit passing an extensive PM should be put in service and loaded with product. In addition to performing proper preventive maintenance, a technician should be able to diagnose problems with the refrigeration system including refrigerant flow problems, leak testing, compressor oil change, and filter drier change.

ENGINE LUBRICATION SYSTEM

Engine oil and filters should be changed at regular intervals according to the manufacturer's recommendations. Because mileage is not an issue with refrigeration units, the intervals will be determined by hours of engine operation, recorded on the engine hour meter.

Engine Oil Change

Before the engine oil is changed, the engine should be warmed up to operating temperature. This will allow the oil to drain faster and more completely than if the engine oil is cold in the oil pan. Once the engine is warm, shut the unit off and remove the battery ground cable (so the engine can't accidentally start). Remove the drain plug from the oil pan or uncap the oil pan extension hose.

Refer to **Figure 17-1** for drain plug and oil filter location.

Have a large enough container to capture all the oil in the engine, approximately 16.9 quarts (16 liters) for a trailer unit. The unit and trailer must be kept in a level position to ensure that all the oil is allowed to drain from the oil pan. It is important to get as much of the oil out as possible because most of the dirt particles are in the last few quarts (liters) of oil to drain from the oil pan. When drainage is complete, reinstall the oil pan plug or cap and refill the oil pan with 16–17 quarts (16 liters) of oil. Then check the engine oil dipstick. Run the unit, check the oil level again, and adjust as necessary.

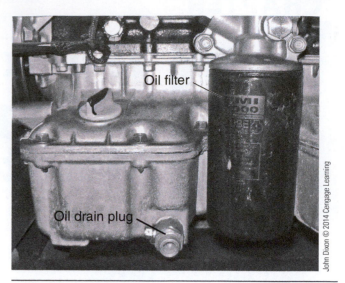

Figure 17-1 The location of the drain plug and oil filter.

Note: Always follow the manufacturer's recommendations for oil viscosity ratings for the ambient conditions to which the piece of equipment will be subjected.

Oil Filter Replacement

When changing engine oil filters, have a catch basin ready because there will be some oil loss. Use an oil filter wrench or large water pump pliers to remove the filter. Before installing the new oil filter, lightly lubricate the filter gasket with clean engine oil. Fill the oil filter with oil from a clean container, making sure no foreign material enters the filter. Failure to prime the filter with oil may allow the engine to operate for a period with no oil supplied to the bearings. Tighten the new filter until the gasket comes in contact with the filter base, and then tighten an additional half turn. Start the engine and check for leaks.

Fuel Filter Replacement

Fuel filters are another item that should be changed while servicing the engine. **Figure 17-2** shows a typical spin-on fuel filter.

- Start by removing the primary and secondary filters and discard. Using clean diesel fuel, lubricate the rubber gasket on the filter.

Note: Do not attempt to prime the fuel filters before installation because unfiltered fuel can damage the injection pump.

Figure 17-2 A typical spin-on fuel filter.

Figure 17-3 A bleeder screw as used on a Thermo King refrigeration unit.

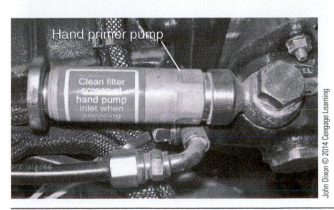

Figure 17-4 A manual hand pump used to prime the fuel injection pump.

- Tighten the secondary fuel filter until the rubber gasket makes contact with the filter base; then tighten an additional half turn.
- Install the primary fuel filter and spin it on, leaving the filter loose (rubber gasket not contacting the filter base).
- Operate the manual hand pump located on the injection pump until fuel bubbles surround the top of the primary filter.
- Tighten the primary fuel filter until the rubber gasket makes contact with the filter base; then tighten it an additional half turn.
- Start the engine and inspect for potential leaks.

BLEEDING THE FUEL SYSTEM

Bleeding of the fuel system will be necessary if the unit is allowed to operate until all the fuel in the tank is depleted. (The bleeding process removes unwanted air from the fuel system.) If this happens:

- Fill up the fuel tank for the refrigeration unit.
- Next, crack the bleeder screw located on top of the injection pump. Refer to **Figure 17-3,** which shows a bleeder screw as used on a Thermo King refrigeration unit.
- Unscrew the pump plunger (counterclockwise) from the manual hand valve. **Figure 17-4** shows

the manual hand pump used to prime the fuel injection pump.

- Operate the manual hand valve by depressing the plunger; spring force will allow the plunger to return for another stroke. This may take a few minutes of operation.
- Operate the hand pump quickly until air bubbles escaping from the bleeder screw are replaced by clean running fuel with no bubbles.
- Tighten the bleeder screw while continuing to operate the hand pump.
- Screw the hand pump plunger back into the pump assembly (clockwise).
- Operate the glow plugs if necessary and start the unit.
- Allow the unit to operate until the engine runs clean (engine not missing). This will ensure that all the air is worked out of the fuel system. Also make sure the alternator is not charging heavily. If it is, allow it to recharge the battery or install a battery charger to bring the battery back up to

John Dixon © 2014 Cengage Learning

Figure 17-5 A typical battery charger with boost capabilities.

full charge state. **Figure 17-5** shows a typical battery charger with boost capabilities.

Note: The battery is usually discharged almost completely due to operators trying unsuccessfully to start the unit, so the battery should be checked and charged as necessary, especially in cold ambient conditions.

For Carrier refrigeration units using a mechanical fuel pump, a bleeder valve is located on the top of the fuel injection pump. The bleeding process for these units is very similar to that of the Thermo King unit. To start:

- Turn the bleed valve (red) counterclockwise until it is completely open.
- Unscrew the pump plunger (counterclockwise) from the manual hand valve.

- Operate the hand pump quickly until a positive pressure (resistance) is felt on the plunger, indicating that fuel is flowing.
- Screw the hand pump plunger back into the pump assembly (clockwise).
- Operate the glow plugs if necessary and start the unit.
- Allow the unit to operate until the engine runs clean (until the engine is not missing); this will ensure that all the air is worked out of the fuel system.
- Close the bleed valve by turning it (clockwise) until it is completely closed.

Bleeding Fuel System with Electric Fuel Pump

If the refrigeration unit is equipped with an electric fuel pump (usually mounted on the fuel tank mounting bracket), proceed as follows:

- Turn the bleed valve (red) counterclockwise until it is completely open.
- Operate the glow plugs if necessary and start the unit.
- Allow the unit to operate until the engine runs clean (until the engine is not missing); this will ensure that all the air is worked out of the fuel system.
- Close the bleed valve by turning it (clockwise) until it is completely closed.

AIR FILTER SERVICE/ REPLACEMENT

Air cleaners filter all air entering the engine for the combustion process. In time the air cleaner will become dirty, causing the air to be restricted from entering the intake of the engine, resulting in a loss of horsepower, increased fuel consumption, and shortened engine life. The speed at which the air cleaner becomes fouled is proportionate to the conditions the unit is operating in. For example, the air cleaner will become dirty faster if the unit is operated in dusty areas or on secondary or gravel roads. Air cleaners should be serviced or replaced at every PM service. There are two styles of air cleaners are used in the transport refrigeration industry, oil bath type and dry type. Older model refrigeration units used air cleaners of the oil bath type while current units use a dry-type air filter.

Oil Bath Air Cleaner

The oil bath air cleaner is a serviceable unit.
Refer to **Figure 17-6** to see the components of an oil bath air cleaner.

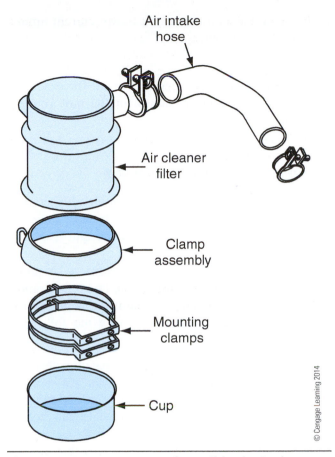

Air intake hose

Air cleaner filter

Clamp assembly

Mounting clamps

Cup

© Cengage Learning 2014

Figure 17-6 Components of an oil bath air cleaner.

All that needs to be done is to simply:

- Remove the lower cup of the assembly and discard the old, dirty oil.
- Wash the cup in solvent to remove sludge that has adhered to the cup assembly.
- Dry the cup of any solvent residue.
- Fill the cup to the oil level mark with clean oil of the same weight as that in the engine's crankcase.
- DO NOT OVERFILL THE CUP.

Dry-Type Air Cleaner

Many manufacturers using dry-type air cleaners will install an air restriction indicator in the air intake elbow. Refer to **Figure 17-7** to see the air restriction indicator. This indicator should be inspected periodically to confirm that the air cleaner is not restricted. Replace the dry air filter cartridge when the red signal remains in view with the engine shut off. To replace the dry-type air filter, stop the engine and remove necessary clamps. Discard the old air cleaner and install the new unit. Be certain all clamps are tight and fit is correct. Any leaks in the air intake will allow foreign material directly into the engine, causing damage and premature wear. **Figure 17-8** shows a typical dry-type

John Dixon © 2014 Cengage Learning

Figure 17-7 The air restriction indicator shows when the air cleaner is restricting the flow of fresh air to the engine's intake and warns when it is time to change the air filter.

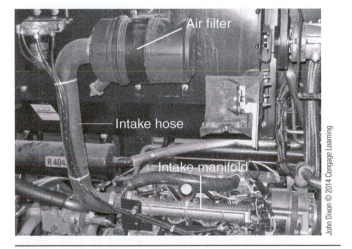

Air filter

Intake hose

Intake manifold

John Dixon © 2014 Cengage Learning

Figure 17-8 A typical dry-type air cleaner and intake system.

air cleaner and intake system. Depress the reset button on top of the restriction indicator after replacement has been performed.

DRIVE BELTS

During every service, the drive belts should be inspected and adjusted or replaced as necessary. It is important for V-belts to be in good condition, with proper tension, to provide adequate air movement across the evaporator and condenser coils. Look for signs of cracking, scuffing, or wear. Belt tension should always be checked during these routine services. A belt that is too loose can whip (long belts) as well as slip, and belts that are too tight create unnecessary strain on bearings and the belt itself. The use of a belt tensioning tool is recommended in order to achieve the recommended belt tension. Follow the manufacturer's recommendations for belt tension specifications.

Note: Always loosen belt adjustments before installing or removing a belt. Trying to pry a belt over a pulley will result in shortened belt life due to internal belt cord damage.

CAUTION *NEVER try to make any belt adjustments with the unit running. Due to the fact that many units today have automatic stop/start features, the best way to service belts on a refrigeration unit is to first disconnect the negative battery terminal. This will prevent accidental starting of the unit and serious personal injury or even death.*

GLOW PLUGS

The glow plugs should be checked whenever the unit comes in for service. The glow plugs preheat the combustion chamber to aid in quick starting, especially in cold ambient conditions. The glow plug circuit is energized whenever the preheat switch/start switch is toggled (in older units) and is shut off when the operator releases the switch. Some new units eliminate the preheat switch and the glow plug circuit is energized by the processor during the startup procedure. If glow plugs become defective, the unit will begin to have trouble starting, especially in cold ambient conditions.

Glow Plug Test

Glow plugs can be tested with the unit amp meter or each plug can be individually tested with an ohmmeter. To test for a burnt-out glow plug, energize the glow plug circuit. There should be a draw of approximately 28–30 amps. If the amperage rating is below 28 amps there could be a defective glow plug (burnt out) or a poor electrical connection at one or more glow plugs. To identify the defective plug or connection, remove the power terminal bar from each of the glow plugs and clean any corrosion as necessary. With an ohmmeter, use the engine block as the ground; test the resistance of each of the glow plugs individually. Each glow plug should indicate a resistance of around 1.55 ohms or a current draw of about 7 amps. A shorted glow plug will be indicated by a full current

discharge on the amp meter or a blown current limiter (fuse) at the amp meter.

Note: Refer to the service manual for exact current and resistance values for the unit you are working on.

ENGINE COOLING SYSTEM

Engine cooling systems for transport refrigeration units are similar to those used by trucks. Refer to **Chapter 8** for complete antifreeze strength tests and tests for antifreeze condition.

All equipment that contains antifreeze requires periodic maintenance and inspection, and refrigeration units are no different. This maintenance is performed to ensure that the antifreeze provides the following benefits for the reefer's cooling system:

- It prevents the coolant from freezing down to −30°F (−34°C).
- It slows down the formation of rust and mineral scale that can cause the engine to run hot or overheat.
- It slows down the corrosion (acid) that attacks the internal components of the cooling system.
- It provides the necessary lubrication for the water pump seal.

A good practice recommended by many manufacturers is to drain, flush, and replace the antifreeze mixture every two years (unless extended life coolant is used) to maintain total cooling system protection. Failure to maintain the condition of the coolant can result in scaling and higher acidity readings. If antifreeze is to be replaced with an ethylene glycol based coolant, it is recommended to use a 50/50 mixture of antifreeze and distilled water. This is true even of refrigeration units that will not be exposed to freezing temperatures. In the summer the accumulator can get cold enough to freeze water as it circulates around the perimeter of the tank. Also, a 50/50 mixture provides the necessary corrosion protection and water pump lubrication. During a PM, always check the coolant level and top up or test for coolant leaks, as necessary. Refer to **Chapter 7** for more information on pressure testing cooling systems.

Coolant Replacement

As previously stated, many manufacturers recommend replacement of the coolant every two years. The following is a basic guide to correctly removing and replacing coolant in a refrigeration unit.

Note: Always comply with environmental regulations in your area for properly disposing of coolant.

- Warm up the engine until it reaches operating temperature.
- Open the drain cock located on the engine block and allow coolant to drain, observing the color of the coolant. If the coolant is dirty, the cooling system should be flushed before replenishing with new coolant. See the section later in this chapter on flushing the cooling system.
- Run clean water into the top of the radiator or coolant expansion tank and allow it to flow through the system until the water draining out of the block is also clear. Refer to **Figure 17-9**, which illustrates the coolant expansion tank with cooling system pressure cap. For radiator cap testing, refer to **Chapter 8** of this text.
- Visually inspect all the coolant hoses for deterioration and check the hose clamps for tightness.
- Remove the tension from the water pump belt and inspect the bearing for looseness and signs of coolant leakage from the seal.
- Test the radiator cap. Please refer to **Chapter 8** for complete radiator cap test procedures.
- Premix a 50/50 solution of antifreeze and demineralized water in a container before adding to

the cooling system. Extended life coolants are premixed.

- Close the drain cock on the engine block.
- Refill the cooling system and warm the engine up with the radiator cap on loosely. This allows trapped air to be removed from the cooling system. Top up the system as necessary and secure the radiator cap.

CAUTION *Avoid direct contact with hot engine coolant.*

Flushing the Cooling System

When the coolant in the system is very dirty, flush using the following sequence:

- Run clean water into the top of the radiator and allow it to flow through the system until the water draining out of the block is also clear.
- Close the drain cock on the engine block.
- Install a commercially available radiator and block flushing agent. Run the unit, following the flushing agent manufacturer's instructions for correct procedures.
- Open the engine drain cock and drain the flushing agent.

Note: If the engine has already developed an overheating problem caused by scaling, the flushing agent will have little effect on the cooling system. In this case, the engine would have to be disassembled and the cylinder block and heads boiled in a soak tank.

Figure 17-9 The coolant expansion tank with cooling system pressure cap.

DEFROST SYSTEM

The defrost system should be checked on every service to check defrost system components. Start by running the unit on high-speed cool until the unit box temperature indicates it is at or below 38°F (3.3°C). Once this has been achieved, depress the manual defrost switch. The unit should now shift from the cool mode to the defrost mode. If the unit fails to go into the defrost mode, consult the service manual for defrost cycle checkout procedures.

Many units use a defrost air switch to initiate the defrost cycle. The defrost air switch senses the air pressure on the inlet side and the outlet side of the evaporator coil. **Figure 17-10** illustrates two different styles of air switches.

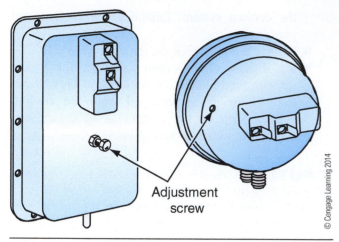

Figure 17-10 Two different styles of air switches responsible for initiating a defrost cycle.

When ice builds up on the coil, the pressure created by the evaporator fan will increase in comparison to the outlet side of the evaporator coil. When this happens, the bellows section within the air switch moves, tripping a microswitch and initiating a defrost cycle. The defrost cycle terminates when the defrost termination switch mounted on the evaporator reaches 56°F (13.3°C). Air switches are preset by the manufacturer and usually do not require recalibration unless the switch does not function properly or you want to test a new switch before installation.

Defrost Air Switch Check

Before testing the defrost air switch, check the sensing tubes into the evaporator section. Make sure tubes are not kinked, crushed, blocked, or split and that there is no low spot within the tube that could trap moisture. On Thermo King units, one tube will be clear and the other black. Examine the probes in the evaporator housing that the ends of the tubes connect to, making sure they are also not blocked. To perform this test, you will need an ohmmeter and a **magnehelic gauge**. **Figure 17-11** illustrates the setup to test a defrost air switch.

Follow these steps to test and set up the air switch:

- Remove both the sensing tubes from the air switch.
- Remove the wires from the terminals of the microswitch (connected to the air switch) and install the leads from the ohmmeter.
- Install the tube from the magnehelic gauge to the air switch that originally had the black hose connected to it. It is the tube that is routed through on the high pressure side of the evaporator.
- Raise the pressure with the magnehelic gauge until the switch closes, as indicated by the ohmmeter.

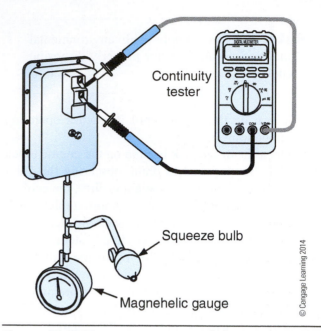

Figure 17-11 The setup to test a defrost air switch.

As the switch closes, read the scale on the magnehelic gauge. This is the set point of the air switch. The reading on the gauge will be indicated in inches or millimeters of H_2O.

- If the switch does not have continuity, pressurize the magnehelic gauge to the specifications found in the service manual and turn the adjusting screw clockwise until the switch closes, as indicated by the ohmmeter.
- Repeat the test procedure several more times to be sure the new setting is correct.
- Remove the test equipment and install the wire leads back onto the switch terminals. Install the sensing tubes back to their original positions, the black hose from the high pressure side of the evaporator to the air switch side stamped BLACK, and the clear sensing tube from the low pressure side of the evaporator to the side of the air switch marked CLEAR.

Note: Thermo King units use color-coded sensor tubes with their air switches. Carriers generally differentiate their sensor tubes by referring to one as the high pressure side and the other as the low pressure side. Testing and making adjustments to the Carrier air defrost switch is identical to the procedure previously described, with the exception of the switch setting. Always refer to the appropriate service manual for correct air switch specifications.

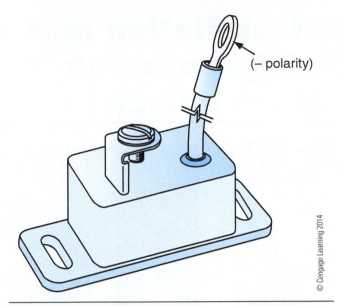

(– polarity)

© Cengage Learning 2014

Figure 17-12 A defrost termination switch.

Defrost Termination Switch

The defrost termination switch is mounted to the evaporator to control the defrost cycle of the unit based on the temperature it senses. **Figure 17-12** is a defrost termination switch used by Thermo King. The switch will close, completing a circuit, whenever the temperature of the evaporator is below approximately 38°F (3.3°C). If the switch is closed and a defrost cycle is initiated, hot gas will enter the evaporator section, quickly melting away ice built up on the evaporator. The water falls down into the drip pan, is expelled outside of the controlled space via the drain tubes, and drops to the ground. During this process, the defrost damper door is closed (Thermo King) or the evaporator fan is effectively turned off (units manufactured by Carrier), preventing the cycling of heat through the cargo area. When the evaporator temperature reaches approximately 56°F (13.3°C), the defrost termination switch will open, terminating the defrost cycle.

REFRIGERATION UNIT PRE-TRIP

Modern microprocessor-controlled refrigeration units incorporate a feature called a pre-trip. When a pre-trip is selected on the control panel by the unit operator, the unit will cycle itself through all different modes of operation. In each mode the processor evaluates the unit's performance and will signal a fault if parameters are not met. This test should be performed by the operator before the unit is loaded with product.

EXTERNAL LEAK CHECKING

If an external refrigerant leak is suspected, refer to **Chapter 7** of this text and read the testing procedures.

TESTING REFRIGERANT LEVEL

If a refrigeration unit is run with an inadequate supply of refrigerant, the suction pressure will be lower than normal, causing the refrigerant in the evaporator to boil well before it reaches the end of the coil. This will cause the box temperature to rise, causing an insufficient cooling complaint by the operator. The suction pressure will continue to drop with refrigerant loss.

Testing of the refrigerant level should be performed every time the unit comes in for routine maintenance. It should also be checked when recharging completely or topping up after major or minor service work on the refrigeration system. If it is suspected that there might not be a sufficient charge of refrigerant in a unit, the following test may be performed.

The following pressures are for units operating on R-12. Check the specific unit manual for actual specifications and for the refrigerant that is being used.

1. When the unit is operating in cool mode, there should be at least 150 psi of discharge pressure. (It might be necessary to cover the condenser to achieve this head pressure if the ambient temperature is low.)
2. Suction pressure and box temperature are about equal.
3. The ball should just be floating in the receiver tank.

The preferred method would be to lower the box temperature to 0°F (−18°C). For trailers loaded with products that must be carried at a temperature above 32° (0°C), it is recommended that an insulated test bag be placed over the evaporator so that a temperature of 0°F (−18°C) can be achieved. If it is impossible to meet these conditions, the unit could be low on refrigerant.

Recharging of the Refrigeration System

When it is necessary to completely recharge the refrigeration unit, evacuate the unit. (See evacuation procedures later in this chapter.) With the manifold gauges installed and the suction and discharge service valves mid seated, place a refrigerant cylinder of the correct type on an accurate scale and note the weight.

Manifold Gauge Installation and Removal on a Refrigeration Unit with Manual Service Valves

PHOTO SEQUENCE 5

P5-1 Before installing the manifold gauge set, remove the service valve caps.

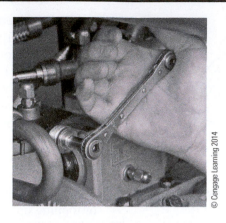

P5-2 Ensure that the stems are fully back seated.

P5-3 Remove acorn nuts from gauge ports.

P5-4 Install manifold hoses, blue to suction, red to discharge.

P5-5 Mid-seat the stems ¼ turn clockwise.

P5-6 Operate unit and take reading from the gauges.

PHOTO SEQUENCE 5 (Continued)

P5-7 To remove manifold set, back-seat the discharge service valve with the unit running.

P5-8 Open both manifold gauges and bleed the discharge side into the suction side to minimize refrigerant loss.

P5-9 Back-seat suction service valve.

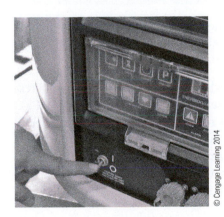

P5-10 Stop unit operation.

P5-11 Remove gauge service lines.

P5-12 Install all caps.

Find the specification for the refrigerant capacity in the service manual (usually marked on the serial number plate as well). Open the valve on the cylinder so that liquid refrigerant is being dispensed. Open the hand valve on the discharge side of the manifold and let the liquid refrigerant enter the system. Watch the weigh scale and shut off the manifold and tank valve when the required weight of refrigerant has been dispensed into the unit. If the refrigerant stops flowing and the required weight has not been installed, shut off the hand valve to the discharge manifold. Change the feed from the refrigerant cylinder so that only vapor can be dispensed. Start up the unit and place it in high-speed cool. Note the pressure on the suction gauge and open the suction manifold hand valve, allowing the vaporous refrigerant to flow into the low side of the system. The compressor will draw the refrigerant from the tank into the system as long as the tank pressure is higher than the suction pressure. Position the suction side manifold hand valve so that the refrigerant vapor is flowing at approximately 30 psi (207 kPa) higher than the base suction pressure. Watch the weigh scale, and turn off the manifold gauges and tank valve when the correct weight of refrigerant has been dispensed into the unit. A refrigerant management center can be used to perform many tasks for the refrigeration system. It can remove and recycle the refrigerant, evacuate the system, and install the correct amount of refrigerant as selected by the operator of the equipment. **Figure 17-13** is a picture of a refrigerant management center.

Partial Recharging of the Refrigeration System

When adding refrigerant, most manufacturers recommend that only vapor be added to the system through the suction service valve when the unit is running. If liquid refrigerant is added directly to the low side, damage to the compressor will result. Once it is determined that the unit is low on refrigerant, the following steps may be taken.

1. Attach the manifold gauge set.
2. Attach the manifold center hose to the refrigerant tank.
3. Open the tank valve in the vapor position and purge the center hose. (This may not be necessary with today's self-sealing refrigerant lines.)
4. If it is not possible to lower the unit or trailer temperature to 0°F (−18°C) due to the type of cargo, be sure to install a test bag over the evaporator.
5. Start the unit with the thermostat set at −20°F (−29°C).

John Dixon © 2014 Cengage Learning

Figure 17-13 A refrigerant management center capable of performing all refrigeration-related servicing.

6. Cover the condenser to help build up a compressor head pressure of at least 150 psi (1034 kPa).
7. Begin adding refrigerant vapor through the low side, suction the service valve until the receiver tank sight glass ball is floating, then close the tank vapor valve and bleed the line into the low side of the system.

COMPRESSOR OIL LEVEL CHECK

The compressor oil level should be checked whenever a refrigerant leak is suspected or when the unit has had refrigeration components serviced or replaced. To check the oil level of the compressor, follow the manufacturer's recommendations for compressor oil level checks. Most manufacturers have very similar procedures but they may vary due to the use of different refrigerants and pressures. Usually the unit is run for 15 to 20 minutes in high-speed cool. This will allow any oil that may have been trapped in the low side of the system to be drawn back into the compressor. (Sometimes oil can get trapped in the evaporator if the unit has last been running in the heat or defrost mode.) The oil level sight glass can then be checked. The compressor oil level sight glass is

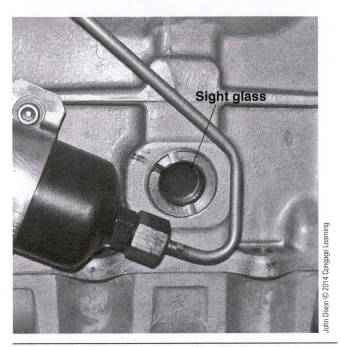

Figure 17-14 The location of the sight glass on a compressor.

positioned in the body of the compressor, close to where the oil pan is bolted to the compressor body. Generally the oil level in the sight glass should be one-quarter to halfway up the window of the sight glass. **Figure 17-14** shows the location of the sight glass on a compressor.

COMPRESSOR PUMP DOWN

Compressor oil should be changed periodically. Check the service manual for the compressor oil change interval. Compressor oil may be changed without removing the refrigerant charge from the unit. Only the compressor itself must be purged of refrigerant. To remove the refrigerant from the compressor, a process called a **compressor pump down** is required. Follow the steps below to perform a compressor pump down.

- Install the manifold gauge set and mid-seat the suction and discharge service valves.
- Run the unit in high-speed cool for 5 to 10 minutes to stabilize the system.
- While the unit is running, front-seat the suction service valve.
- When the suction pressure at the compressor is running in a 15- to 20-inch (51–68 kPa) vacuum, shut the unit off. If the compressor will not hold a vacuum, refer to the service manual for step-by-step procedures to determine what course of action to take next. The compressor must hold a

vacuum before proceeding. If the unit does hold a vacuum, continue on to the next step.

- If the compressor contains residual refrigerant, it may be necessary to restart the unit to allow it to pull down into a vacuum again. Residual refrigerant can be identified by a cool compressor body and bubbling of the compressor oil as observed through the compressor sight glass.
- With the compressor in a vacuum and the unit shut off, front-seat the discharge service valve.
- Using the manifold hand gauge, bleed some refrigerant from the high side to the low side until the low side has a 1- to 2-psi (6.9–13.8 kPa) positive pressure. When the compressor is opened to the atmosphere, it is important to have a positive pressure or foreign material may be drawn into the compressor.

Note: NEVER operate the unit with the discharge service valve closed because severe damage will occur.

- The compressor is now pumped down. Service work on the compressor can now be performed.

Placing Compressor in Service

When putting the compressor back into service after a compressor pump down, follow the steps below:

- With the service valves closed (front seated), install a vacuum pump to the manifold gauge set. Turn on the vacuum pump and open the hand valves of the manifold set. Draw the compressor into a deep vacuum and check to see that it holds for 5 minutes.
- Back-seat the service valves. The unit is now ready to be placed back into service.

COMPRESSOR OIL CHANGE

To change the compressor oil, first perform a compressor pump down. With the compressor at a positive crankcase pressure of 1 to 2 psi (6.9 to 13.8 KPa), place a drain pan under the compressor oil pan and open the fill plug to release the slight positive pressure. Then the drain plug may be removed. Allow the compressor oil to drain completely. Replace the sealing surface on the drain plug and remove the fill plug on the top of the compressor body. Install the quantity of oil as specified

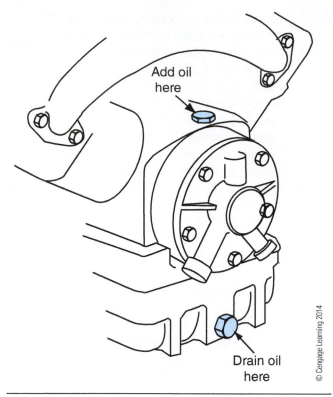

Add oil here

Drain oil here

© Cengage Learning 2014

Figure 17-15 The plugs in the compressor used to remove and refill compressor oil.

in the unit's service manual. Replace the sealing washer (or O-ring) on the oil fill plug and reinstall. Put the compressor back in service. Refer to **Figure 17-15** for oil fill and drain locations.

LOW SIDE PUMP DOWN

A **low side pump down** can be performed whenever service work is required on the low pressure side of the refrigeration system. This procedure allows the technician to service any component downstream of the receiver tank outlet valve all the way to the compressor without removing the refrigerant from the unit. To perform the low side pump down:

- Start by installing a manifold gauge set on the compressor and run the unit for 5 to 10 minutes to stabilize the refrigeration system.
- While the unit is running (high-speed cool), front-seat the receiver tank outlet valve. This will trap all refrigerant between the compressor discharge and the receiver tank.
- When the suction pressure at the compressor is running in a 15- to 20-inch (51–68 kPa) vacuum, shut the unit off. If the unit will not hold a vacuum, refer to the service manual for step-by-step procedures to determine what course of

action to take next. The unit must hold a vacuum before proceeding. If unit does hold a vacuum, continue on to the next step.

- Using the manifold hand gauge, bleed some refrigerant from the high side to the low side until the low side has a 1- to 2-psi (6.9 to 13.8 kPa) positive pressure. When the low side of the system is opened to the atmosphere, it is important to have a positive pressure or foreign material may be drawn into the low side.
- Service work on the low side of the system can now be performed.

Preparing for Back in Service/ Filter Drier Replacement

Once any service work has been completed, it is recommended that the filter drier be replaced. The replacement of the drier should be the last maintenance performed because the drier will start to absorb moisture as soon as the protective caps are removed from the new component. Once the new drier is installed, immediately evacuate the low side of the system. Open the receiver tank outlet valve (back seat). Remove the manifold gauge set. Now the unit is ready to go back in service.

REFRIGERANT REMOVAL

When the refrigeration system requires internal maintenance to repair or replace components, it may become necessary to remove the entire refrigerant charge. If this procedure is required, connect a refrigerant recovery system to the unit to ensure that it is compatible with the refrigerant contained within the unit. Refer to the instructions provided by the manufacturer of the refrigerant recovery equipment for proper operating and installation procedures.

Refer back to **Figure 17-13** to see a typical refrigerant management center. This piece of equipment is capable of performing all aspects of managing the refrigeration system. Refer to **Chapter 7** for further details on refrigerant management centers.

EVACUATION PROCEDURES

An evacuation is performed on the unit whenever the refrigeration system has been opened to the air, whether it be for servicing a component or a severe refrigerant leak. Evacuating procedures are much like the ones discussed in **Chapter 7** but on a bigger scale. Please refer to **Chapter 7,** which details the benefits of evacuation and dehydration of the refrigeration system.

Transport refrigeration equipment must be evacuated to remove air and moisture that was able to enter the system when it was open. Moisture is very lethal to the internal components of a refrigeration system. Moisture is very unfavorable in the system because it can cause copper plating, acid sludge formation, and freezing up of the TXV, as well as the formation of acids, resulting in metal corrosion within the system.

In order to perform a proper evacuation, a good, serviceable vacuum pump is required. This pump must be able to draw 5 CFM. In addition, a vacuum indicator should be used. A vacuum gauge is a much more accurate instrument for reading the vacuum level of the system; it is better to use a vacuum gauge than to rely on the manifold gauge set.

The following is a general process for effectively evacuating and dehydrating a refrigeration system. Always follow the manufacturer's recommended procedures.

- During the evacuation process, try to maintain a 60°F (15.6°C) temperature.
- Before evacuation, ensure that all refrigerant has been removed from the system.
- Attach the refrigerant hoses to the suction and discharge service valves. Also tee in the vacuum gauge. If working on a solenoid control system (Carrier), it is recommended to evacuate from three points due to the closed solenoid SV2 while in the off mode. The third point should be the service valve at the receiver tank outlet.
- Before opening the service valves (back seated) and the port to the vacuum gauge, run the vacuum pump until a deep vacuum is achieved. Then shut off the vacuum pump and observe the vacuum gauge. If the vacuum gauge rises, this indicates a leak in the evacuation equipment. Repair the leak as necessary until the vacuum equipment can hold a vacuum.
- Mid-seat the suction and discharge service valves on the compressor as well as the receiver tank outlet valve if the system is of the solenoid control system type.
- Operate the vacuum pump and evacuate the unit until the vacuum gauge indicates 2000 microns.
- Turn the vacuum pump off and seal it off from the system while monitoring the vacuum gauge. Wait a few minutes to make sure the vacuum holds.
- Restart the vacuum pump, allowing it to evacuate the unit down to 500 microns. Close off the vacuum pump and turn it off. Observe the vacuum gauge to be sure it holds the vacuum for 5 minutes. This will check for residual moisture or leaks.

- Refrigerant can be drawn into the system while it is in a vacuum. Note the weight of the refrigerant cylinder on an accurate weigh scale. Allow the correct amount of refrigerant to enter the unit by observing the scale. Do not overcharge. Refer to the unit service manual for refrigerant capacity and type.

SOLDERING AND SILVER BRAZING

A technician working on mobile refrigeration equipment must become skilled in the techniques of soft soldering and silver brazing to make repairs to the refrigeration system or to change defective components. All components within the refrigeration system are plumbed together with copper tubing that may use mechanical fittings, but due to the possibility of refrigerant leaks, most connections are soldered or silver brazed together. Whenever soft soldering or silver brazing, safe working practices should be strictly adhered to; these include:

- Work in a well-ventilated area with a fan or exhaust fan to vent harmful fumes that are produced from the parts and filler material during the brazing process. These fumes can cause sickness and in some cases could be fatal.
- Wear protective clothing (nonflammable) when performing brazing and soldering techniques.
- Wear proper eye protection at all times while brazing or soldering.
- Maintain all soldering and brazing equipment in good working order.
- Remove all flammable material from the area that will be used to solder or braze. When working on the unit itself, keep the flame away from flammable components such as wires and rubber components. Maintain your presence of mind while working in tight spaces.

INERT GAS BRAZING

During the brazing process, temperatures of 1150°F–1450°F (621°C–788°C) must be reached for some types of brazing alloys to melt and flow properly. At these temperatures, copper will react in the presence of oxygen to form a scale of copper oxide on the internal surfaces of the copper tube. The flow of refrigerant can cause this scale to break off in the form of flakes that will break down into a fine powder that can plug driers, filter screens, and parts of the TXV.

This can be avoided if there is no air (oxygen) inside the copper tubing while it is being brazed. This can be accomplished by allowing dry nitrogen to flow through the copper pipe while the work is being performed. Before starting to braze, allow time for the nitrogen to sweep the air out of the system. This can be done by connecting a hose from the pressure regulator of a nitrogen tank, pushing the nitrogen through the refrigeration system, and allowing it to flow out the open suction service valve into the atmosphere.

Note: Nitrogen cylinders are under extremely high pressure and require a pressure regulator to safely step down the pressure for this procedure. All that is required is a slow steady flow of nitrogen while brazing is being performed.

It is highly recommended that the inert gas method of brazing be used whenever possible. The nitrogen sweeps the line of trapped refrigerant, preventing acid formation caused by refrigerant breakdown as well as preventing copper oxide formation.

Silver Brazing

Silver brazing has become extremely popular in the transport refrigeration industry due to its superior qualities as compared to other soldering techniques. Silver brazing offers high strength, corrosion resistance, and vibration-proof and leak-tight joints. In fact, a silver brazed joint has vibration resistance and a strength safety factor at least 10 times stronger than that of soft solder joints. (From an environmental perspective, silver brazing permits far fewer refrigerant leaks than soldering or mechanical joints.)

One of the other advantages of silver brazing is its ability to join the similar and dissimilar metals used in the refrigeration industry. These metals include copper, brass, steel, malleable iron, etc. Aluminum and magnesium are probably the only metals that cannot be silver brazed with silver brazing alloys. The most common alloys for silver brazing are:

- Sil-Fos 5
- Sil-Fos 6
- Sil-Fos 15
- Phoson +

Silver alloy brazing is a procedure for joining metal when the filler metal has a melting point higher than 800°F (427°C) but below the melting temperature of the metals being joined.

Procedures for silver brazing are similar to those used for soft soldering. There are six steps for producing strong, leak-proof silver brazed joints. These include:

- Tight fitment of the joint
- Clean joint contact area
- Correct use of brazing flux on the joint
- Stress relief at the joint during brazing
- Correct heating and alloy flow techniques
- Final cleaning

Joint Fitment. When preparing a joint for brazing, ensure that the tubing is cut squarely. The use of a tube cutter is highly recommended. Deburring of the cut tube should be performed with the use of a deburring tool or reamer. Joint clearance should not be more than 0.005 inch (0.127 mm) and should be uniform all around the joint.

Joint Cleaning. Joint cleanliness is one of the most important aspects of joint preparation. Joint contact areas must be cleaned completely of oil, grease, rust, or oxides. Oil and grease may be removed with the use of a solvent, making sure to remove all residues. The inside of the joint may be cleaned with a tubing brush. The outside of the tube can be cleaned effectively with 320-grit sanding cloth.

Figure 17-16 illustrates some of the tools used to prepare surfaces for silver brazing.

Note: Once the surface has been cleaned, avoid touching the surface with your hands; oils and acid from the skin can prevent the alloys from sticking to the surface.

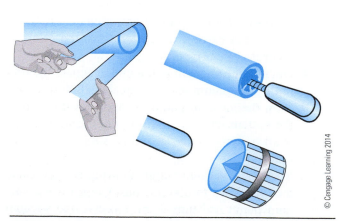

© Cengage Learning 2014

Figure 17-16 Some of the tools used to prepare surfaces for silver brazing.

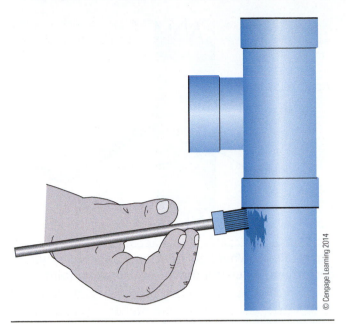

Figure 17-17 The correct procedure used to apply flux to a joint.

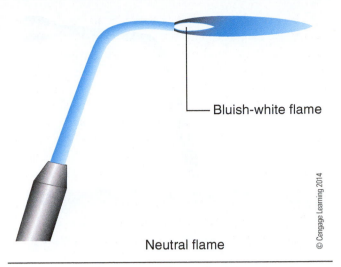

Figure 17-18 The correct flame for silver brazing.

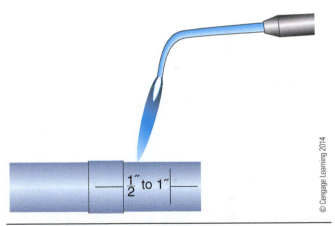

Figure 17-19 Where to start heating the tube to preheat the joint.

Joint Fluxing. The proper use of **flux** is also an important aspect of joint preparation. The flux should be applied so that it does not get inside the joint, causing contamination inside the system. To accomplish this, insert the cleaned tube part way into the fitting and brush the flux around the outside of the joint. The tube is then inserted all the way in. Brush flux evenly around the outside surface of the joint. **Figure 17-17** illustrates the correct procedure for applying flux to a joint.

Joint Stress Relief. Before making a brazing joint, align the assembly so that it is supported in such a way that it will not hinder expansion or contraction and so that there is no strain placed on the joint during brazing and cooldown.

General Joint Heating and Alloy Flow. Heating should be performed with a number 5 oxyacetylene tip for most efficient silver brazing. Air and propane torches and air-acetylene torches can also be used for tubing up to 1 inch (25 mm) in diameter. The flame should be adjusted for a neutral flame (blue flame with small inner cone). **Figure 17-18** illustrates the correct flame for silver brazing. The size of the torch tip should be selected so that the flame surrounds the circumference of the tube. Start heating the tube about 1/2 to 1 inch (13 to 25 mm) away from the end of the fitting. **Figure 17-19** illustrates where to start heating the tube to preheat the joint. Keep the flame moving around the circumference of the tube. Heat evenly so

that the tube expands evenly and carries the heat uniformly to the end inside the fitting. **Figure 17-20** illustrates the sweeping technique used to evenly heat the joint. Once flux on the outside of the tube has melted and has a milky appearance, move the tip of the torch up to the fitting. Keep the flame moving back and forth from the tube to the fitting, keeping it pointed at the tube to maintain temperature. The bulk of the heat should be concentrated at the fitting. The intention is to bring the fitting and tube up to temperature quickly and evenly before applying the brazing alloy. Do not stop moving the torch tip because it can quickly overheat a small section of the fitting. When the flux becomes a clear fluid on both the fitting and the tube, touch the alloy to the seam between the tube and the fitting. If the temperature is correct, the alloy will flow easily into the joint. Once enough alloy has been applied to the joint, one final

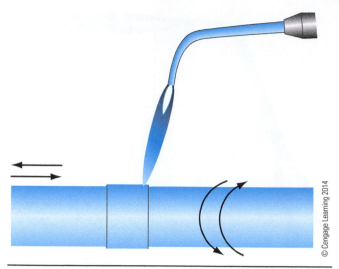

Figure 17-20 The sweeping technique used to evenly heat the joint.

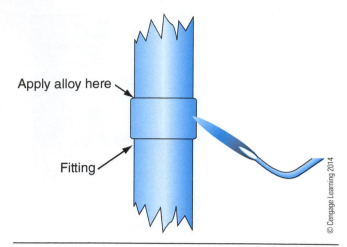

Figure 17-21 Where to heat and apply alloy when performing a vertical down silver brazing joint.

pass of the torch should be made at the base of the joint to expel any entrapped gas and flux.

Final Cleaning. Once the joint has been brazed, immediately apply a wet rag or brush to crack and wash the flux off the outside of the joint. If the pieces that have been joined are a subassembly not yet attached to the unit, flux from the inside of the joint can be removed by flushing with water. This can only be performed if it is possible to completely dry out the components before they become part of the system.

Vertical Down Joint Technique

When making a vertical down joint, heat the entire joint to temperature as described in the general section, heating the tube first and then the fitting. When the joint has reached the melting point of the alloy, concentrate the heat on the fitting while constantly moving the flame. The alloy will follow the direction of the heat, flowing down into the joint. **Figure 17-21** illustrates where to heat and apply alloy when performing a vertical down silver brazing joint procedure.

Vertical Up Joint Technique

When making a vertical up joint, heat the entire joint to temperature as described in the general section, heating the tube first and then the fitting. When the joint has reached the melting point of the alloy, concentrate the heat on the fitting while constantly moving the flame. Avoid heating the tube below the fitting as this will result in the alloy running out of the joint and down the tube. Heating the area slightly above the seam of the joint will cause the alloy to be drawn up

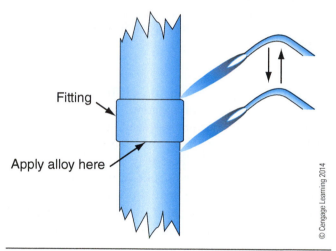

Figure 17-22 Where to heat and apply alloy when performing a vertical up silver brazing joint.

into the entire joint area. **Figure 17-22** illustrates where to heat and apply alloy when performing a vertical up silver brazing joint procedure.

Horizontal Joint Technique

To make a horizontal brazing joint, preheat the entire joint to temperature as described in the general section, heating the tube first and then the fitting. In the case of a horizontal joint, the alloy should be applied at the top of the seam while heating from the lower portion of the joint. The capillary action of the heat and gravity will cause the alloy to run down toward the heat and into the joint. Usually there will be a buildup of alloy at the bottom of the joint. **Figure 17-23** illustrates where to heat and apply alloy when performing a horizontal silver brazing joint procedure. Visually inspect the joint to ensure that alloy is visible all the way around the

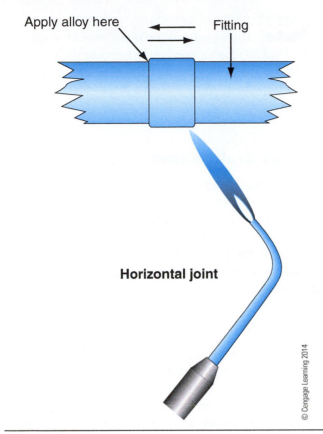

Apply alloy here

Fitting

Horizontal joint

Figure 17-23 Where to heat and apply alloy when performing a horizontal silver brazing joint.

joint; continue to heat and add more alloy to the joint as necessary.

Brazed Joint Disassembly

To disassemble a silver brazed joint, start by fluxing the entire joint. Heat the joint uniformly above the melting point of the alloy. Using a slight twisting action, pull the joint apart. Follow the six steps again to re-braze the joint back together, using additional alloy.

Soft Soldering

Soft soldering, sometimes referred to as low-temperature soldering, is commonly used in the refrigeration industry. The techniques of soft soldering are the same as for silver brazing. **Figure 17-24** illustrates locations to heat and apply solder during soft soldering procedures. Simply follow the six steps that are used for silver brazing. Temperatures for soft soldering are much lower than those required for silver brazing. The recommended filler metal for soft soldering is usually 95/5. This solder consists of 95% tin and 5% antimony and has a melting point of 450°F (232°C). Soft soldering can be performed with an oxyacetylene torch

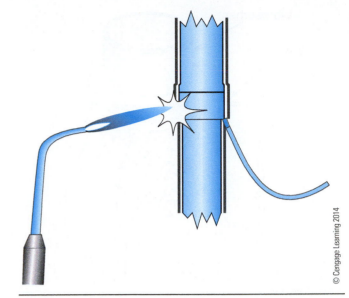

Figure 17-24 Soft soldering techniques for locations to heat and apply solder.

and can also be successfully performed with a simple propane torch. Flux is required for this type of soldering.

Note: When soldering or brazing around components that may be damaged by heat traveling by conduction, a **heat sink** should be used. **Figure 17-25** illustrates a heat sink. A heat sink will absorb the damaging heat that can destroy components such as vibrasorbers and service valves. If a heat sink is unavailable, a wet rag wrapped around the vibrasorber will draw enough heat away to prevent damage to the component. **Figure 17-26** illustrates the use of wet rags as a heat sink.

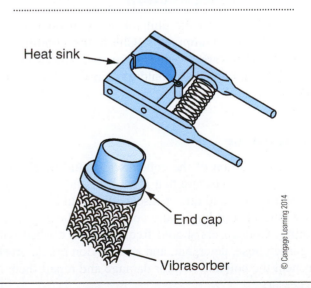

Heat sink

End cap

Vibrasorber

Figure 17-25 A heat sink used to draw away damaging heat from components when brazing or soldering.

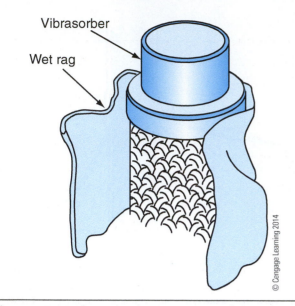

Vibrasorber

Wet rag

© Cengage Learning 2014

Figure 17-26 The use of wet rags as a heat sink.

STRUCTURAL MAINTENANCE

During all PM services, check the following items and service them as required.

Mounting Bolts

The mounting bolts fixing the unit to the trailer and engine mount bolts should be checked and tightened if necessary. Check the unit service manual for mounting bolt specifications.

Unit Visual Inspection

Visually inspect the unit for signs of oil leaks on the engine and compressor. Check the condition of wires and hardware; repair as necessary. Check for any physical damage that could possibly influence the performance of the unit.

Condenser

The condition of the condenser should be checked on every PM to ensure that it is clean and that there is no buildup of road dirt or insects that could restrict air flow and in turn hinder the condenser's heat transferability. Condensers should frequently be washed with a garden hose, detergent, and a soft nylon bristle brush. Inspect the coil and fins for damage and repair them as necessary.

Note: High-pressure washers should never be used to clean a condenser because the high-pressure water can damage the delicate fins and can impede air flow through the condenser.

Defrost Drain Hoses

During the PM, make sure the drain hoses are free of obstructions. This can be accomplished by running water into the evaporator drain pan and making sure water flows from both the drain hoses. (Ensure the unit is level.) If there is an obstruction, it can usually be cleared by blowing compressed air up through the drain tubes.

Evaporator

The condition of the evaporator should be checked on every PM to ensure that it is clean and that there is no debris that can restrict air flow and in turn hinder the evaporator's heat transferability. The evaporator fan can draw shrink wrap as well as particles from cardboard boxes and deposit them on the evaporator service. Evaporators should frequently be washed with a garden hose, detergent, and a soft nylon bristle brush. Inspect the coil and fins for damage and repair them as necessary.

Note: High-pressure washers should never be used to clean an evaporator as the high-pressure water can damage the delicate fins and can impede air flow through the evaporator.

Defrost Damper Door

If the unit is equipped with a defrost damper door, check the damper door bushing for wear as well as wear to the shaft. Ensure that the blade of the shaft seals the flow of air and that the damper door solenoid is bottomed out when energized. Make any adjustments as necessary. Some manufacturers do not use a defrost damper door in their defrost systems. Instead, they may choose to use an electric fan that can be switched off during the defrost cycle. Still others use an electromagnetic clutch to effectively turn on or turn off the fan shaft to the evaporator section of the unit (Carrier).

Summary

- A PM on a reefer unit includes engine service, fuel system service, compressor service, drive belt service, refrigeration system checks, glow plug system checks, and structural maintenance checks.

- Engine oil and filters should be changed at regular intervals according to the manufacturer's recommendations.

- Fuel filters are additional items that should be changed while the engine is serviced.

- The bleeding process removes unwanted air from the fuel system.

- Air cleaners filter all air entering the engine for the combustion process.

- Belt tension and condition should always be checked during every PM.

- Glow plugs preheat the combustion chamber to aid in quick engine starting.

- The condition of the engine coolant should be checked periodically. Coolant not capable of performing can damage the engine as well as the cooling system.

- The cooling system should be flushed when it appears to be very dirty.

- The purpose of the defrost system is to remove accumulated ice from the evaporator coil.

- The defrost air switch is responsible for initiating a defrost cycle. (Electronic timers as well as microprocessors can also initiate a defrost cycle.)

- The defrost termination switch is responsible for ending the defrost cycle.

- Refrigerant level should be checked during every PM. Follow the instructions in the unit service manual for proper procedures for checking the level for that particular unit.

- The compressor oil level should be checked whenever a refrigerant leak is suspected or when the unit has had refrigeration components serviced or replaced.

- A low side pump down can be performed whenever service work is required on the low pressure side of the refrigeration system. This procedure allows the technician to service any component downstream of the receiver tank outlet valve all the way to the compressor without removing the refrigerant from the unit.

- To remove the refrigerant from the compressor, a process called a compressor pump down is required.

- The compressor oil can be changed or service work to the compressor or suction throttling valve may be performed when the compressor is pumped down.

- The evacuation process removes air and moisture that are able to enter the refrigeration system when it has been opened for service.

- Always follow safety precautions when soldering or brazing refrigeration components.

- Causing an inert gas to flow through the joint will prevent copper oxide from forming on the inside of the tube.

- To produce a trouble-free silver brazed joint, always follow the six steps.

- Follow correct heating procedures for the type of joint you intend to braze.

- Soft soldering requires the same preparation as silver brazing; the real difference is the melting point of the solder.

- The structural integrity of the unit should be examined on a PM. This would include examination of the unit engine and fuel tank mounting; visual inspection of the unit; checking cleanliness of the coils; checking evaporator drainage; and inspecting the defrost damper door system.

Review Questions

1. In reference to soft solder 95/5, what are the contents?

   A. 95% antimony, 5% tin

   B. 95% tin, 5% antimony

   C. 95% lead, 5% antimony

   D. 95% lead, 5% tin

2. At what temperature will soft solder 95/5 flow?

   A. 300°F (149°C)

   B. 350°F (177°C)

   C. 400°F (204°C)

   D. 450°F (232°C)

3. Copper oxide on the internal surfaces of a tube being silver brazed can be eliminated by:

   A. Flowing oxygen through the tube while it is being brazed

   B. Flowing helium through the tube while it is being brazed

   C. Flowing nitrogen through the tube while it is being brazed

   D. Flowing air through the tube while it is being brazed

4. A silver brazed joint is _____ times stronger than a soft soldered joint.

   A. Two

   B. Four

   C. Eight

   D. Ten

5. *True or False:* When servicing a diesel engine, it is important to drain the engine oil after the engine has warmed up so that the oil will flow out faster and more completely than if it were cold.

   A. True

   B. False

6. Bleeding the fuel system is necessary to remove unwanted air from the fuel system when the unit has:

   A. Run out of fuel

   B. Had fuel filters replaced

   C. Had maintenance performed to the fuel system

   D. All of the above statements are correct.

   E. None of the above statements is correct.

7. The receiver-drier can be replaced while the unit is in the following state:

   A. A low side pump down

   B. A compressor pump down

   C. A state in which all the refrigerant has been removed from the system

   D. Both statements A and B are correct.

   E. Both statements A and C are correct.

8. Compressor oil can be drained and replaced when the unit is in the following state:

   A. A low side pump down

   B. A compressor pump down

   C. A state in which all the refrigerant has been removed from the unit

   D. All of the above statements are correct.

9. The defrost termination switch senses the temperature of the:

   A. Condenser coil

   B. Evaporator

   C. Compressor discharge

   D. Hot gas line

10. What instrument(s) is/are used to make adjustments to a defrost air switch?

    A. Voltmeter

    B. Ohmmeter

    C. Magnehelic gauge

    D. Both A and B are correct.

    E. Both B and C are correct.

Glossary

absolute zero Absolute zero is the point where no more heat can be removed from a system.

accumulator A storage vessel located between the outlet of the evaporator and the inlet to the compressor. Its purpose is to prevent liquid refrigerant from returning to the compressor.

algorithm A step-by-step problem-solving procedure, especially an established, recursive computational procedure for solving a problem in a finite number of steps.

alternative refrigerants Any approved refrigerant to replace an ozone-depleting substance in an air-conditioning or refrigeration system.

ambient temperature The temperature of the surrounding air. In a refrigeration system, this refers to the temperature outside the controlled space.

antifreeze A chemical specifically designed to increase the boiling point and decrease the freezing point of engine coolant. Antifreeze is generally mixed at a 50/50 ratio with water.

aqueous A solution in which a component(s) is dissolved in water.

atmospheric pressure The weight exerted on the earth's surface at a given elevation, generally expressed as 14.7 psi (101.3 kPa) at sea level.

auto stop/start A feature incorporated into many refrigeration units that saves fuel by allowing the unit to shut down when the set point is reached and that restarts the unit when box temperature differs from the set point.

auxiliary power units Portable, truck mounted systems that can provide climate control (heating and cooling) as well as electricity for electrical appliances.

Battery Council International (BCI) A trade organization that brings together the leading lead acid battery manufacturers in North America and other major players from around the world. Externally, BCI provides information and resources on the industry to numerous outside organizations and researchers.

bidirectional Two-way communication is a form of transmission in which both parties involved transmit information.

binary switch Controls (stops) compressor operation if pressure is too high or too low.

bleeding refrigerant Slowly releasing the refrigerant pressure by allowing it to pass into an area that is lower in pressure.

bleeding the fuel system The act of removing air from a fuel system. Bleeding the fuel system is necessary when the fuel tank has been allowed to run dry. The injection must be brimmed and the air bled (removed) from the fuel system.

blend air system An air-conditioning system that controls temperature by blending heated and cooled air to the selected cab temperature.

boiling point The temperature at which a liquid changes states into a vapor.

box temperature The temperature of the controlled space inside the truck box or inside the trailer.

British thermal unit (BTU) BTU is the measurement for quantity of heat energy. It requires 1 BTU of heat energy to raise the temperature of 1 pound of water 1 degree Fahrenheit.

bypass circuit Term used to describe the flow of engine coolant through the engine before the thermostat opens, allowing the coolant to recirculate through the entire cooling system.

capillary tube The tube with a calibrated inside diameter and length used to transmit pressurized gas from the remote bulb to the power head portion of the thermostatic expansion valve (TXV).

catastrophic Extremely harmful; complete physical ruin.

change of state Rearrangement of the molecular structure of matter as it changes between any two of the three physical states: solid, liquid, or gas.

charging cylinder A container with a visual indicator for use where a critical or exact amount of refrigerant must be dispensed.

check valve A valve in a refrigeration system that allows the flow of refrigerant in one direction and stops the flow in the opposite direction.

chlorofluorocarbons (CFCs) A synthetic compound used in refrigerants such as R-12, more accurately designated CFC-12.

Clean Air Act (CAA) A Title IV amendment, signed into law in 1990 by President Bush, that established national policy relative to the reduction and elimination of ozone-depleting substances.

cold cranking amps The load in amperes that a battery can sustain for 30 seconds at $0°F$ ($-17.8°C$).

combustible Capable of igniting and burning.

compressor A major component used to increase refrigerant pressure and circulate refrigerant throughout the system.

compressor pump down Service procedure used to remove refrigerant from the compressor crankcase so that service work may be performed on the compressor, for example, replacement of compressor oil.

condensation Condensation is the formation of liquid drops of refrigerant from refrigerant vapor.

condenser A major component of the refrigeration system used to change high-pressure refrigerant vapor into high-pressure liquid refrigerant.

condenser pressure bypass valve Used with the three-way valve to ensure condenser pressure does not exceed discharge pressure. Helps the three-way valve shift from the heat mode to the cool mode of operation.

conduction The transfer of heat through solids.

constant discharge temperature control An HVAC system that automatically maintains the selected temperature within the cab regardless of ambient conditions.

continuity Continuous electrical connection.

continuous run mode Mode of reefer operation in which the unit never shuts down once the set temperature has been achieved; used for products requiring constant air circulation.

convection The transfer of heat by the circulation of a vapor or liquid.

conventional starter motor The pinion gear is on the same shaft as the motor armature and thus rotates at the same speed. (No mechanical advantage is obtained.)

cool cycle A refrigeration mode of operation used when the contents inside the controlled space require the removal of heat to maintain the desired set point.

counterflow A radiator design in which the coolant flows into the bottom radiator tank, through the radiator to the top tank, and then from the top tank back down through the radiator to the bottom tank, and back into the engine.

crossflow A radiator in which the coolant flows from one side to the other, as opposed to a vertical flow radiator.

data bus The linking of separate computers together.

de-energize To remove the power supplied to an electrical component.

deep cycle The rapid, continuous, full discharging and recharging of batteries.

defrost cycle A refrigeration cycle the melts ice accumulated on the evaporator coil.

desiccant The medium in an air-conditioning or refrigeration system that removes moisture (water) from the system. The desiccant is located in the receiver-drier, filter drier, or accumulator drier.

discharge line Refrigerant line that connects the discharge side of the compressor to the inlet of the condenser.

disposable refrigerant cylinders Containers for packaging, transporting, and dispensing of refrigerant, designed for one-time use. It is a violation of federal law to refill these cylinders, commonly referred to as DOT-39s.

double pass When coolant flows through the core of the radiator twice before flowing back into the engine.

downflow A radiator in which the coolant flow is from the top tank to the bottom tank, as opposed to a crossflow radiator.

drier Contains a desiccant and usually a filter to remove moisture and contaminants from the refrigerant.

dry charged Batteries that are distributed with no electrolyte in the cells.

electromagnetic clutch A device installed into the drive pulley on refrigeration compressors. A metal core and a coil of wire produce a magnetic effect when electrical current is applied to the coil. This device effectively turns the compressor on and off.

energize To apply electrical power to a component.

EPA Abbreviation for the Environmental Protection Agency.

equalized To be equal or in balance.

equalizing See equalized.

evacuated See evacuation.

evacuation To create a vacuum within a system to remove all trace of air and moisture.

evaporation Evaporation is the process of liquid turning to vapor.

evaporator The component of an air-conditioning or refrigeration system that transfers the heat from the controlled space to the refrigerant within.

externally equalized A style of TXV that receives a direct pressure feed from the evaporator outlet.

fan cycling switch Usually located in the receiver-drier, the fan cycling switch turns the engine cooling fan on or off at predetermined refrigerant pressures. (It will override the engine fanstat to maintain safe refrigerant pressure.)

fanstat A combination temperature sensor and switch (usually pneumatic) used to control the engine fan cycle.

fast charging The rate of charging a battery using high amperage of approximately two hours.

field circuit An electrical circuit that provides the current to set up a magnetic field in the rotor.

fixed orifice tube A device that converts high pressure liquid Refrigerant into low pressure liquid refrigerant.

flash gas The gas resulting from the instantaneous evaporation of refrigerant in a pressure-reducing device such as an expansion valve or a fixed orifice tube.

flooded A term used to indicate that the evaporator is receiving more liquid refrigerant than what can be evaporated.

flushing To remove solid particles such as metal flakes or dirt. Refrigerant passages are purged with a clean dry gas such as nitrogen.

flux A substance used in the joining of metals when heat is applied, to promote the fusion of metals.

four-way valve (reversing system) A valve used in a refrigeration system that reverses the flow of refrigerant so that the evaporator becomes the condenser and the condenser becomes the evaporator while in the heat or defrost mode.

freezing Freezing is a phase in which a liquid turns to solid when its temperature is lowered below its freezing point.

gassing To give off gas.

gear reduction starter motor A starter motor that uses gears to obtain a mechanical advantage to crank the engine.

greenhouse effect A greenhouse is warm because glass allows the sun's radiant heat to enter but not escape. Global warming is caused by some gases in the atmosphere that act like the glass of a greenhouse, hence the term greenhouse effect.

greenhouse gases Gases in the atmosphere that create the greenhouse effect.

heat cycle A refrigeration mode of operation used when the contents inside the controlled space require heat to maintain the desired set point.

heat exchanger An apparatus in which heat flows from one fluid or gas to another fluid or gas, on the principle that heat flows from the warmer substance to the cooler substance.

heat of respiration The heat given off by ripening fruit or vegetables in the conversion of starches and sugars.

heat transfer The flow of heat based upon the principles of thermodynamics.

high-pressure relief valve A mechanical device designed to release dangerously high refrigerant pressures from the system to the atmosphere.

hot load The condition of a load that has not been pre-cooled before being loaded for transportation. (Refrigeration units are not designed to cool the load but are designed to maintain the temperature of the load.)

humidity A term used to describe the amount of water vapor in the air.

HVAC Abbreviation for heating, ventilation, and air conditioning.

hydrochloric acid A corrosive acid produced when water and refrigerants are mixed as within an air-conditioning system.

hydrochlorofluorocarbons (HCFCs) Any of a class of inert compounds of carbon, hydrogen, hydrocarbons, chlorine, and fluorine, used in place of chlorofluorocarbons as being somewhat less destructive to the ozone layer.

hydrofluorocarbons (HFCs) A synthetic compound used in refrigerants, such as HFC-134a.

hydrometer A device used to measure the specific gravity of a coolant used in determining its freeze protection. Specific hydrometers may also be used to test a battery's state of charge.

hydronic heater systems A heating system using fluid as a medium for heat transfer.

hygroscopic Readily absorbing and retaining moisture.

impeller A rotor located in a conduit to impart motion to a fluid.

interface When two or more computers communicate with each other.

internally equalized A TXV that does not sense the evaporator outlet pressure.

kinetic energy Energy in motion.

latent heat The heat required to cause a change in state of a substance without changing its temperature.

latent heat of condensation The energy required to change a substance from a gas to a liquid.

latent heat of fusion The energy required to change a substance from a liquid to a solid.

latent heat of vaporization The energy required to change a substance from a liquid to a vapor (gas).

leak detectors Instruments used to locate refrigerant leaks in an air-conditioning or refrigeration system.

LED Abbreviation for light emitting diode.

liquid line The refrigerant line leaving the condenser before reaching the entering device. Also describes the state of the refrigerant.

litmus test A test for chemical acidity or basicity using litmus paper.

load test Checks a battery's performance rating.

low pressure cut-off switch An electrical switch that is activated at a predetermined low pressure. This switch usually opens the electromagnetic clutch circuit, preventing damage to the compressor.

low-side pump down A term used in a refrigeration system when the receiver tank outlet valve has been front-seated and all refrigerant in the system is trapped between the compressor outlet and the receiver tank. Anything downstream of the receiver can be serviced when the unit is in this state.

magnehelic gauge A tool used to measure both positive and negative low air or gas pressures.

manifold gauge set A device consisting of a manifold (a pipe or chamber with multiple apertures for making connections), calibrated gauges, and charging hoses.

moisture indicator Instrument used to measure moisture content of a refrigerant.

MOSFET An electronic device used to signal the fan actuator circuit to activate the on-off fan drive for the cooling fan.

multi-temp unit A trailer refrigeration unit that has the ability to maintain compartments at different temperatures.

nominal Variations that are insignificantly small.

non-positive Pumps that discharge liquid in a continuous flow are referred to as non-positive displacement pumps and do not provide a positive seal against slippage, as do positive displacement pumps.

orifice tube A refrigerant metering device used at the inlet of evaporators to control the flow of liquid refrigerant into the evaporator.

orifice tube air-conditioning system A type of air-conditioning system that uses a fixed orifice tube instead of a TXV as a metering device to control refrigerant flow.

open circuit voltage test A method of determining a battery's state of charge.

overrunning clutch A device that prevents damage to the starter motor once the engine has been started by disengaging the pinion gear from its drive shaft with the use of spring-loaded wedged rollers.

ozone layer A layer in the stratosphere (at approximately 20 miles or 32 km above the earth) that contains a concentration of ozone sufficient to block most ultraviolet radiation from the sun.

PAG Abbreviation for the synthetic lubricant polyalkaline glycol.

pilot solenoid An electrically activated valve used to shift a three-way valve system into the heat mode of operation.

PM An acronym for preventive maintenance.

potted Mounting of electronic components in an epoxy compound that provides a complete environmental seal, which also enhances heat sinking for electrical components.

pre-cooling The cooling of a truck or trailer box to remove residual heat and humidity from the controlled space before it is loaded with product.

pressure Defined as the force per unit area; refrigerant pressure is commonly measured in pounds per square inch.

pre-trip Checking manually or automatically if a refrigeration unit is functioning properly before it is loaded with product.

pulse width modulation (PWM) Constant frequency digital signal in which on/off time can be varied to modulate duty cycle.

purge To remove all refrigerant from an air-conditioning or refrigeration system.

quench valve Device that monitors the temperature of the compressor discharge line. If the temperature is too high, the quench valve opens to allow a small amount of liquid refrigerant into the suction line, cooling the compressor.

radiation The transfer of heat without heating the medium through which it is transmitted.

ram air effect Air forced into the engine or passenger compartment by the force of the vehicle moving forward.

receiver A container for the storage of liquid refrigerant.

receiver-drier A combination container for the storage of liquid refrigerant and a desiccant.

recirculation To move or flow continuously along a closed path or system. For example, the air in the controlled space of the refrigeration system is recirculated through the evaporator. The coolant is recirculated through the engine block until the thermostat opens.

recovered refrigerant See recovery.

recovery The recovery of refrigerant means to remove it, in any condition, from a system and store it in an external container without necessarily testing or processing it in any way.

rectifier A device, such as a diode, that converts alternating current to direct current.

recycle To clean the refrigerant for reuse by oil separation and pass it through other devices, such as filter driers, to reduce moisture, acidity, and particulate matter. Recycling applies to procedures usually accomplished in the repair shop or at a local service facility.

recycled refrigerant See recycle.

refillable refrigerant cylinders Refrigerant cylinders that can safely dispense refrigerant or be refilled with refrigerant, as with a refrigerant recovery station.

refractometer An instrument that can be used to test the freeze protection of antifreeze; other styles can be used to check the specific gravity of battery electrolyte.

refrigerant hose Same as refrigerant line but usually manufactured with rubber and other materials.

refrigerant identifier A device used to determine the purity of refrigerant.

refrigerant lines Hoses or piping that connect one A/C or refrigeration component to the next.

refrigerants The chemical compound used in a refrigeration system capable of vaporization at a very low temperature and pressure.

remote bulb A sensing device connected to the expansion valve by a capillary tube. This device senses the evaporator outlet temperature and transmits pressure to the expansion valve for its proper operation.

reserve capacity The ability of a battery to sustain a minimum vehicle electrical load in the event of a charging system failure.

residual heat The heat built up in the components making up the controlled space (wall, floor, roof, etc.).

residue Matter remaining after completion of an abstractive chemical or physical process, such as evaporation, combustion, distillation, or filtration.

rotary vane compressor A type of positive displacement air-conditioning compressor.

safety valve Valves used to safely relieve pressure within a refrigeration (A/C) system.

saturated liquid Describes the presence of both liquid and vapor coexisting during a change of states.

scan tool Electronic tool that allows the technician to retrieve trouble codes as well as interface with the vehicle's computer.

scroll compressor A type of refrigerant compressor design.

Section 609 A section of an EPA document that states that any person who services motor vehicle air conditioning must be certified. Servicing motor vehicle air conditioning includes repairs, leak testing, and topping off of air-conditioning systems low on refrigerant, as well as any other repair to the vehicle that requires dismantling any part of the air conditioner.

sensible heat Heat that causes a change in the temperature of a substance, but does not change the state of the substance.

service valves Devices located on the compressor that enable the service technician to check the operating pressure of a system and perform other necessary operations.

set point The selected temperature to which a reefer thermostat is set.

shore power systems Electrical outlets that the truck can plug into to provide power to run HVAC systems or any other appliance the operator wished to plug into the vehicle's outlets (TV, refrigerator, microwave, etc.).

short cycling A term used when a refrigeration unit alternates between heat and cool modes (or turns on or off) more often than normal.

shutterstat A temperature actuated control mechanism located in the coolant manifold. It is used to control the louvers that control the amount of air flow through the front grill of the vehicle.

sight glass A window in the liquid line or in the top of the drier; this window is used to observe the liquid refrigerant flow.

single pass When coolant flows through the core of the radiator once before flowing back into the engine.

slip rings These conduct current to the rotor. Most alternators have two slip rings mounted directly on the rotor shaft; they are insulated from the shaft and from each other. A spring-loaded carbon brush is located on each slip ring to carry the current to and from the rotor windings.

slow charging The rate of charging a battery, using low amperage for a duration of 8 to 10 hours.

SNAP Acronym for Significant New Alternative Policy.

solenoid control valve A valve controlled by an electromagnet, used to perform work. A solenoid is made with one or two coil windings wound around an iron core.

specific gravity Scientific measurement of liquid weight based on the ratio of the liquid's mass to an equal volume of distilled water, commonly used to measure battery state of charge.

stabilize When refrigerant pressure holds steady.

stand-alone systems These systems for HVAC operations are contained in a structure above the truck parking spaces. Ventilation is obtained from a hose attached to the truck's window.

starved A term that refers to an evaporator that has too little refrigerant metered into it, resulting in poor cooling and a low suction pressure.

stepper motor Small devices convert electric signals to a stepped rotating motion that directly moves a door in the air ducting.

strainer A metal mesh located in the receiver, expansion valve, and compressor inlet to prevent particles of dirt from circulating through the system.

stratospheric ozone depletion Damaging effects to the layer above the earth's surface at an altitude of between 7 and 30 miles (11 and 48 kilometers). This layer protects the earth from the ultraviolet (UV) rays from the sun.

subcooling The cooling of liquid refrigerant below its condensing temperature.

suction line The line connecting the evaporator outlet to the compressor inlet.

suction pressure regulator Device designed to limit compressor crankcase suction pressure during the heat-defrost cycle or on startup.

superheat A term used to describe the temperature of a vapor after all liquid has evaporated and the temperature of the vapor is above the boiling point of the liquid.

supplemental coolant additive This is present in antifreeze or can be added to the cooling system periodically to enhance many performance characteristics of a coolant solution.

swash plate compressor A mechanical system that is used for pumping; it has an angle plate attached to a center shaft, and pistons attached along the axis of the shaft. As the shaft is rotated, the pistons move in and out of the cylinders, producing suction and pressure.

sweep To remove moisture, air, and contaminants for a refrigeration system or component by flushing it with a dry gas, such as nitrogen.

temperature Heat intensity as measured with a thermometer.

thermal equilibrium When no temperature change takes place between two objects (heat balance).

thermatic viscous drive A style of engine cooling fan that uses silicone fluid as a drive medium between the drive hum and the fan drive plate.

thermistor A temperature-sensing resistor that has the ability to change values with changing temperature.

thermistor vacuum gauge Device used to accurately measure the last, most critical inch of vacuum during the evacuation process.

thermometer A device used to measure temperature.

thermostat A device that is used in truck cooling systems to maintain optimum engine operating temperature. Or a device to control the heating and cooling operations of a refrigeration unit to maintain a desired box temperature.

thermostatic expansion valve (TXV) A refrigerant metering device used to maintain a constant superheat setting at the evaporator outlet.

thermostatic switch A device used to cycle the clutch to control the rate of refrigerant flow as a means of temperature control. The vehicle operator controls the temperature desired.

three-way valve A valve trademarked by the Thermo King Corporation, used to control the flow of refrigerant for either heating or cooling modes of operation.

ton A ton of refrigeration capacity is equivalent to 12,000 BTUs of heat energy.

top up To add refrigerant to a system that is low on its refrigerant charge.

total dissolved solids (TDS) Dissolved minerals measured in a coolant by testing the conductivity with a current probe (TDS tester). High TDS counts can damage moving components in the cooling system, such as water pumps.

trinary switch Device that controls (stops) compressor operation if pressure is too high or too low. Its third function is as a shutter/fan override switch. This helps to maintain the compressor's discharge pressure within its designed operating range.

truck stop electrification A way to provide electricity to trucks that eliminates the need to idle the truck engine to produce the electrical power, allowing the vehicle operator to use an onboard HVAC system and other creature comforts requiring electricity.

two-piston-type compressor Describe a compressor that uses two reciprocating pistons.

TXV air-conditioning system An air-conditioning system that uses a TXV instead of an orifice tube to meter refrigerant into the evaporator.

ultraviolet That part of the electromagnetic spectrum emitted by the sun that lies between visible violet light and x-ray.

vacuum Any pressure below atmospheric pressure.

vacuum pump A mechanical device used to evacuate the refrigeration system, to rid it of excess moisture and air.

variable displacement compressor A type of air-conditioning compressor that can have its displacement changed, varying the length of the piston's stroke.

ventilation To admit and/or recirculate fresh or conditioned air.

vibrasorber A device used on refrigeration equipment to isolate engine and compressor vibration from being transmitted through to the copper refrigerant lines.

virgin refrigerant New refrigerant that has never been used in a refrigeration system.

water valve controlled system HVAC system that varies the flow rate of coolant through the heater core, thereby controlling the discharge air temperature.

wet charged Batteries that are distributed with electrolyte in the cells.

Index

Note: Page numbers referencing figures are italicized and followed by an "*f.*" Page numbers referencing tables are italicized and followed by a "*t.*"

Absolute zero, 28
AC. *See* Alternating current
Acc-U-Flush, 171
Accumulator, 6–7, *7f,* 71–72
 desiccant in, 73
 function of, 237
 internal piping, *72f*
 mounting, *71f*
 operation of, 237–238
 protective caps, 73
 screen, 73
ACPU. *See* Air conditioning protection
 unit
Activated alumina, 232–233
Air circulation. *See also* Recirculation
 control, 4
 truck-trailer, 220, *221f*
Air cleaner
 dry-type, 285, *285f*
 oil bath, 284–285, *285f*
Air conditioning, 131
 common problems, 178
 components, 4–5, 41–42, *42f,* 61–62
 disabling, 145
 history of, 2
 maintenance, 57
 modern, 2–3
 problematic history, 161
 refrigeration differentiated from, 28
 schematic, 198–202, *199f*
 servicing, 104, 161
 special tools for, 7–9
 switch, 145
 system pressure, 206
 troubleshooting, 210
Air conditioning protection and diagnostic
 systems (APADs), 177. *See also*
 CM-813 controller; CM-820
 controller
 components of, 178
 cost reduction from, 179
 installation of, *182f*
 potted, 178
 self-destruction protection from, 179
 Volvo, *178f*

Air conditioning protection unit (ACPU),
 177
 four pin connector, *186f*
 potted, 178
 six pin connector, *186f*
Air filter service and replacement, 284–285
Air outlet vents, 147
Air restriction indicator, *285f*
Air selection switch, 143, *144f,* 145
Algorithms, 178
Alternating current (AC), 271
Alternator
 components, 270–273
 installation, 272
 output test, 272
 pulley, *272f*
 removal, 272
 turn resistance, *273f*
Aluminum, 120
 radiators, 123
 tubing repair, 98
Ambient air, 3, 76
Antifreeze, 114, 115, *116f,* 286
 freeze protection of, *117f*
 high silicate, 120–121
 recovery, 10–11, *11f*
 recycling, 10–11
 skin contact, 120
 waste, 10–11
APADs. *See* Air conditioning protection
 and diagnostic systems
APU. *See* Auxiliary power units
Aqueous, 117
ATC. *See* Automatic Temperature Control
Atmospheric pressure, 33–34, *34f*
 measuring, 33
Automatic Temperature Control (ATC),
 151
Auxiliary power units (APU)
 mounted, *155f*
 output, 154–155

Batteries, 260–270
 cell operation, 261
 cell voltage, 261

cold weather and, *263f*
components, *260f*
construction, 260–261
conventional, 262–263
cross section, *260f*
deep cycle applications, 262–263
diagram, *261f*
dry charged, 262
freezing, 266
installation of, 269
low maintenance, 263
maintenance, 264–265, *264f*
maintenance free, 263, 264
preventive maintenance,
 276–277
ratings, 263–264
removal of, 269
safety, 261–262
shipping, 262
storage, 264–265
testing, 265–268
types of, 262–263
urban legends, 265
warning label, *261f*
water added to, 262
wet charged, 262
Battery charging
 fast, 268–269
 slow, 268
 systems, 269–270
Battery Council International (BCI),
 264, *264f*
Black light leak detection,
 102, *103f*
Bleeder screw, *283f*
Bleeder valve, 284
Bleeding, 283–284
 with electrical fuel pump, 284
Blend air system, 142
 climate control panel, 150
 ventilation, 150–151
Blink codes
 clearing, 181
 CM-813 controller, 181
 troubleshooting, 182–185

Boiling point
  defined, 32
  lowering, 36
  R-404a, 219
  raising, 36
  of refrigerant, 200
  of water, 33, 35–36
Box temperature, 218
British thermal unit (BTU), 81
  examples of, 30–31
BTU. *See* British thermal unit
Bunk heater, 131
Bunk override switch (BUNK OVRD),
  146
  canceling, 146–147
Bus system
  refrigerant charging, 207
  refrigerant recovery, 207–208
  service procedures, 202–205

Cadillac, 2
CCA. *See* Cold cranking amps
CCOT. *See* Cycling clutch orifice tube
  system
CDTC. *See* Constant Discharge
  Temperature Control
Celsius scale, 30
CFC-12, 17, 20
CFCs. *See* Chlorofluorocarbons
Change of state, 32–33
Check valve, 198
  non-serviceable in-line, 247–248,
    248f
  receiver tank bypass, 231
  serviceable, 247, *248f*
Chlorine, 16
Chlorofluorocarbons (CFCs), 16, 19–20
  CFC-12, 17, 20
  contents of, 19
  ozone layer and, *17f*, 19
Clean Air Act, 95
  Section 608, 17
  Section 609, 16, 17
Clean Water Act, 11
Climate control, 131
  dash controls, *130f*
  general information, 151–152
  general maintenance, 152
  operator maintenance, 151
  supplemental cab, 152–154
Climate control panel, *143f*
  blend air system, 150
  sleeper, 147–150, *147f, 149f*
  Western Star, *151f*
Clutch. *See also* Cycling clutch orifice
  tube system
  electromagnetic, 81, *81f, 82f, 200f*
  overrunning, 274
CM-813 controller, 177
  A/C drive, 189
  blink codes, 181
  compressor control rules, 180–181

diagnostic faults, 181
engine fan control, 181
fault code clearing, 187
fault code table, 182
inputs, 179–180
outputs, 180
rules for, 180–181
testing, 182
CM-820 controller, 177, 179
  DATA+ and DATA-, 189
  diagnostics, 189–190
  engine fan trigger, 188
  example code, 189–190
  fan, 189
  fault code clearing, 193
  fault codes, 190
  functions, 187–188
  high-pressure input, 189
  inputs, 189
  low-pressure input, 189
  outputs, 189
  pinout definition, 188
  Thermostat input, 189
  troubleshooting, 191–193
Cold, 28
  thermostat position, *129f*
  weather, *263f*
Cold cranking amps (CCA), 263
Compressor, *200f, 203f*, 217
  back in service, 293
  binary switch, 85–86, *86f*
  CM-813 control rules for, 180–181
  crankcase, 228
  defective, 170, *171f*
  discharge pressure, 206
  discharge side, 4, 43
  failure, 43
  four-cylinder, *228f*
  freezing, 210
  functions of, 42
  high pressure cut-off switch, 85, *86f*
  high-pressure relief valve, 87, *87f*
  low pressure cut-off switch, 85, *85f*
  lubrication, 50–52
  manifold gauge set and, 170, *171f*
  operating controls, 82
  operation, 228
  plugs, *294f*
  preventive maintenance, 293
  protection devices, 85–87
  pump down, 293
  rotary vane, 46–47, *46f, 47f*
  scotch yoke, 49, *49f, 50f*
  scroll, 49–50, *50f, 51f*
  semi-hermetic, 206
  servicing tools, 12
  suction pressure, 206
  suction side, 4, 43
  swash plate, *4f*, 45–46
  trinary switch, 86, *87f*
  two-piston, 43–45, *44f*
  in vacuum, 211

variable displacement, 47–49, *47f,*
  48f, 49f
variations, 43
Compressor fan
  cycling switch, 86–87, *87f*
  timers, 87
Compressor oil
  adding, 208
  changing, 293–294
  level checking, *52f*, 208–209, 292–293
  preventive maintenance, 292–294
  refrigerant separated from, 10
  removing, 51, 208–209
  in swash plate compressor, *52f*
  type, 51
Condensation
  defined, 32
  latent heat of, 32–33
Condenser, 4–5, *5f*, 217, 230–231
  air flow obstruction, 165, *165f*
  cleaning, 53
  heat transfer principles, 52, 231
  manifold gauge set and, 165, *165f*
  performance of, 52
  pressure bypass valve, 246–247, *247f*
  preventive maintenance, 300
  with service valves, *200f*
  servicing, 52–53
  typical assembly, *52f*
  washing, 300
Conduction, 29
  defined, 114
Constant Discharge Temperature Control
  (CDTC), 142, 147–148
Convection, 29
  defined, 114
Coolant. *See also* Antifreeze; Extended life
  coolant;
    Supplemental coolant additive
  acids forming in, 120
  benefits of, 116
  expansion tank, 127, *127f, 287f*
  flow through coolant filter, 122, *122f*
  flow through radiator cap, *127f*
  high engine temperature switch, 276
  mixing heavy duty, 120
  pH level, 118, 120
  pressure tester, *136f*
  recycler, 122, *122f*
  refraction index of, 117
  replacement, 286–287
  skin contact, 120
  strength testing, 117–118
Coolant filters
  coolant flow through, 122, *122f*
  with shutoff valves, *121f*
  strengths of, 121
Coolant heaters
  fuel-fired, 153–154
  operation of, 154
  output, 153
  shutdown, 154

Cool cycle, 244
    four-way valve, 254–255, *254f*
    solenoid control system (carrier),
       251–252, *251f*
    three-way valve, 245, *246f*, 248, *249f*
Cooling fans, 132–135
    pusher-type, 132
    suction-type, 132
Cooling system, 114
    components, *115f*, 122–136
    flushing, 287
    functions of, 114
    leaks, 135–136
    preventive maintenance, 286–287
    tap water in, 120
    testing, 126
Coperland Corporation, 49
Cycling clutch orifice tube system
    (CCOT), 84

Data bus, 177
DC. *See* Direct current
Defrost air switch, 287
    styles, *288f*
    test setup, *288f*
Defrost cycle, 244
    four-way valve, 256
    solenoid control system (carrier), 253
    three-way valve, 245–246, 249
Defrost system
    damper door, 300
    drain hoses, 300
    preventive maintenance, 287–289, 300
    termination switch, 289, *289f*
Delphi Harrison Thermal Systems, 2
Delta T, 206
Desiccant, 54. *See also specific desiccants*
    in accumulator, 73
Direct current (DC), 271
Discharge valve
    front-seated service, 229
    rotary vane compressor, 46
Distributor tube, 236
Drain plug, *282f*
Drier. *See also* Filter drier;
    Receiver-drier
    liquid line solenoid valve and, *201f*
Drip pan, 248
Drive belts, 285–286
Dry-type air cleaner, 285, *285f*
Dura Flush 141, 171

ECU. *See* Electronic control unit
EG. *See* Ethylene glycol
ELC. *See* Extended life coolant
Electrical I/O definition, 179
Electrolyte, 261
    gassing, 268–269
    level, 262, *263f*
    safety, 262
    specific gravity of, 265

Electromagnetic clutch, *81f*, *82f*, *200f*
    de-energized, 81
    description, 81
    energized, 81
    operation, 81
Electronic control unit (ECU), 87
Electronic leak detection, 8, 103, *103f*, *104f*
Electronic weight scales, 11
    portable, *11f*
Engine, 216–217
    cold weather and, *263f*
    coolant high-temperature switch, 276
    low engine oil pressure safety switch,
       275, *276f*
    lubrication system, 282–283
    maintenance, 224
    oil change, 282
    preventive maintenance, 282–283
Engine fan
    CM-813 controller and, 181
    CM-820 controller and, 188
    control, 181
    trigger, 188
Environmental Protection Agency (EPA),
    10, 16
    penalties, 18
EPA. *See* Environmental Protection Agency
Ethylene glycol (EG), 11, 115
    freeze protection of, *117f*
    toxicity of, 116
Evacuation
    hoses, 203
    improper, 202
    leaks and, 204
    manifold gauge set for, 203
    moisture and, 295
    multiple port, *204f*
    one-time procedure, 205
    preventive maintenance, 294–295
    reliability of, 205
    setup diagram, *205f*
    triple, 204–205
Evaluation procedure, 106–107
Evaporation, defined, 32
Evaporator, 6, *6f*, 217–218
    construction, 70, 71, 237
    design considerations, 71
    driver's, 201
    fin, *70f*
    flooded, 71
    freezing, *82f*, 210
    function of, 6
    heat transfer and, 70
    inlet repair, 98
    parcel rack, *202f*
    plate, *70f*
    pressure regulating device, 238
    preventive maintenance, 300
    service, 71
    temperature control, 82–85
    tube, *70f*
    washing, 300

Extended life coolant (ELC), 115, 117, *121f*
    advantages of, 121
    price of, 121
Eye wash station, *23f*
Eyewear, 8, *8f*
    refrigerant and, 22, 160

Fahrenheit scale, 30
Fan hubs
    on/off, 134
    sections, 134
    thermatic-viscous drive, 134
Fans
    belts, 135
    blade check, 134
    CM-813 controller and, 181
    CM-820 controller and, 189
    compressor, 86–87, *87f*
    cooling, 132–135
    engine, 181, 188
    fiberglass, *133f*
    pulleys, 135
    pusher-type, 132
    shrouds, 134–135, *135f*
    suction-type, 132
    switch, 143, 147
    thermo-modulated, 134
Fanstat controls, 134
Fault code 01 indicator (FMI), 190
Fault codes
    clearing, 187, 193
    CM-813 controller, 181, 182, 187
    CM-820 controller, 190, 193
    table, 187
Field diode, 271
Filter drier, 198, *232f*
    back in service, 294
    functions, 232
    liquid line installation, 233
    low-side installation, 233
    materials, 232–233
    vapor line installation, 233
Fixed orifice tube (FOT), 62, 66–68,
    67f, 76
    close-up, *68f*
    color coding, 67
    filter screen, 67
    parts of, *67f*
    refrigerant flow through, *79f*, *80f*
    restriction in, 168–169, *168f*
    size of, 67
    system overview, 78–79
Flame-type leak detection, *8f*, 102, *103f*
Flash gas, 76–77, 78, 234
Flushing
    catastrophic failures and, 171
    cooling system, 287
    endorsements for, 171
    guidelines, 171–172
    with HFCF-141b, 172
    procedures, 172
    solvents, 171, *172f*

FMI. *See* Fault code 01 indicator
FOT. *See* Fixed orifice tube
Four-way valve
    cool cycle, 254–255, *254f*
    defrost cycle, 256
    heat cycle, 255–256, *255f*
    operation, 253–254
Freezing
    battery, 266
    compressor, 210
    defined, 32
    ethylene glycol protecting from, *117f*
    evaporator, *82f*, 210
    methane, 19
    propylene glycol protecting from, *117f*
    specific gravity and, 266
Fuel filter
    replacement, 282–283
    spin-on, *283f*
Fuel-fired coolant heaters, 153–154
Fuel-fired interior heaters, 152–153
    operation of, 153
    output of, 153
Fuel injection pump, *283f*
Fusion, 32–33

Glow plugs, 286
    testing, 286
Greenhouse effect, 18–19, *19f*
    animated view of, 19
Greenhouse gases, 19, 142

Halide torch, 102
Halogen leak detection, 8
HCFCs. *See* Hydrochlorofluorocarbons
HCl. *See* Hydrochloric acid
Heat, 28–29. *See also* Superheat
    defined, 29
    flow, *29f*
    molecular motion increasing with, *28f*
    sensible, 31
    truck-trailer and residual, *220f*
    types, 31
Heat, latent, 31
    of condensation, 32–33
    of fusion, 32–33
    of vaporization, 32–33
Heat cycle, 244
    four-way valve, 255–256, *255f*
    solenoid control system (carrier), 252–253, *252f*
    three-way valve, 245–246, *247f*, 248–249, *250f*
Heater
    bunk, 131
    control valve, 131
    coolant, 153–154
    core, 130–131, *131f*
    fuel-fired interior, 152–153
    hoses, *131f*
    hydronic, *154f*

Heat exchanger, 115, *234f*
    operation of, 234
Heat sink, 299, *299f*
    wet rags as, *300f*
Heat transfer, 29
    condenser and, 52, 231
    evaporator and, 70
    rate of, 30, 122
Heavy Duty Truck Systems 5, 260
HFC-134a, 17, 20
    SNAP and, 104
HFCF-141b, 172
HFCs. *See* Hydroflourocarbons
High compressor discharge pressure switch, 276, *276f*
High engine coolant temperature switch, 276
High pressure cut-off compressor switch, 85, *86f*
High pressure relief compressor valve, 87, *87f*
Horsepower (HP), 80
Hot load, 219
HP. *See* Horsepower
Humidity, 36–37
    control, 3
    effects of, 37
    relative, 37
    truck-trailer and, *220f*
Hydrochloric acid (HCl), 54
Hydrochlorofluorocarbons (HCFCs), 20. *See also specific hydrochlorofluorocarbons*
Hydrofluorocarbons (HFCs), 20, 219. *See also specific hydrofluorocarbons*
Hydrofluoric acid, 22
Hydrogen gas, 261
Hydrometer
    reading, *265f*
    testing, 265–266
Hydronic heaters, *154f*
Hygroscopic, 54

Idling, 152
Inert gas brazing, 295–296
Interface, 178

Jones, Fred, 215
Jump-starting, 269

kcal. *See* Kilocalorie
Kilocalorie (kcal), 30
KiloPascals, 34
King valve, 198
Lead dioxide, 261
Leak detection
    black light, 102, *103f*
    cooling system, 135–136
    electronic, 8, 103, *103f*, *104f*
    evacuation and, 204
    external, 289
    flame-type, *8f*, 102, *103f*

fluid, 102, *102f*
    halogen, 8
    methods, 102–104
    phosphorus dye, 8–9, 102–103
    problem locations, 102
    refrigerant, *8f*
    refrigerant line, 95
    soapsuds solution, 102
    solution, 8
    ultrasonic, 9, *9f*, 104, *104f*
    ultraviolet, 102, *103f*
    visual inspection, 102
Light emitting diodes (LEDs), 178
Liquid line, 76, 96
    filter drier installation, 233
Liquid line solenoid valve (LLSV), 198
    drier and, *201f*
    parcel rack, 200
Litmus test, 120
LLSV. *See* Liquid line solenoid valve
Load test, 266, 268
    commercial, 266, 268
Low compressor oil pressure switch, 276
Low engine oil pressure cut out switch, 275, *276f*
Low pressure cut-off compressor switch, 85, *85f*
Low-side filter drier installation, 233
Low-side pump-down procedures, 211, 294

Magnehelic gauge, 288
Manifold gauge set, 7, *7f*, 92–94
    calibration, 94–95
    closed, *93f*
    color coding, *92f*
    condenser air flow obstruction and, 165, *165f*
    defective compressor and, 170, *171f*
    defective thermostatic control switch and, 169–170, *170f*
    equalizing, 162
    for evacuation, 203
    excessive air and moisture in, 164, *165f*
    hand valves, 93
    high-side, *93f*
    high-side calibration, *95f*
    high-side open, *94f*
    high side restriction and, 167–168, *167f*
    installation, 290–291
    low charged refrigerant and, 166, *166f*
    low-side, *93f*
    low-side open, *94f*
    overcharged refrigerant and, 165, *165f*
    for R-134 service valve, *100f*
    removal of, 290–291
    service hoses, 95
    sight glass, *95f*
    some air and moisture and, 163–164, *164f*
    thermostatic expansion valve held open and, 169, *169f*

thermostatic expansion valve not
opening and, 168–169, *168f*
troubleshooting with, 162–170
vacuum pump hooked up to, *106f*
very low refrigerant charge and,
166–167, *167f*
Material safety data sheets (MSDS), 22
Maximum operating pressure (MOP), 200
Mercury, 34
Message identifier (MID), 189
Metering devices, 62
Methane, 19
Microprocessor, *218f*
MID. *See* Message identifier
Moisture
evacuation and, 295
excessive, 164, *165f*
manifold gauge set and, 163–164, *164f*,
165f
R-134a and, 233
receiver-drier removal, 54
some, 163–164, *164f*
Moisture indicators, 233–234
center pellet color change, *233f*
receiver-drier, *56f*
Molecular sieve, 233
Montreal Protocol, 17
MOP. *See* Maximum operating pressure
MSDS. *See* Material safety data sheets
Multi-temperature refrigeration units,
219, *220f*

Nitrogen, 296
for purging, 171

ODS. *See* Ozone-depleting substances
Office of Enforcement and Compliance
Assurance, 18
Ohmmeter, 288
Oil bath air cleaner
components, *285f*
preventive maintenance, 284–285
Oil filter replacement, 282
One-time evacuation procedure, 205
Open circuit voltage test, 266
Orifice tube, 5, *5f*
cycling clutch system, 84
fixed, 62, 66–68, *67f*, *68f*, 76, 78–79,
79f, *80f*, 168–169, *168f*
variable displacement, 85
Oxyacetylene tip, 297
Ozone, 16
Ozone-depleting substances (ODS), 18, 19
Ozone layer
CFCs and, *17f*, 19
depletion, 16–17, *17f*
health and environmental concerns
caused by breakdown of, 16

Packard, *2f*
PAGs. *See* Polyalkylene glycols
Performance test, 161–162

PG. *See* Propylene glycol
Phosphorus dye leak detection, 8–9, 102–
103
PM. *See* Preventive maintenance
Polyalkylene glycols (PAGs), 20
safety, 160–161
Popping components dry, 172
Precooling
of controlled space, 220
of product, 219–220
Pressure
air conditioning system, 206
atmospheric, 33–34, *34f*
CM-820 controller input, 189
compressor discharge, 206
compressor relief valve, 87, *87f*
compressor suction, 206
condenser bypass valve, 246–247, *247f*
coolant, *136f*
defined, 33
high head, 210
high suction, 210
low suction, 210
maximum operating, 200
R-12 chart for temperature and, *69f*, *163f*
R-134a chart for temperature and, *69f*,
163f
refrigerant, 80
temperature relationship with, 35–36
tester hand pump, *136f*
thermostatic expansion valve and,
63, 235
three-way valve and, 245
Pressure cycling switch, 84–85
installation of, 84
schematic view, *84f*
Pressure gauges. *See also* Manifold gauge
set
compound, 34, *35f*, 92
per square inch absolute, 34, *35f*
pounds per square inch, 34, *35f*
Pressure regulating devices, 5–7
evaporator, 238
operation of, 238–239
suction, 238, *238f*
Pressure switches
compressor cut-off, 85, *85f*, *86f*
installation, 187
low compressor oil, 276
low engine oil cut out, 275, *276f*
low oil engine cut out, 275, *276f*
mounting locations, *180f*
Pre-trip, 289
Preventive maintenance (PM), 282
air filter, 284–285
batteries, 276–277
bleeding, 283–284
compressor oil, 292–294
compressor pump down, 293
condenser, 300
cooling system, 286–287
defrost damper door, 300

defrost drain hoses, 300
defrost system, 287–289
drive belts, 285–286
dry-type air cleaner, 285, *285f*
engine lubrication system, 282–283
engine oil change, 282
evacuation, 294–295
evaporator, 300
fuel filter replacement, 282–283
glow plugs, 286
mounting bolts, 300
oil bath air cleaner, 284–285
oil filter replacement, 282
refrigerant, 289, 292
structural, 300
unit visual inspection, 300
Prolink 2000, *12f*
Propylene glycol (PG), 115
freeze protection of, *117f*
toxicity of, 116
Pulse modulating water valve, 142, 149
Pulse width modulation (PWM), 142, 149
Purging, 170–171
guidelines, 171–172
nitrogen for, 171
procedures, 172
PWM. *See* Pulse width modulation

R-12, 20
pressure-temperature chart for, *69f*, *163f*
R-22, 17
restrictions on, 20
R-134a
combustion of, 160
moisture and, 233
pressure-temperature chart for, *69f*, *163f*
vapors, 160
R-134 service valve, 100
manifold gauge set for, *100f*
R-404a, 20
boiling point, 219
Radiation, 29
defined, 114
solar, 3
Radiator cap, 125–127, *125f*
coolant flow through, *127f*
functions, 125
removing, 127
testing, 127–128, *127f*
Radiators, 122–124
aluminum, 123
components, 124
construction of, 123
core, 124
core construction, *124f*
counterflow, 123
crossflow, 123, *124f*
double pass, 123
downflow, 123, *124f*
filler neck, 124
hoses, 124
overflow tank, 125, *125f*, 127

Radiators (*Continued*)
  petcock, 123, 124
  plastic, 123
  servicing, 124–125
  single pass design, 123
  tanks, 124
  testing, 125
  tube and fin construction, *123f*
Ram air, *78f*, 122
Receiver, 198
  with service valves, *200f*
Receiver-drier, 6, 53–56
  contaminants, 53
  filter, 53–54
  fitting styles, 55
  installation, 55
  internal components of, *53f*
  location, 54–55
  moisture indicator, *56f*
  moisture removal, 54
  mounting, 55
  pickup tube, 53
  protective caps, 55
  restrictions, 54–55
  saturation point, 54
  service, 55
  sight glass, 55–56, *56f*
  storage filtration provided by, *6f*
  strainer, 53
Receiver tank, 231–232, *232f*
  bypass check valve, 231
  hot gas line, 231
  outlet valve, 198
Recirculation
  button, 145–146
  door, *146f*
  timer, 146
Rectifier diodes, 271, *271f*
Refractometer, 117, *118f*, 265
  accuracy of, 117
  how to use, *118f*, 119
  instructions, 265
  method, *266f*
  operation of, 117–118
  parts of, *117f*
Refrigerant, 19–20, 218–219.
    See also specific types
  alternative, 20
  boiling point, 200
  capacity, 80–81
  capacity warning label, *107f*
  cloudy, 56
  compressor oil separated from, 10
  contaminated, 105
  cycle control valves, 244–256
  eye contact, 22, 160
  first aid, 22
  with flammable hydrocarbon, 12
  health hazards, 22
  liquid, 207
  nonapproved, 105
  one-pound can, *18f*

performance ratings, 80–81
poisonous gas and, 22
pressure, 80
preventive maintenance, 289, 292
purity, *105f*
recycling equipment, 10, *10f*, 109
removal, 294
safety, 22
sale of, 17
skin contact with, 22
states, 80
storage, 54
vapor, 207
weight scale, *108f*
work gloves and, 22
Refrigerant charge
  checking, 206–207
  conditions, 206
  low, 166, *166f*
  manifold gauge set and, 165, *165f*,
      166–167, *166f*, *167f*
  overcharged, 165, *165f*, 206
  very low, 166–167, *167f*
Refrigerant charging, 289, 292
  bus system, 207
  cylinder, 108–109
  full, 207
  with liquid refrigerant, 207
  partial, 108, 207, 292
  procedure, 107–108
  top up, 108
Refrigerant couplers, 95
  flare, *95f*
  O-ring, *95f*
Refrigerant cylinders
  charging, 108–109
  color coding, 20, 21, 22
  cross contamination, 21
  disposable, 20–21, *21f*
  handling, 23
  outlet valves, *108f*
  recovered, 21
  recycled, 21
  refillable, 21, *21f*
  retesting, 21
  storing, 21
  virgin, 21
Refrigerant flow
  through fixed orifice tube, *79f*, *80f*
  through thermostatic expansion valve,
      64f, 76–77, *77f*, *80f*
  three-way valve, 248–250
Refrigerant hoses, 95
  beadlock fittings, 97
  bubble-style crimp, *97f*, *98f*
  color coding, 92
  crimping tools, 98
  double braid, 96
  finger-style crimp, 96–97, *97f*, *98f*
  new-style barrier, *96f*
  repair, 96–98
  size, *97f*

Refrigerant identifier, 11–12, *12f*, 104–105
  display, 12
Refrigerant lines
  construction of, 95
  discharge, 96
  leakage, 95
  liquid, 76, 96, 233
  repair, 96–98
  suction, 96
Refrigerant management center, 109–110,
    207f, *292f*
  installation, 101
Refrigerant recovery
  bus system, 207–208
  equipment, 10, *10f*, 109
Reserve capacity rating, 263–264
Reversing system. *See* Four-way valve
Rotary vane compressor, *46f*, *47f*
  discharge valve, 46
  operation, 46–47
Rotor, 270–271, *271f*

Safety. *See also* Material safety data sheets
  batteries, 261–262
  electrolyte, 262
  eyewear, 8, *8f*, 22, 160
  general workplace, 22–23
  polyalkylene glycols, 160–161
  refrigerant, 22
  work gloves, 22, 160
  workplace, 22–23
Safety switches, 275–276
  water temperature, *276f*
Safety valves, 239
  fusible metal plug, 239, *239f*
  spring loaded piston, 239, *239f*
  types of, 239
Sanden, 49
Saturated liquid, 76
SCA. *See* Supplemental coolant additive
Scaling, 118
Scan tools, 11, *12f*
Schrader-type service valve, 99–100, 230, *230f*
  with hose pin depressor, *100f*
  size of, 100
Scotch yoke compressors, *50f*
  operation, 49
  typical, *49f*
Scroll compressor
  inner section, *50f*
  operation, 49–50, *51f*
Service Pro, 193
Service valves, 98–100, 228–230
  check, 247, *248f*
  condenser with, *200f*
  front-seated discharge, 229
  manifold gauge set and, *100f*
  positions of, *201f*, *230f*
  R-134, 100, *100f*
  receiver with, *200f*
  Schrader-type, 99–100, *100f*, 230, *230f*
  stem-type, 99, *99f*

Set point, 218
SG. *See* Specific gravity
Shop specialty tools, 9–12
Shore power systems, 156
Short cycling, 222
Shutters, 132
Shutterstat, 132
  old and new design, *133f*
SID. *See* Subsystem identifier
Sight glass, 292–293, *293f*
  manifold gauge set, *95f*
  pith ball, 231, *232f*
  receiver-drier, 55–56, *56f*
Significant New Alternatives Policy
    (SNAP), 104
  HFC-134a and, 104
Silica gel, 232
Silver brazing, 295
  advantages of, 296
  alloy flow, 297–298
  common alloys for, 296
  correct flame for, *297f*
  disassembly, 299
  final cleaning, 298
  general joint heating, 297–298
  horizontal joint technique, 298–299, *299f*
  joint cleaning, 296
  joint fitment, 296
  joint fluxing, 297
  joint stress relief, 297
  preparation tools, *296f*
  sweeping technique, *298f*
  vertical down joint technique, 298, *298f*
  vertical up joint technique, 298, *298f*
Skin contact
  antifreeze, 120
  coolant, 120
  refrigerant, 22
SNAP. *See* Significant New Alternatives
    Policy
Soldering, 295
  soft, 299, *299f*
Solenoid control system (carrier),
    250–253
  cool cycle, 251–252, *251f*
  defrost cycle, 253
  heat cycle, 252–253, *252f*
  operation of, 250–251
Solvents. *See also specific solvents*
  flushing, 171, *172f*
Specific gravity (SG)
  electrolyte, 265
  freezing and, 266
  water, 265
Spongy lead, 261
Starter motors, 273–275
  conventional, 273–274, *273f*
  gear reduction, 274, *274f*
  overrunning clutch, 274
  testing, 274–275
  types, 273
Stator, 270, *271f*

Stem-type service valves, 99
  in back-seated position, *99f*
  in front-seated position, *99f*
  in mid-seated position, *99f*
Stepper motor, 150
  placement of, *150f*
Stratosphere, *16f*
Subcooler, 198
Subcooling, 31–32
Subsystem identifier (SID), 189
Suction throttling valve, *238f*
Sulphuric acid, 261
Superheat, 31, 201
  checklist, 209
  determining, 68–70, 236
  example, 236
  test procedures, 209
  thermostatic expansion valve, 235–236
Supplemental coolant additive (SCA),
    115, *116f*
  testing, 118, 120
  test kits, 118
Swash plate compressor, *45f*
  compression stroke, 46, *46f*
  intake stroke, 45, *46f*
  maintenance, 46
  oil check, *52f*
  operation of, 45–46

TDS. *See* Total dissolved solids
Temperature
  absolute zero, 28
  ambient, 216
  box, 218
  Celsius scale, 30
  defined, 30
  engine coolant switch, 276
  Fahrenheit scale, 30
  pressure relationship with, 35–36
  R-12 chart for pressure and, *69f, 163f*
  R-134a chart for pressure and,
    69f, 163f
  scales, 30
  sensors, 151
  water safety switch, *276f*
  wet bulb, 37
Temperature control, 3. *See also*
    Automatic Temperature Control;
    Constant Discharge Temperature
    Control
  evaporator, 82–85
  switch, 145, 147–148
  water-valve controlled system,
    149–150
Thermal equilibrium, 29
Thermistor, 151, *152f*
  vacuum gauge, 107, *107f*
Thermo King, 215, 288
Thermometers, 9, *9f*
  dial-type, 9
  dry bulb, 37
  electronic, 9

Thermostat (TStat)
  by-pass circuit, 129
  closed, *130f*
  CM-820 controller input, 189
  cold position, *129f*
  functions, 128
  hot position, *129f*
  microprocessor controlled, *218f*
  open, *130f*
  operating without, 129
  operation, 129
  rotary dial, *218f*
  set point, 216
  testing, 130
Thermostatic control switch, 82–83,
    82f, 83f
  defective, 169–170, *170f*
  manifold gauge set and, 169–170, *170f*
  mounted, *84f*
  varieties of, 83
Thermostatic expansion valve (TXV),
    202f, 217, 234–236
  assorted, *5f*
  capillary tube, 63
  closed, *65f*
  controlling action, 64–65
  equalizer line, 235
  externally equalized, 65, *65f, 66f*
  fluctuating requirements of, 6
  held open, 169, *169f*
  H valve, 66, *66f, 67f*
  internal components of, *62f*
  internally equalized, 62, *65f, 66f*
  maintenance of, 5
  manifold gauge set and, 168–169,
    168f, 169f
  maximum operating pressure
    equipped, 200
  modulation, 64
  not opening, 168–169, *168f*
  open, *65f*
  operation, 63–65, 235
  pressure, 63, 235
  refrigerant flow though, *64f*, 76–77,
    77f, 80f
  remote bulb, 63
  sensing bulb, 236
  sensing element, 236
  superheat, 235–236
  system overview, 76–78
  throttling, 63–64
Three-way valve, *244f*
  cool cycle, 245, *246f*, 248, *249f*
  defrost cycle, 245–246, 249
  heat cycle, 245–246, *247f*, 248–249, *250f*
  location of, 244–245
  operation of, 245–246
  pilot solenoid, 245, *245f*
  pressure and, 245
  refrigerant flow in, 248–250
TMC. *See* Truck Maintenance Council
Ton, 80–81

Total dissolved solids (TDS), 118
  probe, 120
  testing for, 120
Triple-evacuation, 204–205
Troubleshooting, 160–161
  air conditioning, 210
  blink codes, 182–185
  CM-820 controller, 191–193
  flowchart, *162f*
  manifold gauge set, 162–170
Truck Maintenance Council (TMC), 117
Truck stop electrification, 155–156, *155f*
  combination, 156
  onboard, 156
  shore power systems, 156
  stand-alone, 156
Truck-trailer
  air circulation, 220, *221f*
  auto stop/start, 223
  components, 216–218, *216f*, 228
  construction, 219
  continuous run mode, 223
  electrical system maintenance, 224
  flooring, 219
  front bulkhead spacing, 222
  humidity and, *220f*
  incorrect loading, *222f*, *223f*
  loading factors, 219–221
  loading procedures, 221–222
  maintenance procedures, 223–224
  pallet positioning, 220–221
  performance tasks, 224
  rear door spacing, 221–222
  residual heat and, *220f*
  roof spacing, 221
  side spacing, 221
TStat. *See* Thermostat
Two-piston compressor
  compression stroke, 44, *44f*
  construction of, 43

intake stroke, 43–44, *44f*
  maintenance, 45
  mounting, 43
  operation of, 43–44
TXV. *See* Thermostatic expansion valve

Ultrasonic leak detection, 9, *9f*, 104, *104f*
Ultraviolet (UV), 16
  leak detection, 102, *103f*

Vacuum
  compressor in, 211
  indicator, 295
  micron gauge, 203–204, *203f*
  partial, 35
  perfect, 34, 35
  thermistor gauge, 107, *107f*
Vacuum pump, 9–10, *10f*, 34, *105f*
  capacity, 202
  conditions for, 105
  maintenance, 106
  manifold gauge hooked up to, *106f*
  oil, 105–106
  oil replacement, 202–203
  service, 107
  size, 105
Vapor
  compression, 36, *37f*, *43f*
  line filter drier installation, 233
  R-134a, 160
  refrigerant, 207
Vaporization, 32–33
Variable displacement compressor, 47–49, *47f*
  control valve assembly for, *49f*
  at maximum displacement, *48f*
  at minimum displacement, *48f*
  operation, 49
Variable displacement orifice tube (VDOT), 85

Variable orifice valve (VOV), 68, *68f*
VDOT. *See* Variable displacement orifice tube
Vent door, *144f*, *145f*
Ventilation, 3
  blend air system, 150–151
Vibrasorber, 198, *203f*
  discharge, 230, *231f*
  placement of, *231f*
  suction, 230
  types of, 230
Voltage regulator, 271–272
Volvo, *178f*
VOV. *See* Variable orifice valve

Water
  added to batteries, 262
  boiling point, 33, 35–36
  in cooling system, 120
  specific gravity of, 265
  tap, 120
  temperature safety switch, *276f*
Water pump, 128
  assembly, *128f*
  failure, 128
  inspection, 128
  replacement, 128
Water-valve system, 142, 143–147
  manual, 148–150, *149f*
  manual shutoff, *150f*
  pulse modulated, 142, 149
  solenoid, 149, *149f*
  temperature control, 149–150
  types, 149
Whiteley, William, 2
Winter fronts, 132, *133f*
Work gloves, 160
  refrigerant and, 22
Workplace safety, 22–23